Sea Ice Biota

Editor
Rita A. Horner, Ph.D.
Research Consultant
Seattle, Washington

CRC Press
Taylor & Francis Group
Boca Raton London New York

CRC Press is an imprint of the
Taylor & Francis Group, an **informa** business

First published 1985 by CRC Press
Taylor & Francis Group
6000 Broken Sound Parkway NW, Suite 300
Boca Raton, FL 33487-2742

Reissued 2018 by CRC Press

Library of Congress Cataloging in Publication Data
Main entry under title:

Sea ice biota.

Bibliography: p.
Includes index.
1. Marine biology—Polar regions. I. Homer, Rita, A.
QH95.56.S43 1985 574.92'8 85-3827
ISBN 0-8493-6578-3

A Library of Congress record exists under LC control number: 85003827

ISBN 13: 978-1-315-89744-8 (hbk)
ISBN 13: 978-1-351-07654-8 (ebk)

Visit the Taylor & Francis Web site at http://www.taylorandfrancis.com and the
CRC Press Web site at http://www.crcpress.com

DEDICATION

This book is dedicated to Tom English, friend and teacher, who introduced me to the rewards and frustrations of Arctic research.

PREFACE

Organisms living in sea ice have been known and studied for nearly 150 years. Ehrenberg[1,2] first described diatoms in sea ice from the Arctic and Hooker[3] found diatoms in sea ice from the Antarctic. For nearly 120 years the studies in both polar regions were primarily of a taxonomic and biogeographic nature and over 500 species or groups of species have been reported from sea ice (see Appendix). It was not until the late 1950s that attention turned to the productivity and physiology of this community. The development of new field and laboratory techniques in the last decade has enabled investigators to make more accurate estimates of production within this community, but even with the new methods, comparisons between communities are often difficult. This is partly because of differences in ice structure, but also because of differences in techniques and in reporting data. At the same time, a number of different habitats have been recognized and these may have different species assemblages. In addition, the same habitat in relatively closely spaced geographic areas may have different species assemblages and species succession can occur during the course of an ice algal bloom.

We are just beginning to realize the importance of bacteria in sea ice and relatively little is known about this component of the ice community. Organisms other than diatoms in the size range 2 to 20 μm are just being recognized from the ice. These organisms would be called nanoplankton if they were found in the water column. Cells smaller than 2 μm, other than bacteria, may also be present in the ice, but have not been reported as yet. The importance and often the taxonomic affinities of these small organisms remain unknown.

Sutherland[4] first reported "infusory animalcules" in 1852 and Nansen[5] reported ciliates in sea ice in 1906, but little attention was paid to the ice meiofauna until the last decade. In fact, as recently as the early 1970s, several authors[6,7] reported that no grazing occurred in the ice and net production was released into the water column. We now know that large numbers of invertebrates from several phyla inhabit not only the ice itself, but also the ice-seawater interface, where they graze on the ice algae and bacteria and are in turn utilized by cryopelagic fish species and seals. The ice algae-based food web thus is proving to be an important part of polar ecosystems.

Investigators from a number of countries have been studying the ice community and experimental information is now available from a number of geographic areas. This includes ecological data as well as community and species-specific physiological information. The literature on ice biota is scattered, being found in scientific journals, research and technical reports, symposia proceedings, M.S. theses and Ph.D. dissertations, meeting abstracts, and books on topics ranging from algal ecology to regional oceanography. Much of the material has not been published and some is available only in proprietary or difficult to obtain reports. The purpose of this book is to bring the data and references together in one place and to provide state of the art information on these little-known, but ecologically important, polar communties.

The book begins with a chapter on the history of ice algal studies starting with the earliest reports of ice algae in the literature and extending to the most recent studies, including some not yet in the open literature. An extensive chapter on the physical properties of sea ice follows. This chapter is perhaps more detailed than necessary, but it is these physical properties that make the ice an attractive habitat for the organisms living there. A better understanding of these ice properties may help provide answers to some of the questions biologists have concerning the ability of the organisms to live and grow in the ice.

The three chapters on the ecology, chemical composition and biochemistry, growth, metabolism, and dark survival are generally based on data obtained in the last 20 years and rely on new technology, some specifically designed for the study of ice algae such as specialized sampling techniques and incubators designed to mimic the environment at the bottom of the ice.

The taxonomy of the microalgae found in sea ice is discussed in the next chapter and a preliminary check list of algal species reported from sea ice is given in the Appendix. The early studies on ice algae were mainly lists of species, but unfortunately, except for a few studies primarily in the Arctic, the taxonomy of ice algae has not received much attention in recent work. This situation needs to be remedied.

The chapter on bacteria reviews the sparse data on this important component of the ice community. Most of the information is from studies in the Antarctic and has been obtained since 1980.

The complexities of the faunal communities associated with sea ice are discussed in the final chapter. These organisms range from tiny, colorless flagellates to fishes and marine mammals. They live within the ice or on the undersurface of the ice at the ice-seawater interface; they may be more or less permanent members of the ice community or they may be associated with the ice because food is readily available at the bottom of the ice or because the ice affords a measure of protection from predators. The complex food web based on the ice algae is discussed and pictured in this chapter.

The annual extent of sea ice in polar regions is about 38×10^6 km^2 in winter or about 12% of the total sea surface.[8] Although the areal extent of the ice community is not known, 2×10^6 and 2.6×10^6 km^2 have been suggested for the Arctic[9] and Antarctic,[10] respectively. Thus, if the ice community covers nearly 5×10^6 km^2, it is an important part of the polar ecosystem and deserves more attention in future research.

REFERENCES

1. **Ehrenberg, C. G.,** Einen Nachtrag zu dem Vortrage über Verbreitung und Einfluss des mikroskopischen Lebens in Süd- und Nord-America, *Akad. Wiss., Berlin, Monatsber.*, 1841, 202, 1841.
2. **Ehrenberg, C. G.,** Über neue Anschauungen des kleinsten nordlichen Polarlebens, *Akad. Wiss., Berlin, Monatsber.*, 1853, 522, 1853.
3. **Hooker, J. D.,** *The Botany of the Antarctic Voyage of H. M. Discovery Ships Erebus and Terror in the Years 1839-1843. I. Flora Antarctica*, J. Cramer, Weinheim, 1847 (reprint 1963), 503.
4. **Sutherland, P. C.,** *Journal of a voyage in Baffin's Bay and Barrow Straits, in the years 1850-1851, performed by H. M. ships "Lady Franklin" and "Sophia," under the command of Mr. William Penny, in search of the missing crews of H. M. ships "Erebus" and "Terror,"* Vols. 1 and 2, Longman, Brown, Green, and Longmans, London, 1852.
5. **Nansen, F.,** Protozoa from the ice-floes of the North Polar Sea, *Sci. Res. Norw. N. Polar Exped.*, 5(16), 1, 1906.
6. **Meguro, H., Ito, K., and Fukushima, H.,** Ice flora (bottom type): a mechanism of primary production in polar seas and the growth of diatoms in sea ice, *Arctic*, 20, 114, 1967.
7. **Bunt, J. S. and Lee, C. C.,** Seasonal primary production in Antarctic sea ice at McMurdo Sound in 1967, *J. Mar. Res.*, 28, 304, 1970.
8. **Weeks, W. F.,** Understanding the variations of the physical properties of sea ice, in *Symposium on Antarctic Oceanography*, Currie, R. I., Ed., W. Heffer & Sons Ltd., Cambridge, 1968, 173.
9. **Apollonio, S.,** Chlorophyll in Arctic sea ice, *Arctic*, 18, 118, 1965.
10. **Burkholder, P. R. and Mandelli, E. F.,** Productivity of microalgae in Antarctic sea ice, *Science*, 149, 872, 1965.

THE EDITOR

Rita A. Horner received a B.S. degree from the University of Wisconsin in 1956, a M.S. degree from the University of Minnesota in 1958, and a Ph.D. degree from the University of Washington in 1969. She was a Fulbright student at the University of British Columbia, Canada. She served as an Assistant and Associate Professor in the Institute of Marine Science, University of Alaska, Fairbanks, from 1969 to 1974. Since 1974, she has been a research consultant working on a variety of projects ranging from the biology of ice algae, phytoplankton, and zooplankton to marine fouling organisms. Dr. Horner has authored or co-authored a number of papers primarily on the biology of sea ice biota.

Dr. Horner is a member of the Phycological Society of America, the British Phycological Society, the International Phycological Society, American Association for the Advancement of Science, Sigma Xi, and is a Fellow of The Arctic Institute of North America.

CONTRIBUTORS

Andrew G. Carey, Jr.
Professor
College of Oceanography
Oregon State University
Corvallis, Oregon

Rita A. Horner, Ph.D.
Research Consultant
Seattle, Washington

Gary A. Maykut, Ph.D.
Senior Research Associate
Department of Atmospheric Sciences/
Geophysics Program
University of Washington
Seattle, Washington

Malcolm J. McConville
Plant Cell Biology Research Centre
School of Botany
University of Melbourne
Parkville, Australia

Anna C. Palmisano
Research Associate
Allan Hancock Foundation
University of Southern California
Los Angeles, California

Cornelius W. Sullivan
Director
Professor
Marine Biology Research Section
Department of Biological Sciences
University of Southern California
Los Angeles, California

TABLE OF CONTENTS

Chapter 1

HISTORY OF ICE ALGAL INVESTIGATIONS

Rita Horner

TABLE OF CONTENTS

I. INTRODUCTION

Polar regions have long been a challenge to man's spirit of adventure and his thirst for knowledge. The earliest recorded voyages to Arctic regions were those of the Norsemen starting in 870 A.D. These early adventurers sailed as far west as Labrador and Newfoundland, founded colonies in Greenland, and sailed east to Novaya Zemlya. The search for a Northeast Passage across the top of Eurasia started in the mid-1500s and lasted about a century, but it was not until 1879 that A. E. Nordenskioeld on the *Vega* actually completed the passage. The search for the Northwest Passage to the Orient provided the impetus for much of the Arctic exploration, starting in 1576 with Martin Frobisher's journey to what was later named Baffin Island and lasting until 1906 when Roald Amundsen was finally able to penetrate the northern ice.

The ancient Greeks reasoned that there must be a huge land mass in southern regions to balance the northern land mass. This southern land was called the Antarctic (opposite the Arctic), and it appeared as "Terra Australis Incognita" on many maps during the Middle Ages. It was not until 1820 that an American sealer, Nathaniel Palmer, first sighted the Antarctic Peninsula. A number of countries sent out scientific expeditions in the mid-1830s to search for the South Magnetic Pole, which was not found until 1909. The 1890s saw the first person to land on the Antarctic continent (H. J. Bull of Norway in 1895) and the first person to overwinter on the continent (C. E. Borchgrevink, Norway/U.K.). The first long-term scientific program was the United Kingdom's *Discovery* cruises that lasted from 1923 until 1939.

Naturalists were included on the voyages of exploration in both polar regions and many of the earliest samples of ice algae were collected during these voyages, e.g., Hooker[1] on H. M. S. *Erebus* and *Terror* under the leadership of James Clarke Ross; Sutherland[2] and Dickie[3] during the search for Sir John Franklin; Cleve[4] on the *Vega* Expedition; and Gran[5] and Nansen[6,7] during the First Norwegian North Polar Expedition on the *Fram*.

II. EARLY HISTORY

Diatoms have been known from sea ice for nearly 150 years. They were first described from the Arctic by Ehrenberg[8,9] and from the Antarctic by Hooker.[1] The papers published in the late 1800s and early 1900s were mainly lists of species and it was not until the late 1950s and early 1960s with the development of new field and laboratory techniques that experimental studies began in both polar regions.

A. Arctic

In the Arctic, Sutherland[2] described a "greenish slimy-looking substance" in small cavities on the bottom of the ice and provided observations on the occurrence of the "infusory animalcules" and "minute vegetable forms of exquisite beauty." Dickie[3,10] listed species collected by Sutherland.[2] Cleve and Grunow[11] listed species found in ice from the Kara Sea. Cleve[4,12-15] reported on diatoms from ice collected at Cape Wankarema, Siberia; Baffin Bay and Davis Strait; Franz Josef Land; between Greenland and Spitsbergen; and from north of Jan Mayen Island. The same species were found in all areas leading Cleve[12] to suggest a common origin for these species which he thought was the Bering Strait region. Grunow[16] examined a sample taken from the underside of an ice floe near Novaya Zemlya and found some of the same species that had been found at Cape Wankarema, but the most characteristic Siberian species were missing.

Østrup[17] studied diatoms collected from ice floes off east Greenland and found some of the Cape Wankarema species. He also studied a number of additional forms, several of which he described as new species. Some of the cells contained pigments indicating they were actually living on the ice. Østrup[17] thought the diatoms were carried to the ice by wind or rivers, washed onto the ice by storms, or were frozen into the ice when it formed.

Vanhöffen[18,19] and Gran[20] studied samples from Karajak Fjord on the west coast of Greenland. By March, planktonic species were increasing in the water column and the undersurface of the ice was covered with a brown coating of diatoms. *Nitzschia frigida* Grunow,* *Fragilaria oceanica* Cleve, and *Navicula septentrionalis* Østrup were present and the cells were alive when collected. This is the first study to suggest the seasonal cycle of ice algae.[18] Gran[20] believed that Vanhöffen's investigations indicated the ice attracted a community of diatoms that for "biological reasons", probably light, were able to live on the undersurface of the ice. The community consisted of both planktonic *(Fragilaria, Melosira)* and littoral *(Pleurosigma, Navicula)* species.

The most extensive early description of the ice community is that of Gran[5] and Nansen.[6,7] Most of the samples were collected near 81°N, 125°E during the summer of 1894. Gran[5] distinguished four habitats and described the diatoms from them:

1. The free-floating lumps found in channels between ice floes consisted of large masses of *Melosira* along with free-floating lumps of *Nitzschia* and *Fragilaria*. These were usually found at about 1-m depth floating at the interface between the fresh surface layer and the seawater below.
2. The ice foot at the bottom of ice floes had populations of *Melosira hyperborea* Grunow and *Nitzschia frigida*.
3. Cylindrical holes on the bottom of freshwater melt ponds not connected with the underlying seawater contained cell fragments and spores.
4. Sediment from newly formed ice contained mostly vegetative cells and spores of planktonic species and also some benthic species such as *Pleurosigma stuxbergii* Cleve and Grunow and *Navicula gelida* Grunow.

Gran[5] also distinguished four kinds of diatoms from the ice, including true planktonic forms such as species of *Chaetoceros* and *Thalassiosira* that have a wide geographic distribution; species from plankton and ice, such as *Melosira hyperborea, Nitzschia frigida, Fragilaria oceanica*, and *F. cylindrus* Grunow that are widely distributed in cold seas and occur between ice floes or under ice cover; species that occur only in ice or are littoral and are not present in large numbers; and freshwater species that are also not abundant.

Nansen[6] examined newly formed brownish-red ice collected in October 1893 near 78°N and found that the color was produced by small organisms. In July 1894, Nansen found brown coloration on the ice surface but this was caused by "mineral dust" that he thought originated on land and fell with snow. It then accumulated on the ice surface as the snow melted in summer. At the same time, large amounts of algae were found in leads, on ice floe edges, and down into the water. Nansen said the ice algae were "undeniably most interesting things, a whole new world of organisms that are carried off by the ice from known shores across the unknown Polar Seas, there to awaken every summer and develop into life and bloom."

Also during the summer, yellowish-brown spots developed at the bottom of melt ponds; these increased in size and melted the ice to form round cavities several inches deep. The brown spots consisted of algae and diatoms, but "infusoria and flagellates" and bacteria were also present. In 1906, Nansen[7] described these organisms and their habitats in more detail beginning with the observations he made in the East Greenland Sea in 1882 and Denmark Strait in 1888. Nansen was unable to identify the protozoans to species, but his detailed drawings were reproduced. Nansen thought the "germs" of the organisms were

* Scientific names used in all chapters are those used by the authors cited.

probably frozen into the ice at the sea surface. When the ice melted, the "germs developed in the ponds on the ice in the same way as the diatoms . . . "

Meister[21] listed the flora found in ice floes from the Greenland Sea where *Melosira arctica* (Ehrenberg) Dickie was the dominant species. The silicoflagellate *Distephanus speculum* v. *septenarius* (Ehrenberg) Jørgensen was also present.

Usachev[22] reported on samples collected in the Kara and Laptev Seas. Some of his samples contained freshwater diatoms, green, and bluegreen algae which came from freshwater ice probably formed in the Yenisey River. Most of the diatoms were empty cells, but 10 to 15% of *Melosira* spp. had cell contents. In fact, auxospores of *Melosira granulata* (Ehrenberg) Ralfs germinated and some benthic species were motile when the ice melted. Other samples, containing mostly empty marine species, probably came from the Kara Sea. Some ice floes of mixed origin had freshwater green algae occurring in the middle layers; these floes were probably 2 years old.

In 1949, Usachev[23] reviewed much of the previous work on ice algae, particularly that done by Soviet investigators. Most of the diatom species found frozen in the ice were benthic forms and Usachev considered 24 species to be typical cryophiles, i.e., diatoms developing on the lower surface of the ice and occurring in all collections of Arctic ice. These cryophiles were mostly benthic species that formed mucilage for attachment to the ice and for movement. Winter spores and live cells of benthic and planktonic species began to develop on the lower surface of the ice in early spring because of the small amount of light that penetrated the ice. However, cryophiles were not found in significant quantities or alive in the littoral zone of the Kara Sea and Usachev[23] suggested that these species might develop in the benthos until the ice begins to form and then they choose the underside of the ice as the place for favorable growth to occur.

Usachev[23] described two main types of ice diatom blooms. One consisted primarily of large, mucilaginous masses of *Melosira arctica*, and the other of loose formations of *Chaetoceros septentrionalis* Østrup and benthic cryophiles. He did not find live diatoms in pools on the upper surface of the ice not connected with the underlying seawater as did Gran.[5] The ice blooms, consisting of benthic and littoral diatom species, were different from the phytoplankton which consisted of planktonic genera such as *Chaetoceros*, *Rhizosolenia*, and *Thalassiosira*.

B. Antarctic

In the Antarctic, Hooker[1] was amazed to find "The Waters and the Ice to the South Polar Ocean were alike found to abound with microscopic vegetables . . . " Although they were too small to be seen by the naked eye "they occurred in such countless myriads, as to stain the Berg and the Pack-Ice, . . . , they imparted to the Brash and Pancake-Ice a pale ochreous colour." Diatoms were especially abundant south of about 70°S, with the surface and the sides of every berg being brown. "The Pancake-ice was often seen a few hours after a calm, covering leagues of ocean, and uniformly stained brown from the abundance of these plants." When the pancake ice was collected and allowed to melt, a rich precipitate of diatoms occurred. Brash ice also contained large quantities of diatoms.

El-Sayed,[24] citing Drygalski,[25] suggested that interest in the biology of sea ice dated to the German Antarctic Expedition of 1901 to 1903, when "Vangefen made some interesting observations on the microalgae in the fast ice of the Davis Sea." Heiden and Kolbe[26] studied the diatoms collected during this expedition, including samples taken of ice and at the ice edge. They listed 117 species associated with the ice, but not all species were actually found in ice, some apparently being from water samples collected at the ice edge. "Susswasser-eisdiatomeen" were also found.

Vanhöffen[27] provided the earliest description of a seasonal cycle for ice diatoms in the Antarctic. He found development beginning in early November with a nearly pure culture of *Schizonema*-like forms occurring in the ice by mid-December. These diatoms mixed with the plankton on release from the ice.

Scott[28] was surprised by the large numbers of diatoms caught in the ice, even though their nets were catching vast quantities of cells. He reported the underside of overturned blocks of ice to be colored reddish-yellow by the diatoms.

Hendey[29] and Hart[30,31] reported on diatoms collected during the *Discovery* Expeditions. Hendey[29] suggested that ice contained a polymer of water called trihydrol which stimulated algal growth. A high proportion of trihydrol was said to be maintained in cold waters at high latitudes and this caused the enormous increases in diatoms in areas where ice melted. Hendey[29] found 11 species of diatoms in one sample of melted ice collected from 66°47′S, 69°19′W. This was not considered sufficient to make major comparisons with Gran's Arctic study, except to say that the characteristic Arctic species were missing from the Antarctic. *Fragilaria curta* Van Heurck and *F. linearis* Castracane were the dominant species with *Navicula corymbosa* (Agardh) Cleve and *Amphiprora oestrupii* Van Heurck present in small numbers. The dominant species were neritic and there was no evidence of any freshwater species from the mainland.

Few phytoplankton were found in the Bellingshausen Sea in areas where pack ice was thick leading Hart[30] to suggest that a considerable time interval was necessary after ice breakup before much phytoplankton development could take place. The persistence of the pack ice in spring and early summer strongly retarded phytoplankton development. An abundant flora, primarily of neritic pennate diatoms, developed along the ice edge in late spring. The ice also supported and transported diatom spores which Hart thought were important in maintaining the rich Antarctic flora. Other ways in which the ice influenced the phytoplankton community included increased stability of the water column, increased viscosity, increased trihydrol, and provision of shelter from sea and swells.

Hart[31] found some meroplanktonic species were able to use the pack ice when they needed a solid substrate and could therefore flourish for short periods in the open sea. These neritic and ice edge forms were most important from the beginning to the peak of the main plankton increase. They declined rapidly after the peak and were almost absent from oceanic waters at other seasons. Hart[31] included in this group species of *Thalassiosira, Asteromphalus parvulus* Karsten, *Biddulphia striata* Karsten, *Eucampia balaustium* Castracane, *Chaetoceros flexuosum* Mangin, *C. neglectum* Karsten, *C. sociale* Lauder, *C. tortissimum* Gran, species of *Fragilaria*, and *Nitzschia closterium* (Ehrenberg) W. Smith.

Ice algae were mostly ignored by investigators in both hemispheres until the International Geophysical Year, 1957-58, brought about a renewed interest in polar regions. The development of new or improved techniques to determine plant pigment concentrations and carbon uptake (primary production) led to the beginning of experimental studies on microalgae, including those living in sea ice. However, few geographical regions have been studied, descriptive studies are still needed, and not all of the recent work has been quantitative.

III. STUDIES SINCE 1960

A. Arctic

Bursa[32] reported green-brown films of diatoms on the underside of ice turned over by an icebreaker near Baffin Island. He also reported huge quantities of detritus in the ice that were released as the ice melted and caused reduced transparency in the underlying seawater. Bursa[33] said the ice is a main limiting factor that inhibits metabolic activities for 9 to 11 months but it also provides an ecological niche for organisms living in melt ponds on the ice. The melt pond flora consisted mainly of freshwater green algae and flagellates that occurred on the ice surface as brown-green patches. These algae were said to mix with the seawater and plankton when the ice melted. In addition, the undersurface of the ice acted as a "second sea floor" where diatoms, mostly pennate species, attached.

Apollonio[34] studied the ice algae at Devon Island in the Canadian Arctic where the brown color was mainly in the undersurface of the ice. According to this author, the ice algae were little studied because they are so difficult to collect. This is still a common complaint 20 years later. The ice at the Devon Island site was about 165 cm thick with 20 to 24 cm of snow cover; the ice algal layer was 3 to 4 cm thick and the chlorophyll *a* content ranged from 45 to 150 $\mu g\ \ell^{-1}$* during the first half of June. Phosphate concentrations in melted bottom ice were 1 to 3 μg-at ℓ^{-1} and in the middle of the ice were only about 0.5 μg-at ℓ^{-1}.

After the snow cover was artificially cleared from an area, chlorophyll *a* levels averaged 16 $\mu g\ \ell^{-1}$, or about one fifth the amount when snow was present. During and after snow melt, the chlorophyll *a* concentration dropped to 5 to 27 $\mu g\ \ell^{-1}$. This suggested that light was a major factor in determining production of the ice algae with too much light causing photoinhibition when snow was not present. When snow was present, about 90% of the total radiation was reflected from the snow surface. Apollonio[34] measured 3600 fc (~756 $\mu E\ m^{-2}\ sec^{-1}$) at the snow surface, but only 50 fc (~11 $\mu E\ m^{-2}\ sec^{-1}$) beneath 23 cm of snow or about half the amount of light necessary for the growth of phytoplankton in temperate regions. Thus, the ice algae are adapted for growth under reduced light conditions.

Apollonio[34,35] thought the high concentration of cells on the bottom of the ice provided favorable feeding conditions for zooplankton and in fact, he found small amphipods clustered on the bottom of one of his ice cores. He thought it was unlikely that much of the ice algal community was assimilated into higher trophic levels because it was found at high concentrations throughout the spring, it disappeared as a result of high light intensities, and there was no quantitative information on its utilization. No chlorophyll *a* was found in the water beneath the ice when there was no visible layer of ice algae, so it was not possible to determine whether the ice algae dropped off the ice or remained attached to it, but in a bleached state. The ice algal bloom was clearly separated in time from the normal phytoplankton bloom, however. The extent of the ice algal bloom was not known, but it probably occurred on all 1-year-old ice at the periphery of the Arctic Ocean. Apollonio[34] estimated the total standing crop of ice algae based on the average chlorophyll *a* concentration during the spring. If there are $16 \times 10^6\ km^2$ of sea ice, minus $14 \times 10^6\ km^2$ for which there is no information on the extent of the ice algae, then a standing crop of 25×10^6 kg of chlorophyll *a* is present for 2 months in regions of known production. Using a C:chl *a* ratio of 30, then the ice algae produce a minimum standing crop of 750×10^6 kg C $year^{-1}$.[34]

With regard to grazing on the ice, Mohr and Tibbs[36] found no evidence of amphipods obtaining food from pack ice and they reported cores of ice to be "essentially devoid of microscopic organisms". This is in contrast to the findings of English[37] who reported films of organic material on the bottom and sides of ice floes and amphipods actively grazing in these areas. However, he thought these films probably occurred infrequently and were not very important in the overall Arctic ecosystem.

Meguro et al.,[38,39] in studies done offshore in the Chukchi Sea near Barrow, Alaska, suggested that diatoms were frozen into the ice when it formed in the fall and survived during the winter by forming cysts that started to grow in the spring as soon as light was available. As the light increased, the snow and upper part of the ice disintegrated and became detached from the ice. This release of the ice diatoms from the ice resulted in the sudden blooming of phytoplankton in the seawater. These authors believed that no grazing occurred in the ice community and therefore the net ice production was released into the water column, perhaps inducing local blooming of zooplankton near the ice. This, in turn, attracted sea birds and mammals.

Nutrients in the ice are provided by microbial regeneration in the ice, penetration of

* These are low values being μg per core sample rather than $\mu g\ \ell^{-1}$; the values were later corrected[35] and should be 65 to 645 $\mu g\ \ell^{-1}$.

nutrients from seawater under the ice, and from brine drainage within the ice. The latter is considered the most important because the upper part of the ice can act as a reservoir, gradually supplying nutrients to lower layers as the ice temperature increases.[38,39]

Allen[40,41] suggested heterotrophy as a possible mechanism by which ice diatom cells could survive during the winter. This theory had also been suggested by Rodhe[42] for plankton in water under the ice in subarctic lakes, and by Wilce[43] for sublittoral seaweeds living under Arctic ice.

The most extensive study of the ice algal community and the first in the Arctic to follow the seasonal cycle from fall through winter to ice breakup in spring was that of Horner, Alexander, and their co-workers[44-50] at Barrow, Alaska. These investigators described the ice community in more detail and reported the presence of dinoflagellates, green flagellates, chrysophytes, and euglenoids, as well as diatoms, along with a meiofaunal community consisting of ciliates, turbellarians, nematodes, polychaete larvae, and copepods.[44] They also thought heterotrophy was not important for winter survival.[44] During their studies, it became evident that an *in situ* method was needed to measure productivity in order to prevent physiological damage to cells when they were brought into relatively cold, bright surface conditions when surface coring techniques were used for sampling. They devised a combination sampler-incubation chamber that could be placed in the ice by scuba divers.[45] Inoculation with ^{14}C could be done underwater and cells were not brought to the surface until after the incubation period. The same chambers were also used to collect ice samples for pigment, cell concentration, nutrient, and salinity determinations.

Although cells were present in the ice from the time it formed in the fall, microflagellates were more abundant than pennate diatoms, and the cells were distributed throughout the ice thickness during winter. Measurable metabolic activity did not begin until late March when light reached approximately 2 to 9 $\mu E\ m^{-2}\ sec^{-1}$. Cells concentrated on the bottom of the ice where light levels were optimum and by mid-April, a yellow-brown layer was visible on the undersurface of the ice. Maximum production occurred in late May at light levels of 16 to 65 $\mu E\ m^{-2}\ sec^{-1}$. By early June, the bottom layer of ice was soft and the ice algae were gone from the ice.[46,49] A sharp productivity peak occurred in the water column at this time, but it was short-lived and the dominant species, *Nitzschia grunowii* Hasle, may be common in both the ice and water column. The spring bloom in the water column did not take place until after ice breakup and was composed of different species.

Grant and Horner[50] showed that ice diatoms were capable of growing over a wide range of salinities from 5 to 60‰, with similar growth rates from 10 to 50‰. Greater mortality and longer lag periods occurred at lower salinities, indicating that rapid changes to hyposaline water were more detrimental than changes to hypersaline water. This is not unreasonable because the ice algae are probably living at higher, rather than lower, salinities in the ice.

Studies in the western Beaufort Sea[51] using a redesigned combination incubation chamber-sampler,[52] showed the ice algae providing about two thirds of the spring primary production. As at Barrow, cells were present in the ice throughout the winter, with microflagellates being the most abundant organisms. Cells concentrated at the bottom of the ice in March to April, diatoms became more abundant, and productivity began to increase when the light level was near 8 $\mu E\ m^{-2}\ sec^{-1}$. Maximum production occurred in late May when light levels were 10 to 25 $\mu E\ m^{-2}\ sec^{-1}$. By mid-June, the ice algae were gone from the ice. Species composition was similar to that at Barrow and in other Arctic areas. One major difference from the Barrow study, however, was the finding of a layer of healthy, active cells at the bottom of the ice in November.[53] Species composition and cell concentrations were similar to those found in April.

In the Canadian Arctic, ice algal production in Resolute Bay was considered to be negligible because of high snow cover, but outside the bay, production of this community was higher because snow cover was less.[54] Ice algal production began soon after light was available and continued until early June when ice melt began. Production of the ice algal community was estimated to be similar to that of the shallow water benthos or about 48 g C m^{-2} $year^{-1}$, for inshore and offshore areas.

In Frobisher Bay, nitrate and phosphate accumulated in the sea ice during ice development in winter.[55] The nutrients probably originated in the water when the ice formed and were at least partially utilized by the ice algae in late winter and spring. Chlorophyll *a* also increased in the ice during late winter and spring due to cell division and growth of the ice algae.[56] When the ice melted, nutrients were added to the water column, but probably did not exceed 5% of the total nutrient concentration available in the water. About twice as much chlorophyll *a* as was present in the water column was added to the water when the ice melted. Grainger[56] did not positively connect the release of the ice algae into the water column with the occurrence of the spring bloom in the water column, but he did suggest that the ice algae formed part of an important trophic web. Although rates of photosynthesis were not measured, Grainger[56] calculated production to be in the range of 1 to 10 g C m^{-2} $year^{-1}$ and suggested that ice algal production might be about 10% of the annual phytoplankton production in the water column.

Dunbar and Acreman[57] compared standing crops of diatoms, chlorophyll *a*, salinity, and nutrients in ice from the high Arctic, i.e., Robeson Channel, West Barrow Strait, and West Hudson Bay, with the Gulf of St. Lawrence. Nutrient concentrations were similar in all areas, but highest salinities occurred in the bottom of the ice from the Gulf of St. Lawrence and at the top and bottom of the ice from the northern areas. Chlorophyll *a* concentrations were one to two orders of magnitude higher at the northern stations. More than 96% of the diatoms from the northern stations were benthic, pennate species, while only 57% of the diatoms from the Gulf of St. Lawrence were pennate species. They attributed these differences to differences in the ice regimes. In the Gulf of St. Lawrence, where the ice lasts only 4 to 5 months as compared with 9 to 10 months in northern regions, a special substrate community has evolved to take advantage of the new habitat. In addition, the ice in the Gulf of St. Lawrence is thinner, allowing higher light intensities to occur beneath it. This would tend to favor planktonic species, such as centric diatoms, that normally live at higher light intensities, over benthic or pennate species that normally live at lower light levels.

Also in the Canadian Arctic, Hsiao[58] determined species composition and standing stock in the Eskimo Lakes, Eclipse Sound, and Frobisher Bay, and Grainger and Hsiao[59] studied Frobisher Bay. The ice algal community started to develop at all sites in the late fall when the ice formed. In fact, a relatively large population, indicated by a chlorophyll *a* concentration of 1.18 mg m^{-3}, was found at the bottom of the ice in November in the Eskimo Lakes region.[58] Growth was slow throughout the winter, increasing rapidly in early spring, reaching a peak in late spring to early summer, and declining rapidly as the ice melted. More than 200 species were identified with diatoms being the most abundant organisms during both studies, although other organisms occasionally occurred in relatively large numbers. The dominant species in the bottom of the ice were different from the dominant species found elsewhere in the ice and standing stock was 1 to 2 orders of magnitude greater in the bottom of the ice than in the upper layers.

In southeastern Hudson Bay, Legendre et al.[60] were concerned with the mechanisms controlling phytoplankton blooms under sea ice rather than with the ice algal community itself. However, their study is important because it added information on the fate of the ice algal cells when they leave the ice. Their findings suggest that phytoplankton blooms under the ice result from the seasonal light increase which deepens the photic layer and the increased low-salinity meltwater which increases stratification. The release of the ice algae from the ice may superimpose itself on the phytoplankton bloom, thus increasing the under-ice bloom in this area.

Cross[61] reported on the under-ice biota from Pond Inlet at the northern end of Baffin Island. Distribution of the ice algae was patchy with a well-developed layer of cells present in some areas, but not in others. In some areas, the ice algae were already separating from the ice in mid to late May and a perforated brown sheet was seen floating against and adhering in places to the underside of the ice at one station. By early July, there was a marked decrease in algal density. Some 77 species of algae were recognized with diatoms, chlorophytes, chrysophytes, euglenoids, dinoflagellates, and unidentified microflagellates all being present. *Nitzschia frigida* and *N. grunowii* were codominant, occurring in 40 of 41 samples. Some ten species of pennate diatoms accounted for more than 95% of the total number of algae present. After the peak of the ice algal bloom, thin sheets of algae detached from the ice, but remained partially attached as strands and amorphous masses. High densities of meio- and macrofauna were associated with this algal detritus. Brown or white particles up to 3 mm long were abundant in the water column in June and probably caused the increase in chlorophyll *a* under the ice.

At Disko Island, West Greenland, Andersen[62] tested a surface-operated technique for measuring production by bottom-type ice algae. Blocks were cut from ice no more than 1 m thick and, without turning the ice over, clear plastic chambers were strapped against the bottom of the ice. ^{14}C was injected through a hole in the side of the chamber. Dark uptake was determined by placing the experimental ice and chamber in heavy, black plastic bags. The sample blocks were replaced in the ice to obtain *in situ* conditions. This procedure had the same possible sources of error as the *in situ* techniques of Clasby et al.[45] and Schrader et al.,[52] i.e., possible incomplete mixing of the isotope into the ice and loss of the isotope by upward diffusion into the ice above the sampling chamber. Andersen[62] estimated annual production to be about 0.15 g C m^{-2}, which is considerably less than the 5 g C m^{-2} estimated for Barrow[46] and the 0.7 g C m^{-2} calculated for the nearshore Beaufort Sea.[51] Andersen[62] attributed this low value to the water in Disko Bay being under the influence of the West Greenland Current of partly Atlantic origin, and therefore poor in true ice algae. Previously mentioned possible sources of error in the method, the relatively young age of the ice and therefore the short development time of the ice algal population, the porous nature of the ice that would allow cells to leave the ice, and the possible effects of relatively high light intensities because of the thinness of the ice, all contributed to the low productivity of the ice algae at this site.

In Davis Strait, Booth[63] determined biomass, productivity, and taxonomic composition of the ice algae during April and May. Sampling was done using surface coring techniques, but the bottom of the ice was observed and photographed during coring using a fiberscope. These observations showed that sample loss during coring was not a problem as had been reported by Alexander et al.[46] Productivity was measured by cutting cores in half lengthwise and placing the sections in semicircular clear and opaque plexiglass incubation chambers. The chambers were inoculated with ^{14}C, filled with filtered seawater, and sealed before being suspended beneath the ice with the core sections in their original orientation. The cores were not exposed to surface light conditions during these manipulations.

Ice algal distribution on the underside of the ice floes was highly patchy as determined by visual observation with the fiberscope and by the range of variation of the chlorophyll *a* concentration over a floe. Chlorophyll *a* concentrations decreased from south to north during both April and May, probably in response to variation in available light and increased distance from open water. Most of the chlorophyll was found in the bottom few centimeters of ice with an average of 0.61 mg m^{-2} in the bottom 6 cm. Primary production ranged from 0.003 to 2.36 mg C m^{-2} day^{-1}, with the maximum daily rate being 1 to 2 orders of magnitude less than those from the Antarctic or Chukchi Sea, but similar to those from Godhavn and the Bering Sea. Maximum production occurred at an average light intensity of 1.8 μE m^{-2} sec^{-1}, which was less than the 2.8 to 11.4 μE m^{-2} sec^{-1} needed to initiate production at Barrow.[46]

The dominant species of the bottom ice community were *Navicula transitans* v. *derasa* (Grunow) Cleve and *Nitzschia frigida*, both common components of the bottom ice community throughout the Arctic. In addition to the bottom ice community, an interior or middle ice community consisting primarily of centric diatoms was found once and a surface community consisting of planktonic species was also seen once. These communities have not been reported previously in the Arctic, but are conspicuous features of the Antarctic.

Apollonio[64] has proposed the existence of a "summer ice" algal community that is separated in time and may also be floristically and ecologically separated from the "spring ice" flora. The summer ice community is characteristic of broken, drifting ice that is snow-free and has some surface melt, as opposed to the spring community associated with snow-covered, unbroken ice that is not melting. Grainger[55] first distinguished between intact, stable, first-year ice and broken, drifting ice, but it was Sutherland[2] who first reported a brown, mucilaginous slime on the ice floes or close to the ice in summer. Apollonio[64] believes that the brown coloration often seen in summer on the undersides of ice floes broken and turned over by icebreakers provides additional evidence for the occurrence of a summer ice flora. The summer ice flora is characterized by the high abundance of the centric diatom *Melosira arctica* that forms large, mucilaginous masses, in contrast with the spring ice flora that is largely composed of pennate diatoms.

In the Barents Sea, Mordasova and Fedosov[65] measured chlorophyll and phaeophytin concentrations in ice and water. They found up to 8 mg chlorophyll *a* ℓ^{-1} in "plate drift ice"; an average of 5.75 mg ℓ^{-1} in the lower part of the ice floes, and 4.79 mg ℓ^{-1} in the upper part of ice floes. Of this, chlorophyll *a* averaged 45.4%, chlorophyll *c* 43.5%, and chlorophyll *b* an insignificant 11.1%. Phaeophytin in the ice was 20.1% of the chlorophyll *a* content.

Ice algae are also an important component of the ecosystem in the Bering and southern Chukchi seas. McRoy et al.[66] and McRoy and Goering[67] were the first to study production of ice algae in the Bering Sea where they considered the development of this community to be a feature of late winter to early spring, rather than a continuous occurrence throughout the winter. Ice algal production was found to be 2.2 to 4.8 mg C m^{-2} day^{-1}, with standing stock of 0.34 to 2.97 mg chlorophyll *a* m^{-2}; maximum standing stock occurred just before the disappearance of the ice. These authors believed that the ice algae played an important role in the food webs of sea birds and mammals in the Bering Sea. McRoy and Goering[68] also proposed an annual budget for primary production for the Bering Sea. They said there was still only a fragmentary picture of ice algal development and indicated that productivity of this community was difficult to measure because the cells are temperature and light sensitive and sampling through the ice could damage the cells. Based on their previous estimates of production and a production season of 61 days, they calculated annual ice algal production to be 24 $\times$ 10^6 metric tons of carbon or 9% of the total Bering Sea production.

Alexander and Chapman[69] studied the role of ice communities in the Bering Sea to determine timing, intensity, patchiness, and duration of the ice algal bloom. They wanted to assess the contribution of these communities to the total Bering Sea ecosystem. Chlorophyll *a* from 60 stations in the area between Nome and St. Lawrence Island was patchy on both general and local scales with values ranging from 0 to 213 mg chlorophyll *a* m^{-3}. Closer to the ice edge, they found photosynthesis not inhibited at full daylight intensities and said that by the time the ice breaks up, the ice algae are adapted to surface light and are capable of colonizing the water at the ice edge. This suggests that the ice algae are important in initiating or at least contributing to, the bloom in the water column at the ice edge. Additional support for this hypothesis was the finding of many of the same species in the ice and water column.[69,70] Assuming a 100-day growing season, Alexander and Chapman[69] estimated the ice algae would contribute 0.22 to 1.50 g C m^{-2} $year^{-1}$. Although this may be an underestimate because the ice algal layer may be more than 5 cm thick, the ice algae probably contribute less than 1% of the annual production in the southeastern Bering Sea.

In the northern Bering and southern Chukchi seas, Saito and Taniguchi[71] proposed a scheme of seasonal succession. In late winter to early spring when the ice melts, pennate diatoms are released into the water column. After the ice cover disappears, spring diatoms, such as *Thalassiosira* spp., form an extensive bloom in the shallow surface layer. Both ice diatoms and spring bloom species are thus present in the water. In late spring, the ice diatoms sink, but the spring species remain as a bloom in the surface water. In summer, the spring species are replaced in the surface water by summer species, such as *Chaetoceros* spp. and dinoflagellates. The spring diatoms sink to the bottom and disintegrate, but some of the ice diatoms persist as a patchy layer at depth.

Ice algae have been reported from a number of other areas in the northern hemisphere. *Nitzschia frigida,* a common and often dominant species from ice off northern Alaska and Canada, and recently reported from the Antarctic,[72] was found in ice collected from a Danish fjord by Grøntved,[73] who also gave distribution maps for this species. It is apparently a common, neritic species in the inner Baltic. It is considered to be euryhaline, but stenothermal, not forming dense concentrations at temperatures greater than 3°C.

On the coast of Finland, Häyren[74] reported *Hormidium flaccidum* (Kützing) A. Braun, a green alga, in meltwater on the ice close to shore and a brown color has been reported by Niemi[75] on the underside of ice near Tvärminne. Hickel[76] found a brown-colored layer of pennate diatoms attached to the bottom of ice in Eckenförder Bucht in February and March. He said these cells were "obviously consuming the soluble phosphate of the topmost water layer."

In Japan, Takahashi[77] reported on a floristic study of algae found in sea ice from Lake Saroma, Hokkaido, a lagoon that is connected with the Sea of Okhotsk. Included in 44 taxa from the ice were pennate and centric diatoms, green algae, cryptomonads, dinoflagellates, euglenoids, chrysophytes, craspedomonads, prymnesiophytes, and prasinophytes. Of these, 23 taxa were found only when samples were examined with an electron microscope. Of the cells, 95% were found in the bottom part and only 5% in the upper layers of the ice. Diatoms were the dominant component of the ice community.

At Vostok Bay, Sea of Japan, Propp[78] found a massive growth of microalgae in the bottom of sea ice a month before the phytoplankton bloom in the water column. He thought the total contribution of the ice algae to productive and metabolic processes was small because the ice is thin and occurs over a limited area, but he also said this was one source from which the winter growth of phytoplankton spreads.

Kuznetsov[79] measured chlorophyll and primary production associated with ice in Amur Bay, the Sea of Japan. He said that "ice in moderate latitudes causes an enhancement of productivity representing a substrate for algae and bacteria and that ice thawing stimulates photosynthesis in phytoplanktonic organisms enriching surface waters with biogens." The community consisted primarily of pennate diatoms, especially *Amphiprora paludosa* W. Smith and *Nitzschia* sp., but the silicoflagellate *Distephanus speculum* (Ehrenberg) Haeckel occurred occasionally. Distribution of cells in the ice was not uniform with 1×10^6 cells ℓ^{-1} occurring in the bottom layer and 2.5×10^5 cells ℓ^{-1} occurring in the upper layers. Kuznetsov[79] thought the algae in the upper layers were frozen during ice accumulation which curtailed their metabolic activity. This was indicated by the higher concentration of phaeophytin in the upper part of the ice. Primary production was determined by oxygen production. Ice cores were enclosed in polyethylene bags that were apparently incubated *in situ.* Dissolved oxygen was collected every 8 hr over a 2-day period. In February, production was low, within the limits of the method, but in March when light increased, production was 193 mg C m^{-3} day^{-1}.

B. Antarctic

Bunt and Wood[80] first used the term epontic in referring to diatom communities found on the bottom of sea ice and pointed out that this community was clearly different from the planktonic community. The epontic community consisted of attached species of *Pleurosigma, Nitzschia, Amphiprora,* and *Fragilaria,* as well as of nonattached species living with them as part of that community, including species of *Biddulphia, Coscinodiscus,* and *Asteromphalus.* The epontic community was thought to be important because of its contribution to primary production in any region where it occurs and because it may significantly influence phytoplankton development by reducing light levels under the ice.

The first description of the physical environment of the epontic community was provided by Bunt,[81] who also obtained the first estimates of the concentration of organisms in the ice and their productivity. He found cell concentrations in seepage water ranging from $4.5 \times 10^6\ \ell^{-1}$ in mid-December to $16.8 \times 10^6\ \ell^{-1}$ in early January, with chlorophyll *a* values ranging from 45 to 156 mg m^{-3} over the same time period. These numbers are for seepage (interstitial) water only and do not include values for cells actually attached to ice crystals. At the time of peak chlorophyll *a* concentration in the seepage water, however, the chlorophyll *a* concentration of melted ice crystals was 617 mg m^{-3}. Bunt[81] believed that most of the growth of the ice diatoms was a result of photosynthesis even though light intensity where he sampled was only about 8 fc (~2 μE m^{-2} sec^{-1}).

In a series of papers, Bunt and his co-workers described methods used[82] and reported on seasonal primary production, ecology, and physiology of ice algae from the brash ice zone at McMurdo Sound.[83-89] Bunt[83] thought it was possible that after an initial development period in the ice, certain diatoms might be successful components of the plankton. Although most ice species were considered to be sessile forms and would tend to settle lower in the water column, the planktonic species would be expected to succeed in the plankton. Primary production was light saturated at about 100 fc (~22 μE m^{-2} sec^{-1}) and inhibition occurred at 1100 fc (~240 μE m^{-2} sec^{-1}), indicating extreme shade adaptation.[84] Net carbon fixation was temperature dependent over the range −1.5 to +5°C, with maximum uptake at 10 to 15°C and rapid decline above 15°C.

Bunt,[85] working primarily with *Fragilaria sublinearis* Van Heurck, showed that the temperature most favored for growth was higher than that of the natural habitat regardless of light intensity and suggested that *in situ* conditions must be near the limit for growth. He also showed that chlorophyll *a* per 10^6 cells varied under different conditions and cell carbon varied with temperature and light. In these studies, μg C per 10^6 cells ranged from 90 to 124 and the C:chl *a* ratio varied from 20 to 60 with a mean of 30. This suggests that there is more certainty in predicting cell carbon levels from cell counts than from chlorophyll *a* values.[85] For *F. sublinearis,* the greatest C:chl ratio was at −2°C. Cell volumes can also change with temperature, so that over the temperature range of −2 to +7°C, cell volume was greater at −2°C by a factor of 1.3.

In a series of culture experiments designed to determine dark survival, Bunt and Lee[86] found a mean C:chl ratio of 37 for the four test organisms. They found no growth in controls or cultures containing organics and concluded that there was probably no heterotrophic growth for these species. These authors further suggested that seeding of the ice should be possible with cells exposed to continuous darkness for at least 3 months. Additional support for this theory was provided by Bunt et al.,[87] who found respiratory oxygen consumption in *F. sublinearis* to be depressed at 3°C as compared with 10°C and they assumed respiration would be further depressed at −1.8°C.

Bunt and Lee[88] described seasonal primary production and found production to be limited by the amount of light available to the cells in the brash ice on the undersurface of the ice. The brash ice consists of a layer of unconsolidated crystals with a maze of interconnected small water channels that are continuous with the brine of the hard ice above and the water column below. The ice algae live in the interstitial water between ice crystals, as well as attached to the ice crystals, which makes sampling and study of the diatoms extremely difficult. Cell numbers in the interstitial water and attached to ice crystals increased from mid-July until early December with a more or less constant population from late August through October and a sharp increase from early November to early December with a peak of $>10^6$ cells ℓ^{-1}. The July to September increase may have been caused by cells in the water being carried up to the brash layer on ice crystals formed as a result of supercooling of part of the water column. Chlorophyll *a* concentrations followed the same time course as cell numbers. Near the peak of the growing season in November to December, organic carbon in the ice ranged from 1076 mg m^{-2} at a shaded station to 520 mg m^{-2} at a well lighted station where growth was probably light inhibited. This was considered to be net annual production because there was little detritus and little or no grazing. Some grazing may occur, however, because Bunt and Lee[88] found pteropods swimming just under the ice in winter and small fish with their heads up in the ice apparently feeding.

Bunt[89] reviewed some of the data then available on ice algae. He thought it was reasonable to expect that ice algae might provide an inoculum to the water column if the ice broke up early in the season. He further considered that primary production in the water might be related to the potential of the ice layer to be a center of development for photosynthetic organisms. He questioned the distinction between the phytoplankton and the ice algae, suggesting that more information was needed on changes in taxonomic composition and relative abundance of individual species in the water column and in the ice. As possible sources of ice algae, Bunt[89] considered benthic species being trapped in ice as ice crystals rise from the bottom and cells associated with ice crystals formed in the water column. In addition, cells that are able to remain viable in the dark for long periods of time may be important in initiating the ice algal bloom.

Meguro[90] studied what he called "plankton ice." This is a layer 15 to 20 cm thick consisting of snow that has been infiltrated with seawater containing algal cells. It is located beneath a 30 to 40 cm thick layer of snow and is just above and just below sea level. Sea ice, 50 to 100 cm thick, occurs below the plankton ice. In summer, the cells respond to increased light and grow. As more growth takes place, warming occurs from light absorption by the cells and the plankton layer melts, while the snow and ice layers above and below it remain frozen. This causes the snow and ice layers to separate and the cells in the plankton layer to be released into the water column. Zooplankton may be present in the holes formed by the decaying plankton layer.

This surface plankton layer was considered to be a good biological environment[90] because the temperature is stable at approximately $-1.7°C$ and the snow protects the organisms from colder air temperatures. In addition, the layer is probably oxygen saturated, nutrients are supplied by the penetrating seawater, and plenty of light is available to the cells. The plankton layer was not an accidental or local phenomenon, but was present over a wide area and a large part of the photosynthesis of the pack ice region occurred in the ice.

Bacteria were also found in the sea ice[91] and it was thought that they might be "ecologically connected" with the diatoms. The dominant diatom species in the plankton ice was *Fragilariopsis cylindrus* (Grunow) Krieger.

Fukushima and Meguro[92] believed that primary production in the Antarctic depended on the formation of plankton ice, growth of plankton in sea ice, and the release of plankton into the water column when the ice melted. They used the term "plankton ice" to refer to the brown layer located at either the snow-ice or ice-water interface. The bottom-type plankton ice contained 100 times as much biomass based on diatom volume as the surface-type plankton ice and 560 times as much as the seawater. The dominant ice species were also found in the water column and these authors suggested that cells produced in the ice formed the larger part of the plankton.

Hoshiai and Kato[93] studied ice collected at Lützow-Holm Bay and found flagellates and ciliates in addition to diatoms, but only studied the diatoms. They found eight communities vertically distributed in the ice. The majority of species were the same in the ice and water, but the ice community might change with time.

At Syowa Station, Hoshiai[94-96] studied the seasonal growth and succession of the ice community. Freezing began in mid-February and by late March, a brown layer was present on the bottom of the ice. The autumn layer was apparently a normal event and developed on the bottom of the ice by a process similar to the growth of the spring ice community. Ice continued to grow in winter below the autumn brown layer. Ice growth slowed after mid-September and a second, or spring, brown layer formed at the bottom of the ice in November and remained through December. Chlorophyll *a* peaks were present in late April with approximately 30 mg m^{-2} and in late October to early November with approximately 35 mg m^{-2}. Assuming a C:chl ratio of 50, Hoshiai[96] calculated that the ice algae provided $<10\%$ of the annual primary production in the Antarctic.

Burkholder and Mandelli,[97] studying floating brown slush ice, found large numbers of a "very small diatom", moderate numbers of *Fragilaria curta* and *Nitzschia seriata* Cleve, and small numbers of many other diatom species. They also found dinoflagellates, small green flagellates, and *Phaeocystis pouchetii* (Hariot) Lagerheim in the ice. These authors found 30 times as much chlorophyll *a* and ^{14}C uptake in the ice as in the water column with production in both the ice and water increasing at light intensities up to 18 Klux. This is in contrast to Bunt,[84] who found maximum production at 1 Klux, but Bunt[84] was working with the bottom, brash ice community that would be expected to be shade-adapted, rather than with a surface community that might not be shade-adapted. Production for microalgae trapped in the sea ice was calculated to be about 0.19 g C m^{-2} day^{-1}, or an estimated *in situ* production of approximately 0.5×10^6 tons C day^{-1} for 2.6×10^6 km^2 of sea ice.

Several distinct ice habitats were studied at Signy Island in the South Orkney Islands.[98] The most intensively studied habitat was the coastal tide-crack overflow region. This habitat occurs where the fast ice sheet meets land or a grounded iceberg. The rise and fall of the tide makes a series of parallel cracks through which seawater passes. Pools form in these cracks at spring tides. An accumulation of snow and frozen seawater depresses the ice and seawater penetrates into the slush or infiltration ice at the snow-ice interface. *Navicula glaciei* Van Heurck was the most abundant diatom in this habitat. Diatoms were first detected in late May, but there was little growth until light levels increased and heavy snowfall increased the depth of the habitat. Growth became rapid in early September and a chlorophyll *a* peak of 148 ± 89 mg m^{-2} was reached in early November; the standing crop of carbon was about 5.5 g m^{-2}. The population declined until ice breakup in early December. When air temperatures were above 0°C for long periods, low salinity meltwater penetrated the infiltration ice causing local salinities of <5‰ and diatom pigments changed from the normal yellow-brown color to pale bluish-green.

Other habitats studied in this area included the fast ice sheet, simple pack ice, offshore tide-cracks, and icebergs. All of these habitats were further subdivided into "habitat divisions". Chlorophyll concentrations in these areas varied, as did species composition. Grazing by adults and juveniles of the amphipod *Pontogeneia antarctica* Chevreux occurred in tide-cracks in fast ice.[98]

High diatom concentrations in the tide-crack overflow habitats were thought to be a result of low loss rates rather than of high carbon fixation rates, although ciliates were found to be feeding on ice diatoms late in the season. Major population loss was associated with periods of melting and habitat destruction. Maximum carbon fixation of the inshore ice habitats at Signy Island was estimated to be 1 to 2 g m^{-2} $year^{-1}$,[98] a relatively insignificant amount when compared with the minimum inshore phytoplankton productivity of 80 to 130 g m^{-2} $year^{-1}$ estimated by Horne et al.[99]

In the Weddell Sea, Ackley et al.[100-102] studied ice properties and ice-associated algae and suggested a model for the formation of an interior-type ice algal community. The ice community in the Weddell Sea region is located at intermediate depths within the ice rather than at the ice surface or bottom and has been termed the interior community.[100-102] This community depends on air temperatures at or slightly below freezing to initiate, but not complete, brine drainage. Release of the algae from the interior community occurs along the open ocean-sea ice interface where floes are disintegrating and is a continuous process as opposed to the pulsed release of cells from surface and bottom communities that occur when the ice breaks up. Fluorescence measurements indicated the interior ice algal community was strongly associated with salinity maxima within the ice. Maximum chlorophyll *a* concentration ranged from 0.31 to 4.54 mg m^{-2}.

Garrison and Buck,[103,104] and Ackley[105] have suggested that physical, rather than biological, factors are responsible for the abundance of algae in sea ice. Garrison and Buck[103] found silicate and nitrate concentrations in newly formed ice to be correlated with salinity rather than biomass which suggested to them that accumulation rather than growth was responsible for higher chlorophyll *a* concentrations in ice than in seawater. Chlorophyll *a* levels were 2 to 60 times higher in ice than in the water column and the same species were found in both ice and seawater which Buck and Garrison[104] suggested were evidence for physical accumulation by ice particle scavenging. Growth of the ice over short time scales (approximately hr) suggested that biological factors alone could not account for the increased concentration of algae in the ice as opposed to the water column. Concentration of cells by nucleation can occur in water at its freezing point when a frazil ice crystal nucleates on an algal cell. In addition, scavenging can occur when ice crystals traveling upward in the water column collide with algal cells and more than one algal cell can be collected by each ice crystal. Both these mechanisms are in operation because microscopic examination of thin sections of ice show algal cells within ice crystals (nucleation) and between crystals (scavenging). Scavenging was considered to be more important because of the small size of the ice crystals and the concentrations of algal cells found in the ice.[105]

Another kind of bottom ice community has been reported from east Antarctica.[72] This community, found at the Australian stations of Casey, Davis, and Mawson, developed during September and consisted of strands of diatoms suspended from the bottom surface of annual sea ice. The strands, often extending up to 15 cm into the water column, were dominated by an unidentified tube-forming diatom of the *Amphipleura/Berkeleya* group and chains of *Entomoneis* sp. Cell densities in the strands ranged from 10^6 to 10^7 cells $m\ell^{-1}$. No brash ice layer was present at these sites and a typical bottom-type ice community, consisting primarily of *Nitzschia frigida* and *Entomoneis* spp., occurred in the bottom 3 cm of ice with cell densities reaching concentrations of 3×10^5 cells $m\ell^{-1}$. Maximum production rates of 81 μg C ℓ^{-1} hr^{-1} occurred in November and minimal rates of 2.8 μg C ℓ^{-1} hr^{-1} occurred in mid-January. The phytoplankton bloom in the water column occurred at the same time as the breakout of the fast ice and was dominated by *Phaeocystis* sp. Production reached 5.8 μg C ℓ^{-1} hr^{-1} in early December.

As the bottom ice community disintegrated, a surface type community developed in melt pools located just below the surface of consolidated snow. The melt pools formed by thawing in a layer up to 15 cm deep that occurred 10 cm below the surface. This community is not the same as the snow community described by Meguro[90] because it occurs on fast ice with insufficient snow cover to depress the ice layer and the algae develop in melt pools within the sea ice rather than in the snow layer that has been infiltrated by seawater. The community was dominated by small diatoms, flagellates, and *Phaeocystis* sp. that were apparently transported up brine channels to the melt pools by the flow of brine and seawater during brine drainage and flushing. Large aggregates of cells formed on the bottom of the pools by mid-January and productivity within the aggregates reached 363 μg C ℓ^{-1} hr^{-1}.[72]

Bacteria were also found to be common in the bottom ice community. They were attached to the mucilage sheath of *Amphipleura* sp. and to the valves of *Entomoneis* spp. Bacterial densities ranged from 4×10^5 to 2×10^7 cells $m\ell^{-1}$ and reached 8×10^6 cells cm^{-2} on the mucilage sheath of *Amphipleura* sp. Turnover times of ^{3}H-labeled amino acids suggested that appreciable heterotrophic activity occurred, associated mainly with size fractions greater than 2 μm. Heterotrophic activity of the 0.2 to 2 μm size fraction was only 1 to 6%, suggesting that small, free-living bacteria were not very important in recycling organic material in this community.[72]

IV. CONCLUSIONS

We are left with a number of unanswered questions concerning the growth, physiology, and ecology of ice algae, including questions about growth rates, dark survival, salinity tolerances, light requirements, species succession during ice algal blooms, vertical and horizontal distributions of organisms in the ice, nutrient concentrations in the ice, the presence of meio- and epifauna and the effects of grazing, as well as questions on the origin and fate of the ice algal cells. Some of these questions will be answered in other chapters, some will have to wait for future research.

REFERENCES

1. **Hooker, J. D.,** *The Botany of the Antarctic Voyage of H. M. Discovery Ships Erebus and Terror in the Years 1839-1843,* Vol. 1, J. Cramer, Weinheim, 1847 (reprint 1963), 503.
2. **Sutherland, P. C., Ed.,** *Journal of a Voyage in Baffin's Bay and Barrow Straits, in the Years 1850-1851, Performed by H. M. Ships "Lady Franklin" and "Sophia," Under The Command of Mr. William Penny, in Search of the Missing Crews of H. M. Ships "Erebus" and "Terror,"* Vol. 1 and 2, Longman, Brown, Green, and Longmans, London, 1852.
3. **Dickie, G.,** Notes on the algae, in *Journal of a voyage in Baffin's Bay and Barrow Straits, in the years 1850-1851, performed by H. M. ships "Lady Franklin" and "Sophia," under the command of Mr. William Penny, in search of the missing crews of H. M. ships "Erebus" and "Terror,"* Vol. 2, Sutherland, P. C., Ed., Longman, Brown, Green, and Longmans, London, 1852, cxci.
4. **Cleve, P. T.,** Diatoms collected during the expedition of the *Vega, Vega-Exped. Vetenskap. Iakttag.,* 3, 455, 1883.
5. **Gran, H. H.,** Diatomaceae from the ice-floes and plankton of the Arctic Ocean, *Sci. Res. Norw. N. Polar Exped.,* 4(11), 3, 1904.
6. **Nansen, F.,** *Farthest North,* Vol. 1, Harper and Brothers, New York, 1897.
7. **Nansen, F.,** Protozoa on the ice-floes of the North Polar Sea, *Sci. Res. Norw. N. Polar Exped.,* 5(16), 1, 1906.
8. **Ehrenberg, C. G.,** Einen Nachtrag zu dem Vortrage über Verbreitung und Einfluss des mikroskopischen Lebens in Süd- und Nord-Amerika, *Akad. Wiss. Berlin Monatsber.,* 1841, 202, 1841.
9. **Ehrenberg, C. G.,** Über neue Anschauungen des kleinsten nördlichen Polarlebens, *Akad. Wiss. Berlin Monatsber.,* 1853, 522, 1853.
10. **Dickie, G.,** On the algae found during the Arctic Expedition, *J. Linn. Soc. London Bot.,* 17, 6, 1880.
11. **Cleve, P. T. and Grunow, A.,** Beiträge zur Kenntniss der arctischen Diatomeen, *K. Sven. Vetenskapsakad. Handl.,* 17(2), 1, 1880.
12. **Cleve, P. T.,** Diatoms from Baffins Bay and Davis Strait, *Bih. K. Sven. Vetenskapsakad. Handl.,* 22(4), 1, 1896.
13. **Cleve, P. T.,** Diatoms from Franz Joseph-Land collected by the Jackson - Harmsworth Expedition, *Bih. K. Sven. Vetenskapsakad. Handl.,* 24(2), 3, 1898.
14. **Cleve, P. T.,** Mikroskopisk undersökning af stoft, funnet på drifis i Ishafvet, *Oefv. K. Vetenskapsakad. Foerh.,* 1899(3), 123, 1899.
15. **Cleve, P. T.,** Microscopical examination of dust from drift-ice north of Jan Mayen, *Oefv. K. Vetenskapsakad. Foerh.,* 1900(4), 393, 1900.
16. **Grunow, A.,** Die Diatomeen von Franz Josephs Land, *Akad. Wiss. Wien Denkschr. Math. -Naturwiss. Kl. B,* 48, 53, 1884.
17. **Østrup, E.,** Marine Diatomeer fra Østgrønland, *Medd. Groenl.,* 18, 395, 1895.
18. **Vanhöffen, E.,** Frühlingsleben in Nordgrönland, *Verh. Ges. Erdk. Berl.,* 1893. (Reference not seen, cited from Hooker[1].)

19. **Vanhöffen, E.,** Fauna und Flora Grönlands, *Gronland-Exped. Ges. Erdk. Berl., 1891-1893,* B, II, 1897. (Reference not seen, cited from Hooker[1].)
20. **Gran, H. H.,** Bacillariaceae vom kleinen Karajakfjord, *Bibl. Bot.,* 42, 13, 1897.
21. **Meister, F.,** Mission du "Porquoi-Pas?" en 1929 sous le Commandement du Dr. J.-B. Charcot. Diatomes récoltées par R.-Ph. Dollfus un glace flottante, *Bull. Mus. Nat. Hist. Nat.,* 2nd ser., 2, 329, 1930.
22. **Usachev, P. I.,** Biological analysis of ice-floes, *C. R. Dokl. Acad. Sci. URSS,* 19, 645, 1938.
23. **Usachev, P. I.,** The microflora of polar ice, *Tr. Inst. Okeanol., Akad. Nauk SSSR,* 3, 216, 1949 (in Russian).
24. **El-Sayed, S. Z.,** Biological aspects of the pack ice ecosystem, in *Proc. Symposium on Antarctic Ice and Water Masses,* Deacon, Sir G., Ed., Scientific Committee on Antarctic Research, W. Heffer & Sons Ltd., Cambridge, 1971, 35.
25. **Drygalski, E.,** *Zum Kontinent des eisigen Suden,* G. Reimer, Berlin, 1904. (Reference not seen, cited from El-Sayed[24].)
26. **Heiden, H. and Kolbe, R. W.,** Die marinen Diatomeen der deutschen Südpolar - Expedition 1901-1903, *Dtsch. Süd-Polar-Exped. 1901-1903,* 8(5), 450, 1928.
27. **Vanhöffen, E.,** Biologischer Bericht, in *Die Deutsche Südpolar-Expedition auf dem Schiff Gaus,* Drygalski, E., *Veroeff. Inst. Meeresk. Univ. Berl.,* 5, 1, 1903.
28. **Scott, R. F.,** *The Voyage of the Discovery,* Vol. 1, Smith, Elder, and Co., London, 1907, 95.
29. **Hendey, N. I.,** The plankton diatoms of the Southern Seas, *Discovery Rep.,* 16, 151, 1937.
30. **Hart, T. J.,** On the phytoplankton of the south-west Atlantic and the Bellingshausen Sea, 1929-31, *Discovery Rep.,* 8, 1, 1934.
31. **Hart, T. J.,** Phytoplankton periodicity in Antarctic surface waters, *Discovery Rep.,* 21, 261, 1942.
32. **Bursa, A. S.,** Phytoplankton of the *Calanus* expeditions in Hudson Bay, 1953 and 1954, *J. Fish. Res. Board Can.,* 18, 51, 1961.
33. **Bursa, A. S.,** Phytoplankton in coastal waters of the Arctic Ocean at Point Barrow, Alaska, *Arctic,* 16, 239, 1963.
34. **Apollonio, S.,** The chlorophyll content of Arctic sea-ice, *Arctic,* 14, 197, 1961.
35. **Apollonio, S.,** Chlorophyll in Arctic sea ice, *Arctic,* 18, 118, 1965.
36. **Mohr, J. L. and Tibbs, J.,** Ecology of ice substrates, in *Proc. Arctic Basin Symposium,* Tidewater Publ., Centreville, 1963, 245.
37. **English, T. S.,** Some biological oceanographic observations in the central North Polar Sea, Drift Station Alpha, 1957-1958, Sci. Rep. No., 15, The Arctic Institute of North America, Washington, D.C., 1961.
38. **Meguro, H., Ito, K., and Fukushima, H.,** Diatoms and the ecological conditions of their growth in sea ice in the Arctic Ocean, *Science,* 152, 1089, 1966.
39. **Meguro, H., Ito, K., and Fukushima, H.,** Ice flora (bottom type): a mechanism of primary production in polar seas and the growth of diatoms in sea ice, *Arctic,* 20, 114, 1967.
40. **Allen, M. B.,** Metabolic activities of phytoplankton associated with Arctic sea ice, Sci. Rep., R70-1, Institute of Marine Science, University of Alaska, Fairbanks, 1970.
41. **Allen, M. B.,** High latitude phytoplankton, *Ann. Rev. Ecol. Syst.,* 2, 261, 1971.
42. **Rodhe, W.,** Can plankton production proceed during winter darkness in subarctic lakes?, *Verh. Internat. Verein. Limnol.,* 12, 117, 1955.
43. **Wilce, R. T.,** Heterotrophy in Arctic sublittoral seaweeds, *Bot. Mar.,* 10, 185, 1967.
44. **Horner, R. and Alexander, V.,** Algal populations in Arctic sea ice: an investigation of heterotrophy, *Limnol. Oceanogr.,* 17, 454, 1972.
45. **Clasby, R. C., Horner, R., and Alexander, V.,** An in situ method for measuring primary productivity of Arctic sea ice algae, *J. Fish. Res. Board Can.,* 30, 835, 1973.
46. **Alexander, V., Horner, R., and Clasby, R. C.,** Metabolism of Arctic sea ice organisms, Rep. R74-4, Institute of Marine Science, University of Alaska, Fairbanks, 1974.
47. **Matheke, G. E. M. and Horner, R.,** Primary productivity of the benthic microalgae in the Chukchi Sea near Barrow, Alaska, *J. Fish. Res. Board Can.,* 31, 1779, 1974.
48. **Alexander, V.,** Primary productivity regimes of the nearshore Beaufort Sea, with reference to potential roles of ice biota, in *The Coast and Shelf of the Beaufort Sea,* Reed, J. C. and Sater, J. E., Eds., The Arctic Institute of North America, Arlington, 1974, 609.
49. **Clasby, R. C., Alexander, V., and Horner, R.,** Primary productivity of sea-ice algae, in *Assessment of the Arctic Marine Environment: Selected Topics,* Hood, D. W. and Burrell, D. C., Eds. Institute of Marine Science, University of Alaska, Fairbanks, 1976, 289.
50. **Grant, W. S. and Horner, R.,** Growth responses to salinity variation in four Arctic ice diatoms, *J. Phycol.,* 12, 180, 1976.
51. **Horner, R. and Schrader, G. C.,** Relative contributions of ice algae, phytoplankton, and benthic microalgae to primary production in nearshore regions of the Beaufort Sea, *Arctic,* 35, 485, 1982.
52. **Schrader, G. C., Horner, R., and Smith, G. F.,** An improved chamber for in situ measurement of primary productivity by sea ice algae, *Can. J. Fish. Aquat. Sci.,* 39, 522, 1982.

53. **Schell, D. M.,** personal communication, 1980.
54. **Welch, H. E. and Kalff, J.,** Marine metabolism at Resolute Bay, Northwest Territories, in *Proc. Circumpolar Conference on Northern Ecology,* Vol. 2, National Research Council of Canada, Ottawa, 1975, 69.
55. **Grainger, E. H.,** The annual nutrient cycle in sea-ice, in *Polar Oceans,* Dunbar, M. J., Ed., The Arctic Institute of North America, Calgary, 1977, 285.
56. **Grainger, E. H.,** Primary production in Frobisher Bay, Arctic Canada, in *Marine Production Mechanisms,* Dunbar, M. J., Ed., Cambridge University Press, London 1979, 9.
57. **Dunbar, M. J. and Acreman, J. C.,** Standing crops and species composition of diatoms in sea ice from Robeson Channel to the Gulf of St. Lawrence, *Ophelia,* 19, 61, 1980.
58. **Hsiao, S. I. C.,** Quantitative composition, distribution, community structure and standing stock of sea ice microalgae in the Canadian Arctic, *Arctic,* 33, 768, 1980.
59. **Grainger, E. H. and Hsiao, S. I. C.,** A study of the ice biota of Frobisher Bay, Baffin Island 1979-1981, *Can. Manuscr. Rep. Fish. Aquat. Sci.,* No. 1647, Ottawa, 1982.
60. **Legendre, L., Ingram, R. G., and Poulin, M.,** Physical control of phytoplankton production under sea ice (Manitounuk Sound, Hudson Bay), *Can. J. Fish. Aquat. Sci.,* 38, 1385, 1981.
61. **Cross, W. E.,** Under-ice biota at the Pond Inlet ice edge and in adjacent fast ice areas during spring, *Arctic,* 35, 13, 1982.
62. **Andersen, O. G. N.,** Primary production associated with sea ice at Godhavn, Disko, West Greenland, *Ophelia,* 16, 205, 1977.
63. **Booth, J. A.,** The epontic algal community of the ice edge zone and its significance to the Davis Strait ecosystem, *Arctic,* 37, 234, 1984.
64. **Apollonio, S.,** Arctic marine phototrophic systems: functions of sea ice stabilization, *Arctic,* in press.
65. **Mordasova, N. V. and Fedosov, M. V.,** On the chlorophyll and phaeophytin content of the water and ice of the Barents Sea, *Tr. Vses. Nauchno-Issled. Inst. Morsk. Rybn. Khoz. Okeanogr.,* 75, 80, 1972 (in Russian, English summary).
66. **McRoy, C. P., Goering, J. J., and Shiels, W. E.,** Studies of primary production in the Eastern Bering Sea, in *Biological Oceanography of the Northern North Pacific Ocean,* Takenouti, A. Y., et al., Eds., Idemitsou Shoten, Tokyo, 1972, 199.
67. **McRoy, C. P. and Goering, J. J.,** The influence of ice on the primary productivity of the Bering Sea, in *Oceanography of the Bering Sea with Emphasis on Renewable Resources,* Hood, D. W. and Kelley, E. J., Eds., Institute of Marine Science, University of Alaska, Fairbanks, 1974, 403.
68. **McRoy, C. P. and Goering, J. J.,** Annual budget of primary production in the Bering Sea, *Mar. Sci. Commun.,* 2, 255, 1976.
69. **Alexander, V. and Chapman, T.,** The role of epontic algal communities in Bering Sea ice, in *The Eastern Bering Sea Shelf: Oceanography and Resources,* Vol. 2, Hood, D. W. and Calder, J. A., Eds., University of Washington Press, Seattle, 1981, 773.
70. **Schandelmeier, L. and Alexander, V.,** An analysis of the influence of ice on spring phytoplankton population structure in the southeast Bering Sea, *Limnol. Oceanogr.,* 26, 935, 1981.
71. **Saito, K. and Taniguchi, A.,** Phytoplankton communities in the Bering Sea and adjacent seas. II. Spring and summer communities in seasonally ice-covered areas, *Astarte,* 11, 27, 1978.
72. **McConville, M. J. and Wetherbee, R.,** The bottom-ice microalgal community from annual ice in the inshore waters of east Antarctica, *J. Phycol.,* 19, 431, 1983.
73. **Grøntved, J.,** Phytoplankton studies. I. *Nitzschia frigida* Grun., an Arctic-Inner-Baltic diatom found in Danish waters, *K. Dan. Vidensk. Selsk. Biol. Medd.,* 18(12), 1, 1950.
74. **Häyren, E.,** Zwei Notizen über das Meereseis und die Algen, *Memo. Soc. Fauna Flora Fenn.,* 5, 134, 1929.
75. **Niemi, Å.,** Ecology of phytoplankton in the Tvärminne area, SW coast of Finland. I. Dynamics of hydrography, nutrients, chlorophyll *a* and phytoplankton, *Acta Bot. Fenn.,* 100, 1, 1973.
76. **Hickel, W.,** Planktologische und hydrographisch-chemische Untersuchungen in der Eckenförder Bucht (westliche Ostsee) wahrend und nach der Vereisung im extrem kalten Winter 1962/1963, *Helgol. Wiss. Meeresunters.,* 19, 318, 1969.
77. **Takahashi, E.,** Floristic study of ice algae in the sea ice of a lagoon, Lake Saroma, Hokkaido, *Mem. Nat. Inst. Polar Res. Ser. E,* 34, 49, 1981.
78. **Propp, M. V.,** Exchange of energy, nitrogen and phosphorus between water, bottom and ice in a near-shore ecosystem of the sea of Japan, *Helgol. Wiss. Meeresunters.,* 30, 598, 1977.
79. **Kuznetsov, L. L.,** Chlorophylls and primary production associated with ice in Amur Bay, Sea of Japan, *Sov. J. Mar. Biol.,* 6, 297, 1980.
80. **Bunt, J. S. and Wood, E. J. F.,** Microalgae and Antarctic sea-ice, *Nature (London),* 199, 1254, 1963.
81. **Bunt, J. S.,** Diatoms of Antarctic sea-ice as agents of primary production, *Nature (London),* 199, 1255, 1963.

82. **Bunt, J. S. and Lee, C. C.,** Observations within and beneath Antarctic sea ice in McMurdo Sound and the Weddell Sea, 1967-1968, Methods and Data, Tech. Rep. 69-1, Institute of Marine Sciences, University of Miami, Miami, 1969.
83. **Bunt, J. S.,** Primary productivity under sea ice in Antarctic waters. I. Concentrations and photosynthetic activities of microalgae in the waters of McMurdo Sound, Antarctica, *Antarct. Res. Ser.*, 1, 13, 1964.
84. **Bunt, J. S.,** Primary productivity under sea ice in Antarctic waters. II. Influence of light and other factors on photosynthetic activities of Antarctic marine microalgae, *Antarct. Res. Ser.*, 1, 27, 1964.
85. **Bunt, J. S.,** Some characteristics of microalgae isolated from Antarctic sea ice, *Antarct. Res. Ser.*, 11, 1, 1968.
86. **Bunt, J. S. and Lee, C. C.,** Data on the composition and dark survival of four sea-ice microalgae, *Limnol. Oceanogr.*, 18, 458, 1972.
87. **Bunt, J. S., Owens, O. van H., and Hoch, G.,** Exploratory studies on the physiology and ecology of a psychrophilic marine diatom, *J. Phycol.*, 2, 96, 1966.
88. **Bunt, J. S. and Lee, C. C.,** Seasonal primary production in Antarctic sea ice at McMurdo Sound in 1967, *J. Mar. Res.*, 28, 304, 1970.
89. **Bunt, J. S.,** Microalgae of the Antarctic pack ice zone, in *Symposium on Antarctic Oceanography*, Currie, R. I., Ed., W. Heffer & Sons Ltd., Cambridge, England, 1968, 198.
90. **Meguro, H.,** Plankton ice in the Antarctic Ocean, *Nankyoku Shiryo*, 14, 1192, 1962.
91. **Iizuka, H., Tanabe, I., and Meguro, H.,** Microorganisms in plankton-ice of the Antarctic Ocean, *J. Gen. Appl. Microbiol.*, 12, 101, 1966.
92. **Fukushima, H. and Meguro, H.,** The plankton ice as a basic factor of the primary production in the Antarctic Ocean, *Nankyoku Shiryo*, 27, 99, 1966.
93. **Hoshiai, T. and Kato, M.,** Ecological notes on the diatom community of the sea ice in Antarctica, *Bull. Mar. Biol. Stn. Asamushi, Tohoku Univ.*, 10, 221, 1961.
94. **Hoshiai, T.,** Ecological observations of the coloured layer of the sea ice at Syowa Station, *Nankyoku Shiryo*, 34, 60, 1969 (in Japanese, English summary).
95. **Hoshiai, T.,** Seasonal change of ice communities in the sea ice near Syowa Station Antarctica, in *Polar Oceans*, Dunbar, M. J., Ed., The Arctic Institute of North America, Calgary, 1977, 307.
96. **Hoshiai, T.,** Proliferation of ice algae in the Syowa Station area, Antarctica, *Mem. Natl. Inst. Polar Res.* Ser. E, 34, 1, 1981.
97. **Burkholder, P. R. and Mandelli, E. F.,** Productivity of microalgae in Antarctic sea ice, *Science*, 149, 872, 1965.
98. **Whitaker, T. M.,** Sea ice habitats of Signy Island (South Orkneys) and their primary productivity, in *Adaptations within Antarctic Ecosystems*, Llano, G. A., Ed., Gulf Publ., Houston, 1977, 75.
99. **Horne, A. J., Fogg, G. E., and Eagle, D. J.,** Studies *in situ* of the primary production of an area of inshore Antarctic sea, *J. Mar. Biol. Assoc. U.K.*, 49, 393, 1969.
100. **Ackley, S. F., Taguchi, S., and Buck, K. R.,** Sea ice and ice algae relationships in the Weddell Sea, *Antarct. J. U.S.*, 13, 70, 1978.
101. **Ackley, S. F., Taguchi, S., and Buck, K. R.,** Primary productivity in sea ice of the Weddell Region, Report 78-19, Cold Regions Research and Engineering Laboratory, Hanover, N.H., 1978.
102. **Ackley, S. F., Buck, K. R., and Taguchi, S.,** Standing crop of algae in the sea ice of the Weddell Sea region, *Deep-Sea Res.*, 26A, 269, 1979.
103. **Garrison, D. L. and Buck, K. R.,** Sea-ice algae in the Weddell Sea. I. Biomass, distribution and the physical environment, *EOs*, 63, 47, 1982.
104. **Buck, K. R. and Garrison, D. L.,** Sea-ice algae in the Weddell Sea. II. Population comparisons between sea-ice and water column, *EOs*, 63, 47, 1982.
105. **Ackley, S. F.,** Ice scavenging and nucleation: two mechanisms for incorporation of algae into newly-formed sea ice, *EOs*, 63, 54, 1982.

Chapter 2

THE ICE ENVIRONMENT

Gary A. Maykut

TABLE OF CONTENTS

I. EXTENT AND MORPHOLOGY OF THE POLAR ICE PACK

Sea ice interacts thermodynamically with both the atmosphere and ocean, giving rise to a climate over the polar oceans that is more characteristic of the continental ice sheets than of a marine environment. Melting and freezing of the ice, together with the damping of surface waves and reduction in solar energy input, impact both the temperature and salinity structure of the upper ocean in the polar regions. By area, sea ice roughly accounts for two thirds of the earth's permanent ice cover, but only about 0.1% of its volume. Thus, unlike the polar ice caps, sea ice is only a thin veneer whose thickness and extent can undergo large changes in response to small changes in climate or oceanic heat transport. Because of the positive feedback between ice extent and surface albedo, sea ice is generally considered to be an important element in the climate system.

A. Geographical Distribution

At maximum extent sea ice covers some 8% of the Southern Hemisphere and 5% of the Northern Hemisphere. The nature of the ice pack in the two hemispheres is quite different due primarily to differences in the distribution of land masses. In the south the Antarctic Continent occupies most of the area pole-ward of 70°S, so that sea ice forms a seasonally varying annulus in the region between 55 and 70°S (Figure 1)[1]. Meridional variations are relatively small, the greatest northward extent occurring in the Weddell Sea sector. Most of the Antarctic ice pack is seasonal, covering 20×10^6 km^2 at maximum extent and only 4×10^6 km^2 at the end of summer.[2]

In the Northern Hemisphere the area pole-ward of 70°N is in large part a landlocked ocean basin which is ice-covered during most of the year. Seasonal ice forms in many of the peripheral seas surrounding the Arctic Basin (Figure 2). At minimum extent (August) the ice covers about 7×10^6 km^2, almost double that in the Antarctic, while at maximum extent (March) the area expands to 14×10^6 km^2.[3] Variations in oceanic circulation and the complex configuration of the land masses cause meridional variations in winter ice extent of up to 35 degrees of latitude. Cold air flowing over the Sea of Okhotsk from the Asian Continent results in the formation of a seasonal ice cover which typically extends to the north coast of Japan (44° to 45°N). At the other extreme is the area west of Spitsbergen (0°, 80°N) where warm Atlantic water flowing into the Arctic Basin and the southward drift of the ice combine to maintain the ice edge at a nearly constant position throughout the year. The economic, military, and scientific importance of the region, together with its relative accessibility, have stimulated extensive research efforts and we now have a good general picture of the ice-ocean-atmosphere system in the Arctic. Such is not the case in the Antarctic. Although some satellite-derived data are available regarding the drift, extent and physical characteristics of Antarctic pack ice, (e.g., Ackley[4]), conditions in most ice covered portions of the Southern Ocean are poorly known.

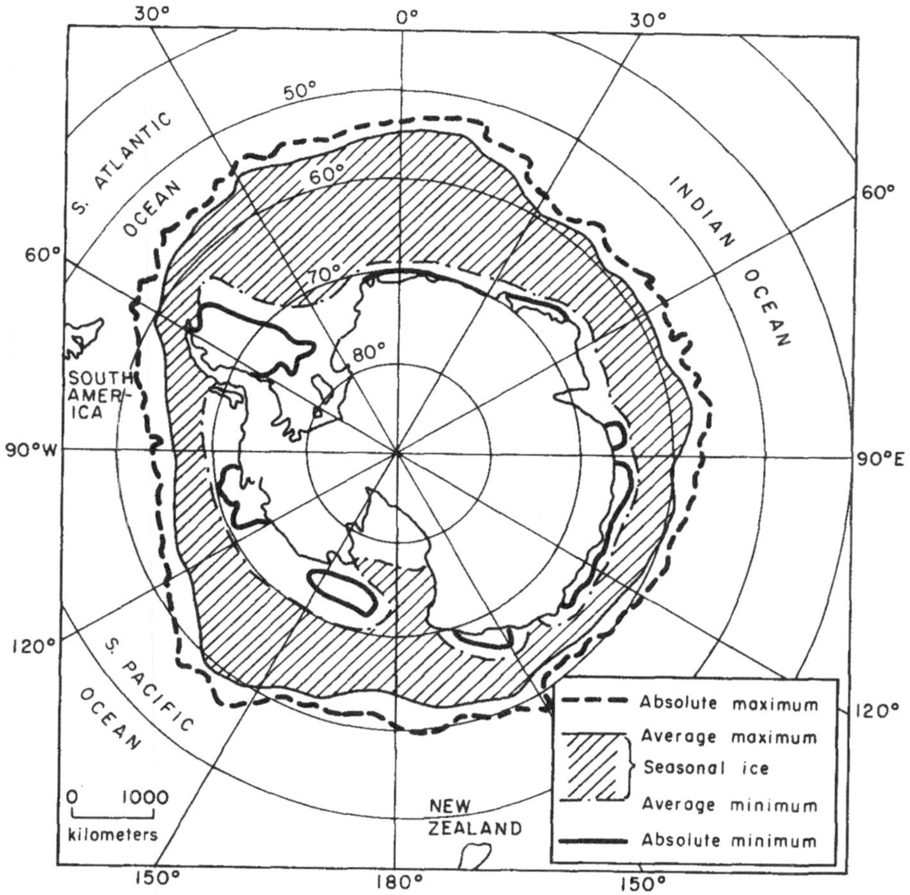

FIGURE 1. The extent of sea ice in the Southern Hemisphere with greater than 1/8 concentration (after CIA[1]).

B. Ice Movement

Sea ice, driven principally by winds and currents, is in almost continual motion. As will be discussed later in the chapter, these motions affect not only the extent of the ice, but also its physical properties and the way in which it interacts with the atmosphere and ocean. Ice motion in the Arctic Basin has been well documented from the trajectories of manned drifting stations and automatic data buoys. Most references describe the two major drift features identified by Gordienko:[5] (1) the Transpolar Drift Stream which transports ice from the western side of the basin across the pole and through the Fram Strait between Greenland and Spitzbergen and (2) a clockwise gyre in the Beaufort Sea region. However, a recent analysis of the mean drift field by Colony and Thorndike[6] shows a somewhat different pattern with anticyclonic motions throughout the Amerasian Basin and southward drift toward Fram Strait in the Eurasian Basin. Average displacements are approximately 7 km day^{-1}. Ice movement in the Greenland and Bering seas tends to be southerly with average velocities on the order of 15 km day^{-1}.[7] While some ice is lost from the Arctic Basin through the Bering Strait, this is small compared to the export through Fram Strait. On an annual basis roughly 10% of the ice in the basin drifts into the Greenland Sea where it eventually melts away, providing a significant heat gain for the basin. Ice movements in the Antarctic are much simpler, being generally clockwise around the continent. A major drift feature exists in the Weddell Sea in the form of a cyclonic gyre which advects ice along the Antarctic Peninsula and allows multi-year ice to develop in the region.

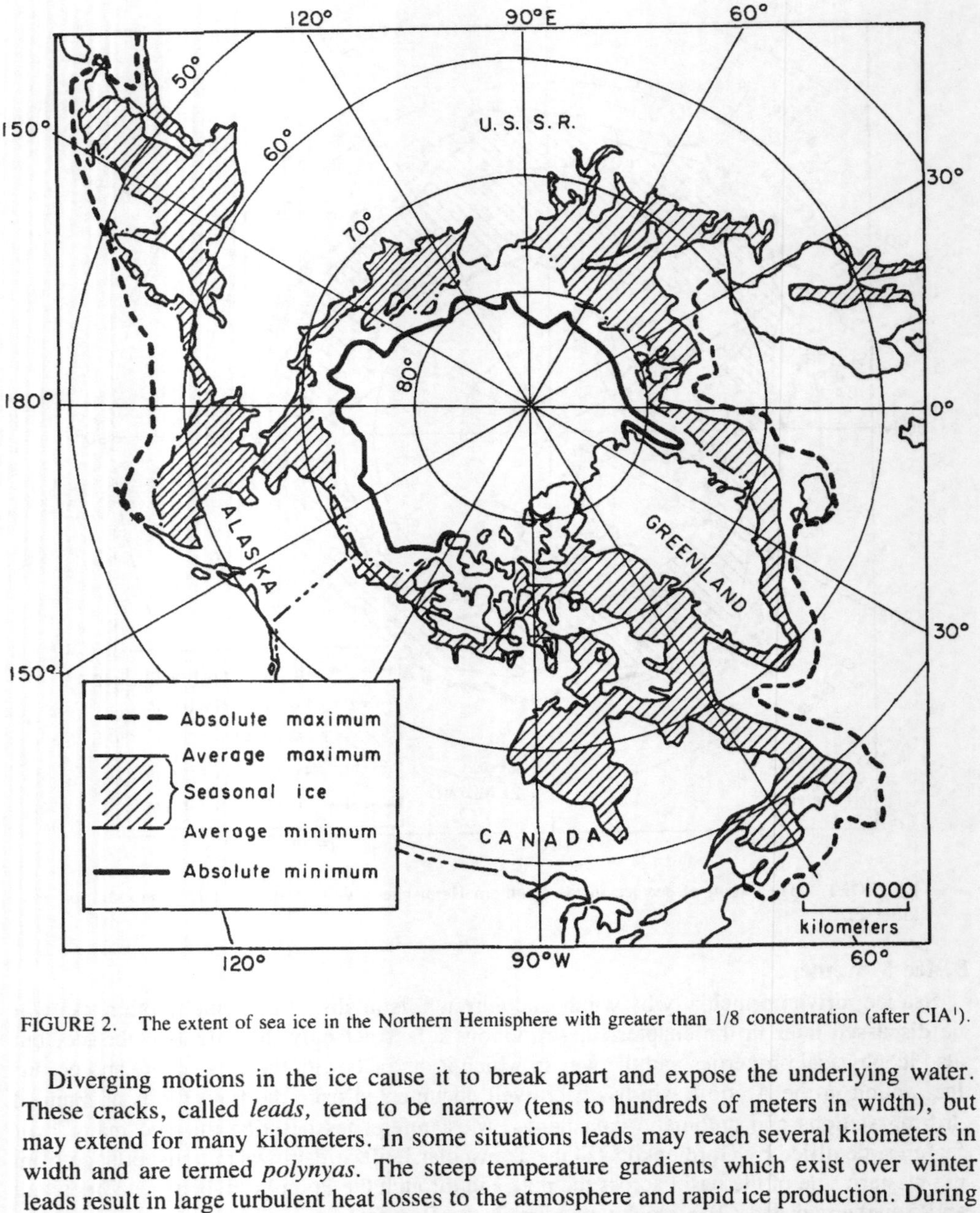

FIGURE 2. The extent of sea ice in the Northern Hemisphere with greater than 1/8 concentration (after CIA[1]).

Diverging motions in the ice cause it to break apart and expose the underlying water. These cracks, called *leads,* tend to be narrow (tens to hundreds of meters in width), but may extend for many kilometers. In some situations leads may reach several kilometers in width and are termed *polynyas*. The steep temperature gradients which exist over winter leads result in large turbulent heat losses to the atmosphere and rapid ice production. During the summer leads act as moving windows which admit substantial amounts of solar energy to the upper ocean. Leads are common in most regions. In the Arctic Basin winter leads occur with an average spacing of about 300 m. Leads tend to freeze over quickly and open water rarely makes up more than 0.5% of the winter ice cover. Young ice ($h < 1$ m) in refrozen leads typically accounts for 5 to 10% of the area. During the summer decay cycle the area of open water increases greatly, reaching values of 10 to 20% in the Central Arctic.

In marginal seas where the ice edge is not confined by land, ice concentration patterns are complex and can vary rapidly with time. Over the interior of the Bering Sea ice pack, for example, Cavalieri et al.[8] found winter concentrations ranging between 85 and 100% with cold northeast winds, but these values dropped to as low as 45% following the onset of warm southerly winds. Ice concentrations in the Antarctic are also fairly low, averaging

60 to 80% throughout most of the year.[2] The large amount of open water suggests that Antarctic ice undergoes substantial divergence. In fact, analysis of satellite imagery indicates that the Antarctic ice pack is everywhere divergent in its mean motion and that northward advection of the ice must result in the continual production of new leads and polynyas.[9,10]

Convergence and shearing motions cause the ice to fracture and pile up into a variety of deformation features, the most notable being pressure ridges. Pressure ridges form linear features up to several kilometers in length which consist of a *keel* beneath the ice and a *sail* on the surface of the ice. Keels are typically about five times thicker than sails. Most ridges in the Central Arctic appear to be built out of ice that was originally less than 1 m in thickness.[11] This may not be true in all areas, however, as detailed morphological studies north of Prudhoe Bay, Alaska, indicated that nearly 50% of the ridges there were composed of blocks between 1.0 and 1.6 m in thickness.[12] Ridge frequency in the Arctic Basin varies with season and location, but is generally in the range of 2 to 6/km. Keel depths are usually less than 20 m. It has been estimated[13] that as much as 40% of the total mass in the Arctic ice pack may be contained in deformation structures. In contrast to the Arctic situation, ridges are generally believed to be uncommon in most parts of the Antarctic due to the divergent nature of the velocity field there. A notable exception is the Weddell Sea where ridged ice may account for more than 15% of the area.[4]

Without continued deformations the polar ice pack would assume a much more uniform character. In areas of seasonal ice the thickness would increase smoothly inward from the edge, while in areas of perennial ice the thickness within a particular climatic regime would be essentially constant. The reality is that a typical sea ice cover contains a great range of thicknesses, all temporarily coexisting under similar thermal forcing. Thicknesses tend to be distributed chaotically with large variations occurring on horizontal scales of only a few meters. Ice motion enhances the transfer of momentum from the atmosphere to the ocean and causes thermodynamic interactions between the air and water to be much more vigorous than would otherwise be the case. Mechanical breakup and advection also act to accelerate the decay and retreat of the ice cover, allowing larger portions of the polar oceans to become ice free each summer.

C. Ice Zones

Because of differences in the properties of the ice, in the dominant ice-ocean interaction processes and in the nature of the scientific problems encountered, the polar ice pack is frequently subdivided into different zones. The most basic distinction is between regions which are perennially ice covered and those with seasonal ice. The *perennial ice zone* in the Arctic, located largely pole-ward of 75 to 80°N (Figure 2), contains about two thirds of all the multi-year sea ice found in the World Ocean. Thickness averages 3 to 4 m in the central part of the basin, but increases significantly north of Greenland and Ellesmere Island as the motion of the ice toward the land causes extensive ridging. The age of this ice is uncertain, but may reach tens of years in the Beaufort sector and at least several years on the Soviet side of the basin.

Multi-year ice in the Antarctic is restricted to the Weddell Sea and parts of the Ross, Amundsen, and Bellingshausen seas (Figure 1). The age of this ice is probably much less than in the Arctic. What is known of drift patterns in the Weddell Sea suggests that the residence time of the ice is no more than 2 to 3 years. Nevertheless, ice thicknesses of up to 5 m have been observed in undeformed ice there.[14] The structure of this ice was quite different from that in the Central Arctic. Some 60 to 70% of the thickness was made up of layers of small ice crystals a few millimeters across rather than the organized columnar structure commonly observed in most Arctic ice. These so-called "frazil" crystals could originate in the water column beneath the ice or in nearby leads, but the exact mechanisms involved have not yet been positively identified. In addition to the more complex internal structure, weak surface melting during the Antarctic summer allows the salinity of the ice near the surface to remain high. As a result, microwave techniques which rely on differences in near-surface salinity to distinguish between first-year and multi-year ice are much less successful in the Antarctic than in the Arctic (see Section IV.E).

The *seasonal sea ice zone* (SSIZ) covers about 25×10^6 km^2, roughly 7% of the World Ocean. The area covered by the SSIZ in the Arctic is about the same as that of the perennial ice. While undeformed ice in the arctic SSIZ is generally less than 2 m in thickness, large amounts of ridged ice are present so that average thicknesses can be much larger, particularly near the coasts. As was pointed out previously, 80% of the highly mobile Antarctic ice pack is seasonal. It has been reported[15] that seasonal ice in the Antarctic can grow to thicknesses at least 50% greater than in the Arctic. This ice has frequently been observed to possess not only layers of frazil crystals, but also a layer of infiltration ice formed by flooding of the ice surface in response to snow loading in the early stages of growth. The generality of these observations has not been established and it is not known whether they apply throughout the southern SSIZ or only to certain regions near the continent. The relative importance of various ice decay processes also differs in the two hemispheres. Surface melting is a major factor in the summer disappearance of Arctic ice. Antarctic ice, on the other hand, is characterized by minimal surface melting and an absence of melt ponds, probably due to lower relative humidity in the atmosphere.[16] This means that heat from the water must be largely responsible for seasonal ice retreat in the Southern Ocean. In addition, higher biological activity in Antarctic ice enhances internal melting, weakening the ice and accelerating the breakup.

Ice-ocean-atmosphere interactions in the shelf regions of the SSIZ appear to have an important effect on the large-scale structure and circulation of the World Ocean.[17] It has long been suspected that salt fractionation during freezing leads to the formation of cold, dense water on the shelves. Such water appears to be the precursor to the Antarctic Bottom Water which fills much of the deep World Ocean. Few details are known regarding production sites, rates, or mechanisms for the formation of this water. Offshore winds produce a number of large recurrent polynyas around the Antarctic Continent, especially in the Ross, Weddell, and Amundsen seas.[18,19] Rapid ice growth in such polynyas and the resulting increase in density leads to vertical haline convection, possibly all the way to the sea floor in some areas. It is probable that much of the Antarctic Bottom Water originates in this way. Similar processes on certain of the Arctic shelves also appear to produce dense water which is critical in maintaining the thermohaline structure of the Arctic Ocean.[20]

Other ice regimes frequently mentioned in the literature include: the *fast ice zone*, the *shear zone*, and the *marginal ice zone*. Shorefast ice occurs along most coasts in the Arctic and Antarctic. It forms early in the winter in shallow water where the water column can cool rapidly to the freezing point. It is anchored to the bottom at the shoreline and frequently at depths of up to 20 m by grounded pressure ridge keels.[21] Although rigidly anchored, fast ice may move tens of meters in response to thermal and mechanical stresses.[22] Such motions are important because of their potential impact on offshore structures. Salinities beneath ice in lagoons and basins where circulation is restricted can reach values in excess of 100‰. The extent of fast ice is determined primarily by sea bottom and shoreline topography and is highly variable. Near Point Barrow, for example, the zone is about 15 km in width while at nearby Harrison Bay it extends some 60 km offshore. Fast ice development is most extensive in areas where the shelf slope is gentle, (e.g., Mackenzie Delta and Laptev Sea). Also usually classified as landfast is much of the ice which fills constricted bays and channels in regions like the Canadian Archipelago. A review of landfast ice and biological characteristics of the underlying water has recently been given by Newbury.[23]

Aircraft and submarine sonar observations indicate the existence of a zone of highly deformed ice in coastal regions along the Alaskan North Slope and Canadian Archipelago.[14,24]

The origin of this so-called *shear zone* is landward movement of the ice pack, causing the ice to converge, shear, and ridge. The frequency of ridges in the shear zone is several times greater than in the central pack, but keel depths are not appreciably larger. The size of the shear zone depends on definition. Wadhams[13] defines the shear zone as the region where ice characteristics differ significantly from those in the Central Arctic. His analysis of submarine data suggests that the shear zone extends some 400 km off the coast of Ellesmere Island where ice motion is normal to the coast and about 160 km off Alaska where the motion is more parallel to the land. Microwave imagery[25] shows that the landward part of this zone is made up largely of first-year ice and that the outer part contains a substantial amount of multi-year ice. The shear zone is clearly a boundary layer between the fast ice and central pack and, as such, understanding its behavior is of particular importance in efforts to develop large-scale dynamic models of sea ice.

One of the most complex and scientifically interesting regions, the *marginal ice zone* (MIZ), is located near the boundary between the ice and open ocean. Here, penetration of surface waves into the pack breaks the ice into numerous small floes whose average size increases rapidly with distance from the edge. In regions like the Bering Sea, where floes are thin (<0.5 m), ice in the outer 5 to 15 km is heavily rafted and many reach thicknesses of up to 5 m.[26] The MIZ is usually taken to include that part of the ice which is affected by the presence of the open ocean, a region some 150 to 200 km in width. The outer part of the MIZ is characterized by large horizontal gradients in the properties of the ice, ocean, and atmosphere. Advection of air across the ice edge can produce large changes in turbulent heat transfer, surface stress, cloudiness, and the radiative fluxes. Comparable changes also occur in the mechanical properties of the ice, surface roughness, and solar input to the ocean. Horizontal variations in density structure and surface stress can give rise to a variety of mesoscale phenomena in the ocean (eddies, jets, upwelling) which impact biological activity and acoustical properties across the MIZ. During the summer, floe breakup and increasing stratification beneath the ice affect the response of the ice to winds and currents and the rate at which heat is transferred from the water to the ice. Biological activity in the MIZ is high.[27] Substantial research efforts are currently underway in the Bering and Greenland seas to investigate not only the properties of the ice cover, but also oceanographic, atmospheric, biological, and acoustical characteristics of the MIZ.[28,29]

II. FORMATION AND GROWTH OF SEA ICE

A. The Freezing of Seawater

The presence of salt causes seawater to respond somewhat differently to atmospheric cooling than freshwater. A density vs. temperature plot of freshwater (Figure 3A) shows the existence of a density maximum at 3.98°C. Surface cooling in a freshwater body whose temperature is above this value causes the density to increase, generating vertical convection that continues until all the water reaches the temperature of maximum density (T_m). Further cooling then produces a density decrease in the surface layer, allowing the water to become thermally stratified. Since conduction rather than convection is then the dominant heat transport mechanism in the water, there is a sharp decrease in the upward flux of heat to the surface and cooling proceeds rapidly. Ice formation in freshwater can thus occur when most of the water column is significantly above the freezing point.

Addition of salt to the water depresses the freezing point of the solution (T_f) according to the relation:[30]

$$T_f \simeq -0.003 - 0.0527S_w - 0.00004S_w^2 \tag{1}$$

where temperature is in °C and the salinity of the water (S_w) in ‰. For many purposes Equation 1 can be approximated simply by $T_f \simeq -0.055S_w$. Salt also reduces T_m, but at a

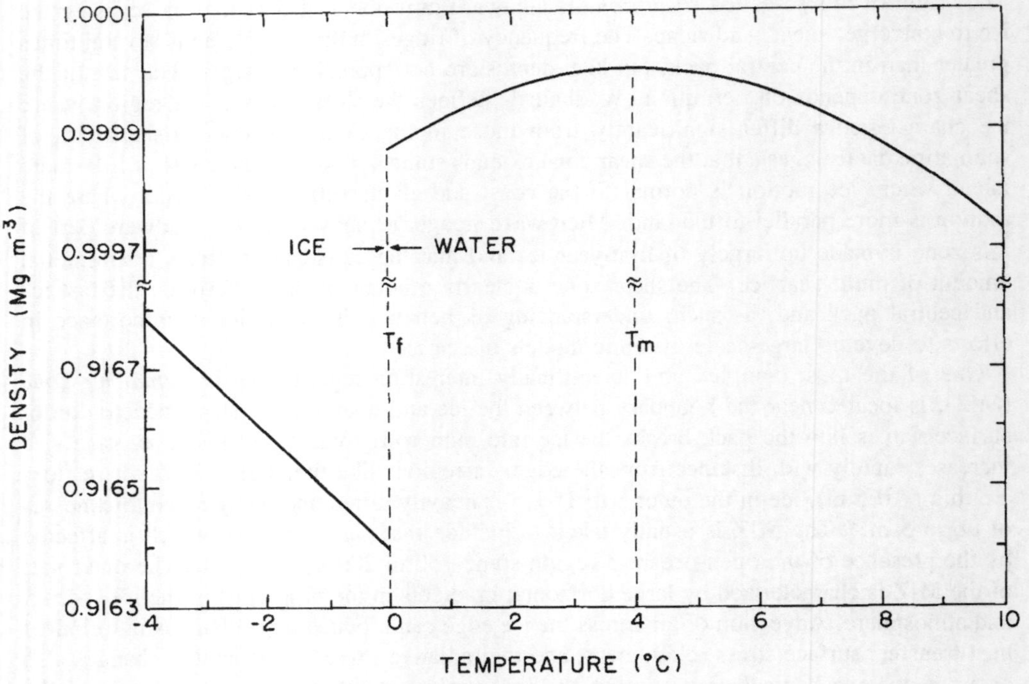

FIGURE 3A. Density vs. temperature diagram for freshwater and ice at atmospheric pressure.

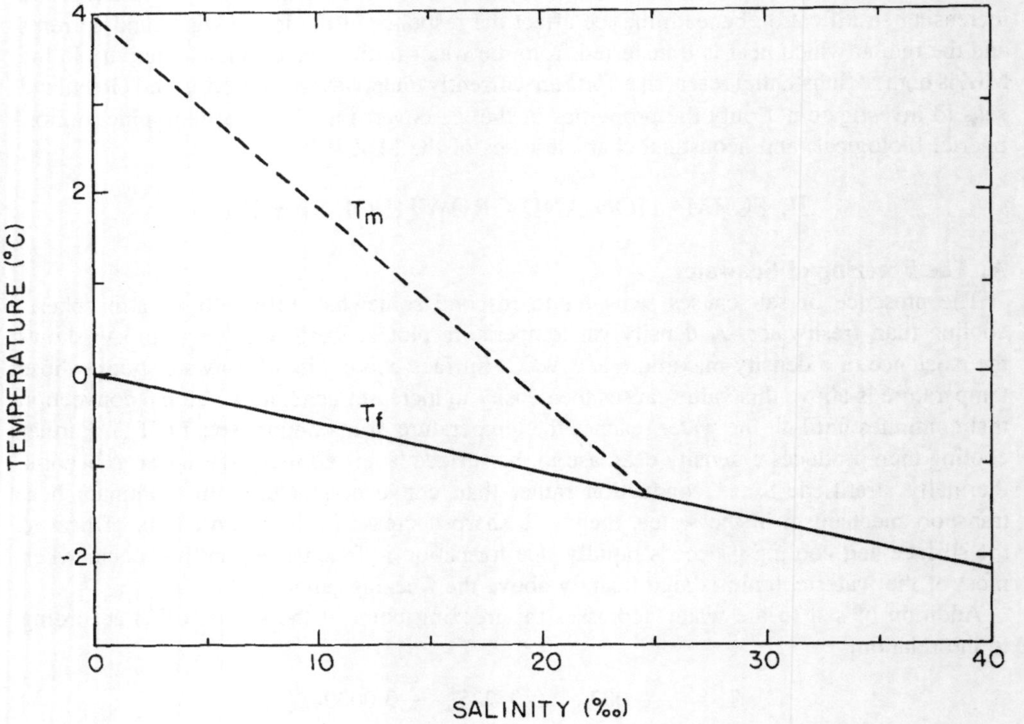

FIGURE 3B. Effect of salinity on the temperature of maximum density (dashed curve) and the freezing point temperature (solid curve).

more rapid rate than T_f (Figure 3B). If $S_w > 24.7‰$, $T_m = T_f$. As this is the case in normal seawater, surface cooling results in a density increase and vertical mixing which continues until the water reaches T_f. Unlike freshwater lakes density gradients in the upper ocean limit the depth (z_c) to which the water must be cooled before freezing can commence. Soviet data[31] indicate that z_c is usually in the 10 to 40 m range, although values in excess of 70 m have been observed. Heat brought to the surface through convective mixing can have a large effect on the onset of freezing and observations[31] show a delay of as much as 2 months in the date of initial ice formation in regions where $z_c = 50$ m compared to those where $z_c = 10$ m.

B. Initial Ice Formation

Once the upper ocean reaches the freezing point, additional heat loss produces slight supercooling of the water and ice formation. The amount of supercooling necessary to initiate ice growth is usually small, probably on the order of a few hundredths of a degree Celsius, although observations in some Antarctic polynyas and near Greenland have shown a supercooling of 0.2 to 0.4°C down to depths of tens of meters.[31] Initial ice formation occurs at or near the surface of the water in the form of small platelets and needles which are termed *frazil ice*. Frazil crystals do not generally exceed 3 to 4 mm in diameter. Continued freezing results in the production of *grease ice*, a soupy mixture of unconsolidated frazil crystals. Because grease ice suppresses capillary wave action, it appears as a smooth, grayish layer whose albedo is only slightly above that of open water. Under quiescent conditions the frazil crystals quickly freeze together to form a solid, continuous ice cover with a thickness of 1 to 10 cm. More typically, wind waves induce turbulence in the water and abrasion between the crystals which inhibit development of a solid cover. Wind and wave action advect frazil crystals downwind where accumulations of up to 1 m in thickness can form in front of obstacles, (e.g., floe edges). Laboratory experiments[32] indicate that when the ice volume is low, grease ice behaves as a fluid whose viscosity increases as the proportion of ice increases. Once the ice fraction exceeds 30 to 40%, sintering and regelation create sufficient bonding between individual crystals that their mobility is drastically reduced and the transition to a solid cover begins. In the presence of a wave field, this transition is usually marked by the formation of *pancakes*, rounded masses of semiconsolidated slush 0.3 to 3.0 m in diameter. Pancakes often display irregular ice rims around their perimeters built up as a result of continual bumping into one another. With time, the pancakes consolidate and are welded together into a composite ice sheet by the freezing of grease ice which forms between them. A detailed review of initial freezing processes, together with photographs of different stages of ice development, appear in Weeks and Ackley.[14]

C. Young Ice Growth

In the early stages of growth (grease ice, pancake ice) the surface of the ice is wet and its temperature stays close to the freezing point. With air temperatures often in the −10 to −20°C range, turbulent and longwave radiation losses are large and ice formation rapid. However, once the ice becomes consolidated, free water is no longer available at the surface and its temperature begins to approach that of the atmosphere. Additional growth then occurs by accretion on the underside of the sheet, the rate depending on how rapidly heat can be conducted from the ice-water interface toward the surface, i.e., on the thermal conductivity and temperature gradient in the ice.

While winds, cloudiness, relative humidity, snowfall, and oceanic heat flux, all affect the rate of ice growth, field measurements suggest that reasonable growth estimates can be obtained from air temperature data alone. Figure 4 shows ice thicknesses observed by Anderson[33] near Thule, Greenland, plotted against the cumulative number of freezing-degree days θ, where:

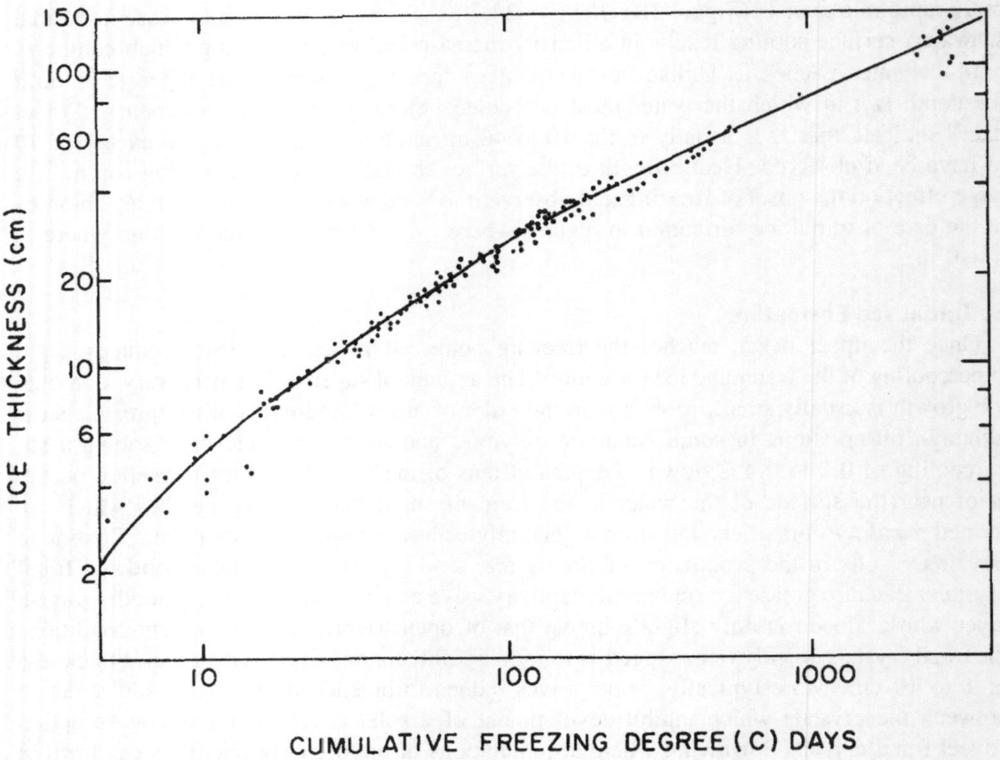

FIGURE 4. Relationship between the thickness of young sea ice with minimal snow cover and the cumulative number of freezing-degree days.[33]

$$\theta \equiv \int_0^t (T_f - T_a)dt \tag{2}$$

t is time and T_a is the air temperature. Anderson found that in the thickness range $10 < h < 80$ cm his data could be fit by a simple power law:

$$h^2 + 5.1\ h = 6.7\ \theta \tag{3}$$

As the ice becomes thicker, the close relationship between θ and h begins to break down. Figure 4, for example, shows that when $\theta = 3000$ degree-days, the observed h varies from 90 to 140 cm. Factors contributing to this breakdown are the increasing thermal mass of the ice, the decreasing magnitude of the heat flux in the ice relative to that in the ocean, and increasing snow cover. The term *young ice* has been somewhat narrowly defined in the *Glossary of Snow and Ice*[34] as ice in the 10 to 30 cm category. In the context of this chapter, the term will be used in a broader sense to refer to ice whose temperature profile is nearly linear and whose growth can usually be described by some simple function of θ, i.e., $h <$ 80 to 100 cm.

Alternate expressions for h (θ) have been derived by investigators working in other regions (see Bilello[35]). Implicit in such formulations is the idea that the exact form of $T_a(t)$ is unimportant in determining h, rather it is only the area under the curve which needs to be known. This is equivalent to saying that, for example, the ice would attain the same thickness in 10 days at −20°C as it would in 20 days at −10°C (assuming that $T_f = 0$°C). Differences in growth predicted by the various empirical formulas presumably reflect differences in environmental parameters (snowfall, heat content of the underlying water column, etc.)

which are not explicitly treated in the formulas. As a result such equations must be considered at least partially site-specific and their utilization approached with caution. Applying an equation obtained from data in the Central Arctic to coastal conditions or to conditions in the Southern Ocean could lead to substantial errors.

To establish a more general relationship, it is necessary to consider the physics governing the situation. The usual approach is to write down energy balance equations at the top and bottom of the ice (see Section V.B.) which are coupled by an equation describing the rate of heat transport between the two boundaries. The problem is greatly simplified in the case of young ice because the conductive flux is directly proportional to the temperature difference between the boundaries. It can easily be shown, (e.g., Maykut[36]) that the form of Anderson's equation follows from the assumption of a linear temperature gradient in the ice and negligible heat flux in the water. The theoretical expression can be written

$$h^2 - h_0^2 + \left(\frac{2k_i}{k_s} h_s + \frac{2k_i}{C_t}\right)(h - h_0) = \frac{2k_i}{\rho_i L}\theta \tag{4}$$

where h_o is the initial thickness of the ice, k_i is the average thermal conductivity of the ice, k_s is the thermal conductivity of the snow, ρ_i is the density of the ice, L is the latent heat of fusion, C_t is a bulk transfer coefficient describing turbulent heat transfer between the ice and atmosphere, and h_s is the snow depth which is assumed to remain constant over the period of integration. The net radiation balance is assumed to be zero. When $h_s = 0$ and the surface temperature of the ice (T_o) is equal to T_a, Equation 4 yields the classic result that h is a simple function of $\theta^{1/2}$. Taking into account the difference between T_a and T_o gives rise to the linear term in the equation. With no snow, the linear term rapidly decreases in importance once h exceeds about 20 cm. A small amount of snow, however, greatly extends the range over which the linear term dominates the growth.

With nominal values for the thermal properties and $h_o = 0$, Equation 4 can be written as: $h^2 + (13.1h_s + 16.8)h = 12.9\theta$. Comparison with Equation 3 shows that the theoretical coefficients multiplying the h and θ terms are substantially larger than the empirically derived ones. If $h_s = 0$, Equation 4 predicts much too rapid a growth. With a few centimeters of snow, the theoretical growth is too slow intitially and too rapid once the ice becomes thicker. One factor contributing to this problem is that, contrary to what was assumed in the derivation of Equation 4, heat conduction in the ice is not usually the dominant heat transfer mechanism in the early stages of ice formation. Strictly, then the equation should be applied only after a solid ice cover has been established. Another complication is the fact that h_s cannot be expected to remain constant over extended periods of time. It is possible to take into account $h_s(t)$ by applying Equation 4 in a stepwise fashion between periods of snow accumulation. Such an approach was taken by Nakawo and Sinha[37] in analyzing ice growth data from the Canadian Arctic. Surprisingly, they found that growth predictions made by including observed variations in h_s were less accurate than those obtained using a large constant snow depth. The reason for the apparent failure of the more rigorous method was probably due to the oceanic heat flux at the bottom of the ice (F_w). Allison[38] reports values of F_w under young, rapidly growing ice of as high as 50 W m^{-2}, which would be sufficient to decrease growth rates by more than a centimeter per day.

Unfortunately, a simple analytic solution to the heat balance equations cannot be obtained when $F_w \neq 0$, making it necessary to employ numerical techniques to estimate the ice growth. In the case of young ice the numerical procedures are relatively simple and growth calculations straightforward. Using a numerical model to describe the thermodynamics of young ice, (e.g., Maykut[39]) has the advantage that many of the simplifying assumptions (e.g., zero radiation balance) that were needed to obtain an analytical expression are unnecessary. Moreover, such a model is general in that it is not restricted to a particular location or season, provided that sufficient information on the thermal forcing is known. If actual data on snowfall, T_a, F_w, wind speed, and the radiative fluxes are available, very accurate growth calculations can be made. Even if direct observations are lacking, it should be possible to improve on empirical predictions through the use of climatological data and suitable parameterizations for the radiative fluxes.[36] The most serious source of error in such a calculation would probably be due to uncertainties in F_w. If only a crude estimate of ice growth is needed, then the formula given by Zubov[40] ($h^2 + 50h = 8\theta$) would probably yield acceptable results as it represents an average of many years of observation from the Soviet Arctic. Because it contains implicit assumptions regarding $F_w(t)$ and $h_s(t)$, its applicability to the Southern Ocean is doubtful.

While estimating the thickness of the ice is of primary importance in many practical problems, the *rate* at which it grows affects the level of convective activity and the structure of the underlying ocean. This is because freezing removes freshwater from the surface of the water column producing a salt flux at the bottom of the ice. The rate of ice production is extremely sensitive to ice thickness, roughly dropping by an order of magnitude in the first 50 cm of growth. Figure 5, based on Anderson's observations,[33] shows growth rate (f) as a function of h for several different air temperatures. Note that the sensitivity of f to T_a tends to decrease as h increases. Because of its low thermal conductivity, snow insulates the ice and decreases the rate at which it can lose heat to the atmosphere. With a numerical model it is possible to quantify the response of young ice to changes in h_s (Figure 6). Increasing h_s decreases f, the effect being much more pronounced for the thinner ice. For the case illustrated, f is essentially constant if $h_s > 15$ cm. The reason for this behavior is best explained in terms of the *thermal conductance* (γ) of the combined ice-snow slab, where $\gamma = k_i k_s/(k_i h_s + k_s h_i)$. The magnitude of f depends on the rate at which heat is conducted through the slab (F_c), which for young ice can be written:

$$F_c = \gamma(T_s - T_f) \tag{5}$$

where T_s is the surface temperature of the slab. Values of $\gamma(h,h_s)$ are presented in Section IV (see Figure 14). When h_s is small, γ (and f) decreases rapidly with increasing h; when h_s is large, γ is small and changes slowly with both h and h_s.

D. Multi-Year and Thick First-Year Ice Growth

The principal simplification used in treating the growth of young ice was the assumption of a linear temperature gradient in the ice, causing f to respond immediately to changes in thermal forcing at the surface. However, measurements of the temperature field in multi-year ice[41] indicate significant departures from linearity, even during mid-winter when temperature fluctuations are relatively small. Figure 7 shows theoretical predictions of f (h) as a function of season for thicker ice in the Central Arctic. The curves demonstrate that differences in thermal mass strongly affect how the ice responds to conditions at the upper surface. In November, over 2 months after freeze-up, ice thicker than about 3 m continues to ablate because the fall cooling has had insufficient time to reach the lower part of the ice. By January it is evident that cooling has penetrated all but the thickest ice. Spring warming is reflected in decreasing values of f in the thinner ice during April and May, while ice thicker than 4 m continues to feel the effects of the winter cooling. In June accretion can occur at the bottom of the ice while vigorous melting is taking place at the surface. Clearly the assumption of a linear temperature profile is invalid for thicker ice whose growth depends more on its thermal history than on the immediate heat balance at the surface.

Heat transport and temperature changes in thicker ice can be predicted by utilizing the standard second-order partial differential equation describing heat conduction in a solid. Combining this equation with heat balance equations at the upper and lower boundaries of the ice results in a model which can simulate the growth of any thickness of undeformed ice (see Section V. B). A detailed model of this type which takes into account effects of

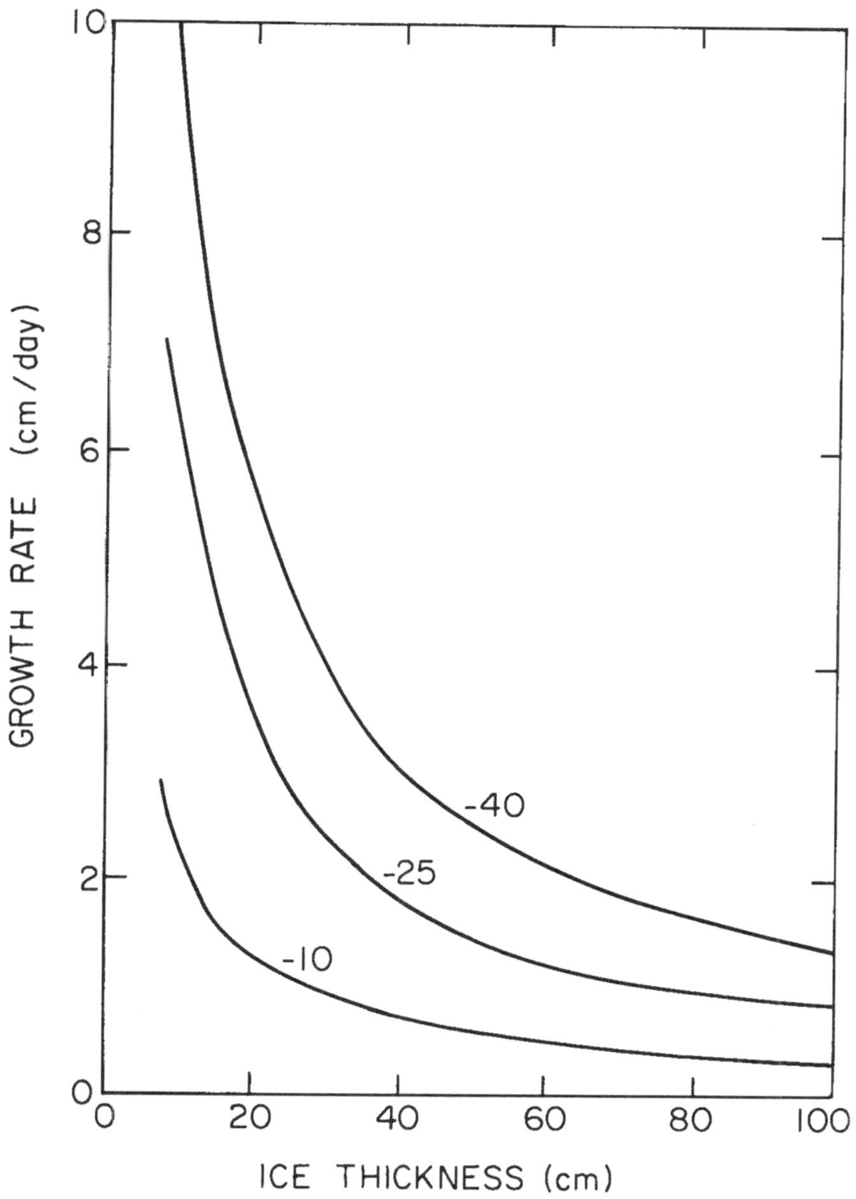

FIGURE 5. Dependence of growth rates in young sea ice on ice thickness and air temperature (°C).[36]

salinity and internal solar heating has been formulated and successfully tested under arctic conditions by Maykut and Untersteiner.[42] A serious drawback to incorporating the heat conduction equation into an ice model is that an accurate solution involves strong numerical constraints and substantial computer resources. The problem can, to a large extent, be overcome by only crudely resolving temperatures within the ice. Numerical experiments[36,43] indicate that results of sufficient accuracy for climate studies or biological applications can be obtained with as few as three or four internal temperature points. Numerical difficulties in this case are minimal.

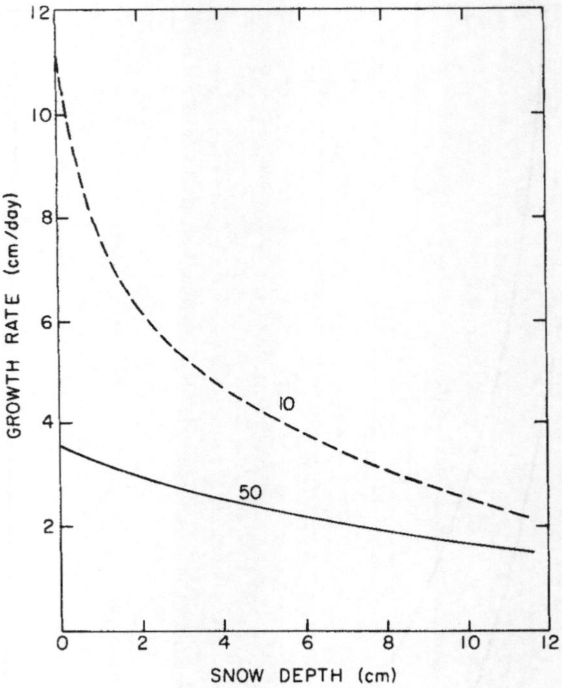

FIGURE 6. Growth rates as a function of snow depth for ice thicknesses of 10 and 50 cm.

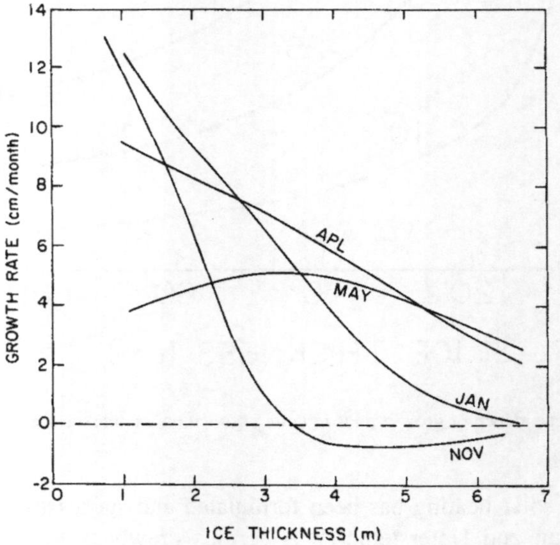

FIGURE 7. Growth rates in multi-year and thick first-year sea ice as a function of season in the Central Arctic.[36]

E. Equilibrium Growth

The concept of an *equilibrium thickness* (h_e) is frequently useful in discussing perennial ice in the Arctic. Such ice loses mass each summer by ablation at the upper boundaries and gains mass the rest of the year by accretion at the bottom. The amount of ice accreted each year depends strongly on h. If interannual variations in atmospheric and oceanic forcing are small, the ice will ultimately attain a steady-state thickness pattern where summer ablation essentially balances the net annual accretion. The ice is then in thermodynamic equilibrium with its environment and ice thicker than equilibrium will experience net annual thinning while ice thinner than equilibrium will undergo net growth. Model simulations (Section V. B) indicate that the magnitude of h_e is especially sensitive to factors which affect the amount of surface melting: air temperature, ice albedo, incoming radiation, and turbulent heat exchange. Despite small year to year changes in the present climate and advection of ice between different climatic regimes, observations show that undeformed older ice in the Arctic Basin almost invariably has a thickness in the 3 to 4 m range, in agreement with theoretical expectations for h_e in the region.[42] Although equilibrium ice in the Arctic may conceivably exist for any length of time, it is composed of ice whose age does not generally exceed 7 to 10 years. Over an annual cycle some 40 to 50 cm of ice are lost from the surface and replaced by an equal amount of new ice at the bottom. Thus, an ice particle will necessarily move upward with time at a velocity of 40 to 50 cm per year until eventually melting away at the surface.

F. Underwater Frazil Production

To this point the discussions have focused on *congelation ice* (sometimes called *columnar ice)* which grows at the ice-water interface in response to conductive heat losses along the temperature gradient in the ice. It has been widely accepted that such ice makes up the vast majority of the polar ice pack, however, recent data suggest that there are some regions where other ice formation mechanisms may be important.As mentioned earlier, multi-year ice in the Weddell Sea appears to be substantially thicker than would be expected on the basis of congelation growth alone. Ice core data from the region indicate that congelation ice accounts for only about 20% of the multi-year ice and 47% of the first-year ice, with fine grained layers of frazil crystals making up most of the remainder. This means that 3 to 4 m of frazil crystals must accumulate and be incorporated into the ice cover over 1 or 2 annual cycles. The higher percentage of frazil in thicker ice presumably reflects a decrease in the rate of congelation growth relative to the frazil ice, suggesting that frazil production is not strongly dependent on ice thickness. The fact that frazil layers are often sandwiched between layers of congelation ice indicates that frazil production is episodic.

Underwater frazil ice also appears to be common along the Antarctic coast. For example, Dayton et al.[44] observed 1 to 4 m thick billows of frazil crystals beneath the ice in McMurdo Sound. On some occasions ice crystals appeared throughout the water column accompanied by the formation of anchor ice on the sea bottom and on lines hanging in the water. Similar conditions were also noted by Morecki[45] near Mirny. It is suspected that periodic incursions of supercooled water generated at nearby ice shelves may be responsible for the coastal observations, but it does not seem likely that this mechanism could explain the extensive distribution of frazil in the Weddell Sea.

Because frazil crystals usually nucleate at some depth within the water column and then float to the surface, frazil accumulations tend to concentrate sediment and algae contained in the water. Weeks and Ackley[14] mention that concentrations of algae in young Weddell Sea ice can be several times that in the water. Microscope examination of the ice revealed considerable numbers of cells within individual crystals suggesting that such cells are an effective nucleating agent for frazil ice. At the same time, 1 to 2 orders of magnitude more cells were located at grain boundaries than within crystals. Since the ice was too young for significant biological growth to have taken place within the ice, it follows that the algae loading was primarily the result of scavenging by the rising crystals.

Analogous phenomena also appear to be widespread in coastal portions of the Arctic where divers frequently encounter accumulations of slush on the underside of congelation ice. This "slush ice" can reach thicknesses of at least 2.5 m but over 50% of the layer appears to consist of unfrozen water and divers can easily reach into it with their hands.[46] While frazil crystals form the bulk of this slush, it often contains large numbers of silt particles located in interstices between the frazil crystals. The source of many of these particles is likely to be scavenging from the water column, but direct entrainment by ice crystals forming on the sea floor could also be a factor. With time, upward loss of heat through the overlaying congelation ice will cause interstitial water in the slush to freeze producing a solid layer. Normal congelation ice may then form below the frozen slush. This was clearly the case in many of the Weddell Sea cores. Biological activity is severely impacted by the presence of silt-laden ice, light transmission and photosynthesis being negligible beneath it. Moreover, advection and subsequent melting of such ice after the spring breakup could represent an important sediment transport mechanism in the Arctic. Because most of the under-ice diving has taken place within a few kilometers of the coast, it is not known whether such intense frazil generation also occurs in deeper water where little if any sediment would be entrained. In the Central Arctic frazil production takes place in leads and in water below the ice, but there is little evidence from core data to suggest a significant effect on ice thickness.

Numerous mechanisms have been proposed to explain the occurrence of frazil ice in the ocean.[14,23,46] All depend on having slight supercooling and some degree of turbulence in the water column. Wind induced mixing is one such mechanism. Its effects are only felt near the surface and are rapidly damped out as the grease ice accumulates. In large leads and polynyas wind and wave action herd the frazil crystals to the downwind side where it can accumulate to depths of up to 1 m.[47] Currents and local circulations can carry some of these crystals beneath the surrounding ice but, except for regions with semipermanent polynyas, the process is selflimiting and does not seem capable of explaining the widespread distribution of frazil in areas like the Weddell Sea.

Another source of underwater frazil ice is the upward movement of water at its freezing point. Decreasing pressure as the water rises raises the freezing point producing supercooling and the potential for frazil formation. This mechanism is believed to be responsible for frazil generation near ice shelves, icebergs, and perhaps even pressure ridge keels. Likewise, contact between water masses of different salinity, which are each at their freezing point, can cause ice to form at their interface through the process of double diffusion. Such ice is known to occur where rivers empty into the sea and during summer runoff of meltwater from the ice pack. In some areas like the Beaufort Sea coast, permafrost extends beneath the sea floor promoting downward conduction of heat and the formation of anchor ice. Periodic release of anchor ice could thus contribute to sediment loading and ice accumulation at the bottom of the ice cover. While most of the above mechanisms are restricted to very special situations, underwater frazil ice can also form as a result of thermohaline convection, a process associated with sea ice throughout the polar regions. Drainage of cold dense brine from the interior of sea ice occurs during much of the year and produces descending plumes of brine that cause supercooling in the adjacent water. Some of the resulting ice forms as hollow stalactites around the brine where it exits the ice, but laboratory studies indicate that as much as half of the ice production is in the form of frazil crystals.[48] Similar plumes also exist under freezing leads. Weeks and Ackley[14] feel that there is a greater potential for frazil production in the Antarctic than in the Arctic and suggest that thermohaline convection could account for much of the frazil observed in the Weddell Sea. None of the above processes have yet been studied in sufficient detail to allow quantitative calculations. It appears, however, that underwater frazil ice is much more widespread than previously thought and that no single mechanism is likely to explain all of the observed phenomena.

G. The Summer Melt Cycle

Existing data indicate significant differences in summer melt progression between regions of perennial ice, coastal ice, and marginal ice and between the Arctic and Antarctic. In the Arctic, melting of the snow cover begins as the air temperature approaches the freezing point and typically takes 2 to 3 weeks. Accompanying the snow melt is a dramatic decrease in albedo due to the puddling of meltwater on the surface. Melt pond coverage on multi-year ice may briefly reach values of 50 to 60%, but quickly decreases to about 30% once the snow cover has been removed. With the development of melt ponds, area-averaged albedos drop from about 0.7 to 0.5. Deepening and consolidation cause a continual decrease in the areal extent of the ponds throughout the summer. Values as low as 10% are common by the end of August. Albedos increase from about 0.50 to 0.55 during this period. Because turbulent heat fluxes and longwave radiation act to remove heat from the surface during the summer, the energy needed to drive the melt cycle in regions of perennial arctic ice comes almost entirely from shortwave radiation.

Initial melting on seasonal ice near land produces nearly complete pond coverage due to the lack of surface topography. Even when the melt cycle is well advanced, it is not unusual for shallow ponds to occupy 60% or more of the surface. In addition, dust deposited from nearby land areas tends to lower the albedos even further. Areally averaged albedos as low as 20% have been reported from such regions.[49] Air temperatures in coastal areas are typically warmer than the ice and the turbulent fluxes make a substantial contribution to the melting. Thus, even when the incoming radiation fluxes are similar, ice decay proceeds more rapidly in coastal regions than in the interior pack. Due to the lower albedos and thinner ice, there will also be greater energy input to the underlying ocean from solar radiation penetrating through coastal ice.

The pattern of ice decay in the Southern Ocean appears to differ strikingly from that in the Arctic. Katabatic winds, higher latitudes, and lower oceanic heat fluxes cause melt rates near the continental boundaries to be small compared to those near the free boundary of the Antarctic ice pack. Moreover, the melt ponds which are so characteristic of ice melting in the Arctic have rarely been observed in the Antarctic and surface melting appears to be minimal. The reason is related to differences in the meteorological variables that control the surface heat balance. Surface melting in the Arctic often takes place with air temperatures significantly below freezing. In the Antarctic, Andreas and Ackley[16] have found that lower relative humidities and stronger winds enhance turbulent heat losses from the ice and allow surface melting to occur only with air temperatures significantly above 0°C, a condition rarely encountered over the ice cover in the Southern Ocean. In spite of the near absence of surface melting, most of the Antarctic ice pack disappears each summer due to melting at the ice-water interface. Gordon[50] estimates that no more than half of the necessary energy could be derived from the relatively warm deep water below the Southern Ocean pycnocline. The remainder must come from shortwave radiation absorbed by the upper ocean, either through leads in the ice pack or near the ice edge.

An open lead absorbs over 90% of the incident shortwave radiation, about half of which is absorbed in the upper 2 m. Part of this energy goes to lateral melting on floe edges, causing an increase in the amount of open water (A_w). This represents a positive feedback process which affects both the rate of retreat of the ice edge and the overall heat and mass balance of the ice pack. In sheltered bays and fjords along the Arctic coast surface melting is generally inadequate to remove the 2 to 2.5 m of ice which grows each winter. The annual disappearance of this ice can only be explained by taking into account the accelerated decay produced by solar heating in leads.[51]

Both ice dynamics (diverging motions) and thermodynamics (lateral melting on floe edges) affect A_w, though field data are not yet adequate to identify their relative importance in the different regions. Except in areas like the Greenland Sea, divergence of ice in the Arctic is believed to be much smaller than in the Antarctic. Nevertheless, substantial amounts of open water exist in the Arctic where the summer ice cover undergoes a dramatic increase in the number of leads and in the total amount of open water contained within the pack. Even in regions of perennial ice, summer values of A_w can reach 10 to 15%.[52] Clearly, the effects of $A_w(t)$ on the input of solar energy and meltwater to the upper ocean can have a substantial impact on biological processes there.

Various attempts have been made to relate the summer decrease in ice thickness to T_a through empirical equations analogous to those developed for young growing ice. Bilello[35] found a correlation of 0.93 between his data and thickness changes given by the equation:

$$\Delta h = 0.55\theta' \quad (6)$$

where Δh = the total decrease in ice thickness and θ' = the accumulated degree-days above freezing. Bilello also mentions a similar relationship [$\Delta h = 0.51\ (\theta'' - 32)$, $\theta'' \equiv$ accumulated degree-days above $-5°C$] developed by Karelin for the Soviet Arctic. The high degree of correlation between T_a and Δh is surprising because shortwave radiation is the dominant term in the summer energy balance. Variations in cloud cover, surface albedo, ice concentration, and F_w would seem to be as important as T_a. The probable explanation is that the observations were taken close to shore and were strongly influenced by warm air from nearby land masses. It is unlikely that such relationships could be employed away from coastal areas where summer values of T_a are invariably close to the freezing point.

III. STRUCTURE AND SALINITY

A. Crystal Orientation

Ice crystals are hexagonal in form and are characterized by one principal hexagonal axis of symmetry (the c-axis) which is the optic axis of the crystal. Perpendicular to the c-axis is the *basal* (0001) *plane* defined by three crystallographic axes (the a-axes) of equal length with angles of 120° between the positive ends. *Prism* faces are 6 crystal faces parallel to the c-axis. The arrangement of atoms in an ice lattice is such that fracture along the basal plane requires the breaking of only two bonds, while fracture along any plane normal to this requires the breaking of four or more bonds. Slip or cleavage are thus most likely to occur parallel to the basal plane.

In natural sea ice the orientation of individual crystals tends to be highly organized. Only near the surface where the ice is composed of frozen slush or grease ice, are crystal orientations random. Once congelation growth begins, crystals whose c-axes are parallel to the ice-water interface quickly begin to dominate the structure of the ice sheet. The explanation for this phenomenon is anisotropic growth on different crystal faces which results in a geometric selection of crystals with favorable orientations. The structure of the ice lattice is such that the density of potential bonding sites is greater on a prism face than on a basal plane. This means that for a given degree of supercooling, growth parallel to the c-axis will be much slower than growth perpendicular to the c-axis. Observations[53] show growth rate differences of as much as two orders of magnitude. Crystals whose c-axes are closer to the horizontal grow downward into the water faster than those with a more vertical orientation, wedging out the slower growing crystals. This process proceeds quite rapidly until only those crystals with c-axes parallel to the freezing interface remain. As we shall see later, this turns out to be an important factor in determining the structure and properties of sea ice.

The region over which the crystal orientation shifts from being random to having essentially all c-axes horizontal is termed the *transition zone* and typically occurs over a vertical distance of only 5 to 10 cm. The transition zone begins at the bottom of the frozen frazil layer and hence may be located anywhere in the upper meter of the ice cover, depending on the initial thickness of the frazil layer. Beneath the transition layer there is little interference between the growth of adjacent crystals and continued development of these crystals occurs primarily in the vertical. The region between the bottom of the transition zone and the ice-water interface is the so-called *columnar zone* and is characterized by crystals strongly elongated in the direction of the temperature gradient. Crystal diameter in the upper half meter of the columnar zone is typically on the order of 1 to 2 cm. There is a definite tendency for average crystal diameter to increase with depth in a roughly linear fashion.[14,54] The reason appears to be associated with the decreasing growth rate of the ice as it thickens.

Until recently it was generally accepted that horizontal c-axis alignments were random throughout the columnar zone. However, several studies (e.g., Cherepanov,[55] Weeks and Gow[56]) now indicate that strong horizontal alignments exist over large portions of the Arctic. Nearshore observations suggest that the c-axes are closely oriented in the direction of the average current at the bottom of the ice. Since the microscale roughness on the underside of the ice is greater when the current is parallel to the c-axis (see next section), Weeks and Gow[56] argue that the resulting increase in turbulence decreases the thickness of the stable boundary layer at the growing interface, giving such crystals a slight growth advantage. Although this explanation is plausible, additional field and laboratory work is needed to verify it.

B. The Skeletal Layer and Brine Entrapment

Examination of the underside of growing sea ice reveals the presence of a delicate layer of thin ice platelets (Figure 8a) arranged perpendicular to the c-axes of the individual crystals. The many platelets growing on the bottom of each exposed crystal are generally parallel (Figure 8b) with a spacing (a_0) of 0.5 to 1.0 mm. The zone of platelet occurrence has a vertical dimension on the order of 1 to 3 cm and is frequently referred to as the *skeletal layer*. It has been found both theoretically and experimentally that a_0 is proportional to $f^{-1/2}$. Clearly, flow parallel to the rows of platelets generates the least turbulence, allowing a slightly thicker boundary layer to develop beneath crystals with c-axes normal to the flow and possibly explaining the apparent growth advantage of crystals more closely aligned with the flow.[56]

Absence of a skeletal layer beneath freshwater ice suggests that the complex microstructure found at the growing interface of sea ice is somehow related to the presence of salt in the water. When seawater freezes, most of the impurities (salts) are rejected from the ice lattice resulting in a thin boundary layer of very salty water ahead of the advancing interface. The continual addition of salt causes the salinity at the top of this layer to be greater than the salinity beneath the layer. Because of the higher salinity, temperature at the ice-water interface will be lower than that in the underlying water. These conditions produce a downward flux of salt and an upward flux of heat through the layer. If we assume that the transport of heat and salt is primarily the result of diffusive processes, it follows that *constitutional supercooling* can occur in this boundary layer. Due to large differences in the rate at which heat and salt diffuse, the flux of heat through the boundary layer is much more rapid than the flux of salt and salinity undergoes an exponential type decrease below the interface, while the temperature increase is more nearly linear. Since it is the salinity which determines the freezing point of the solution, the actual temperature will typically be below the freezing point (Figure 9). Under such conditions, laboratory observations[57] reveal that the ice-water interface initially develops an array of parallel knife-edged cells on which scallops form and deepen until the cells separate into a line of individual platelets. The length and spacing of these platelets act to minimize the amount of supercooling ahead of the interface. An approximate theory has been worked out to describe criteria for constitutional supercooling and the geometry of the interface.[14] Efforts are underway to develop a more exact theory by allowing for free convection near the interface instead of only diffusion.

The significance of the skeletal layer is that it provides a mechanism for including sig-

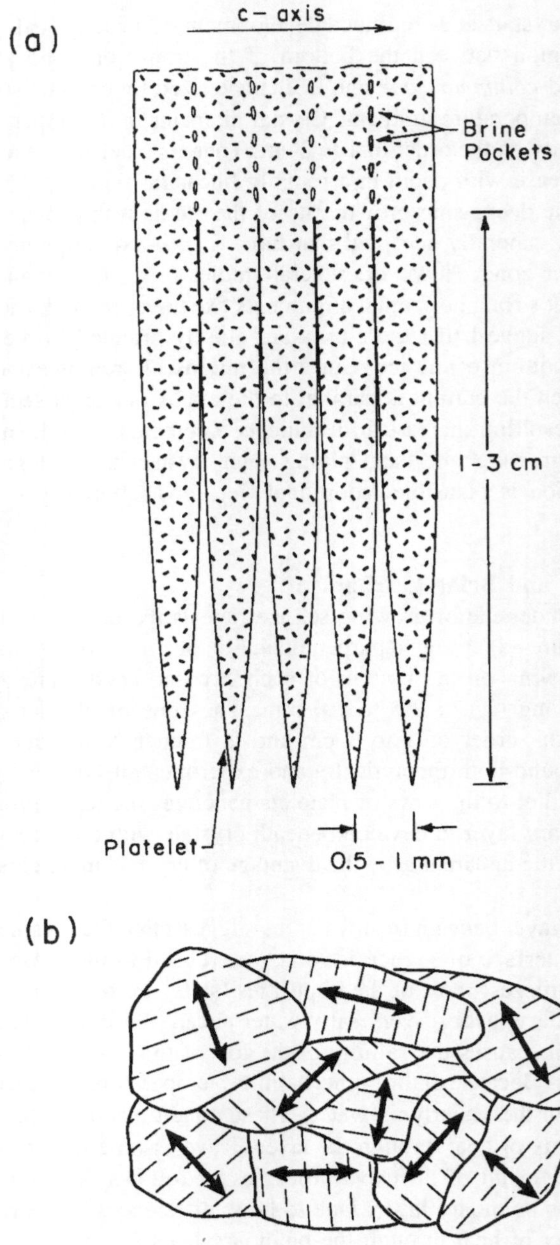

FIGURE 8a. Schematic illustration of the platelet structure on the bottom of a growing ice crystal. Figure 8b. Plan view showing platelet arrangement at the underside of growing sea ice. Curved lines represent crystal boundaries and short straight lines the platelets. Note that platelets on an individual crystal tend to form in parallel sets normal to the horizontal direction of the c-axis (double ended arrow).

nificant amounts of salt in the ice. As the platelets lengthen, they grow wider and periodically small ice bridges form between adjacent platelets, entrapping pockets of brine. These brine pockets are typically long and narrow, being on the order of 0.05 mm in diameter. Sea ice structure is thus characterized by not only the columnar crystals, but also by many vertical

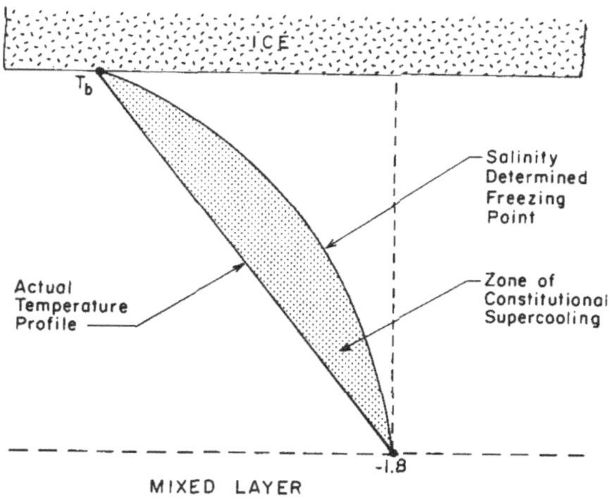

FIGURE 9. Constitutional supercooling in the boundary layer ahead of an advancing ice interface.

strings of liquid inclusions that mark the original boundaries between the platelets. As discussed in Section IV, these brine pockets dictate the bulk thermal and mechanical properties of the ice. Without this entrapment mechanism, there would be little to distinguish sea ice from freshwater ice.

Initially, the composition of the entrapped brine should have the same relative concentration of ions as normal seawater. However, as temperature decreases in the ice, solid salts begin to precipitate out of the solution. While mirabilite ($Na_2SO_4 \times 10H_2O$) begins to form at $-8.2°C$, the most important precipitate is sodium chloride ($NaCl \times 2H_2O$) which occurs below a temperature of $-22.9°C$. Appearance of the $NaCl \times 2H_2O$ causes the ice to take on a milky color and alters its optical properties. Other solid salts which may be present in small amounts include $CaCO_3 \times 6H_2O$, $MgCl_2 \times 8H_2O$, KCl, $MgCl_2 \times 12H_2O$, and $CaCl_2 \times 6H_2O$.[14]

In general, the salinity of the brine in a pocket is independent of the salinity of the original water because of mass changes taking place at the walls of the pocket. The temperature of the brine (T_b), which in the normal state of thermal equilibrium is the same as that of the surrounding ice, must be at the freezing point of the solution (T_f). If $T_b > T_f$, melting will occur on the ice walls, lowering the salinity of the brine (S_b) until $T_f = T_b$. Similarly, if $T_b < T_f$, freezing will occur in the brine, raising S_b until T_f and T_b coincide. This being the case, salinity in a brine pocket can be related to T_b through the following freezing point equations which are based on the work of Assur:[58]

$$S_b = \frac{1000\ T_b}{T_b - 54.11} \qquad T_b > -8.2°C \tag{7}$$

and

$$S_b = \frac{62.4 - 10.31\ T_b}{1.0624 - 0.01031\ T_b} \qquad -22.9 \leq T_b \leq -8.2°C \tag{8}$$

where S_b is in ‰ and T_b in °C. Also fitting the data reasonably well are the approximate relations given by Morey et al.[59]: $S_b = 9.65 - 14.8\ T_b$ when $-8.2 \leq T_b \leq -2°C$ and $S_b = 78.11 - 6.6\ T_b$ when $-22.9 \leq T_b \leq -8.2°C$. The change in the $S_b(T_b)$ relationship below $-8°C$ is due to the precipitation of $Na_2SO_4 \times 10H_2O$ from the solution. Below $-23°C$ the data are somewhat ambiguous and proper freezing point relations have not def-

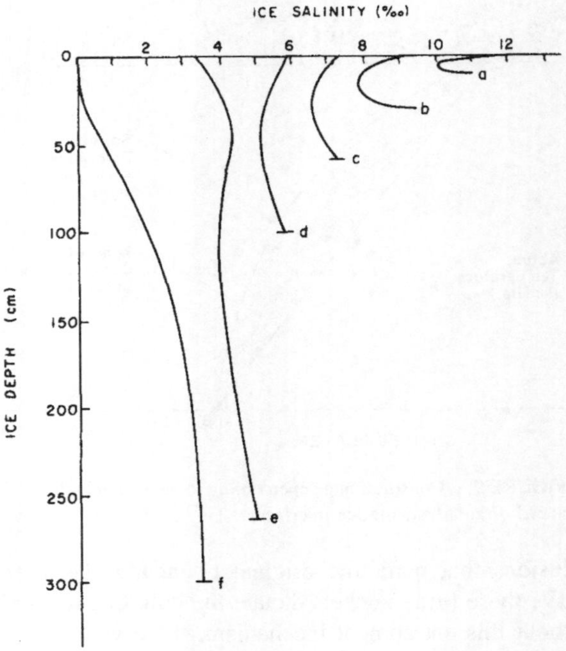

FIGURE 10. Idealized salinity profiles in Arctic sea ice of various thicknesses. Curves a to d represent first-year ice. The remaining curves illustrate the 2 classes of profiles found in multi-year ice: curve f shows S_i beneath hummocks and other elevated areas, while curve e shows S_i beneath low areas where the surface is close to the freeboard level.

initely been established. Since the temperature of the surrounding ice (T_i) will normally equal T_b, it is therefore possible to calculate S_b by simply measuring T_i and applying Equation 7 or 8.

C. Ice Salinity

The salt content of sea ice is usually described in terms of a *bulk salinity* (S_i) defined as:

$$S_i = \frac{\text{mass of salt}}{\text{mass of ice + mass of brine}} \cdot 10^3 \tag{9}$$

where S_i is in ‰. The amount of salt initially entrapped by the growing ice depends strongly on growth rate and water salinity. Laboratory measurements by Cox and Weeks[60] show two different regimes. When $f < 2 \times 10^{-5}$ cm sec^{-1} ($\sim$1.7 cm day^{-1});

$$S_i = S_w(0.8439 + 0.0529 \ln f) \tag{10}$$

while for $f > 2 \times 10^{-5}$ cm sec^{-1}

$$S_i = 0.26\, S_w[0.26 + 0.74 \exp(-7243f)]^{-1} \tag{11}$$

The flux of salt to the water beneath the growing ice can thus be written as $F_{salt} = \rho_i f(S_w - S_i)$ and estimated using Equation 10 or 11.

If the brine did not migrate in the ice, we would expect to see stationary salinity profiles with a maximum near the surface and a continuous decrease with depth, reflecting the reduction in f as the ice thickened. Instead we find that $S_i(z)$ is highly time dependent with a shape that suggests substantial vertical movement of brine after entrapment. Figure 10 shows idealized salinity profiles for different ice thicknesses based on data from a variety of sources.[14,37,61-63] Actual salinity profiles exhibit considerable vertical scatter, presumably correlated with changes in thermal forcing (see Nakawo and Sinha[37]). Likewise, the magnitude of S_i can be quite different for ice of equal thickness because of differences in growth rate. Nevertheless, Figure 10 illustrates several important points. Profiles in thinner ice tend to be C-shaped indicating salt enrichment in the lower half of the slab. The average salinity also decreases as the ice ages, (i.e., becomes thicker) suggesting loss of brine to the ocean. Finally, there is a dramatic change in $S_i(z)$ in multi-year and melting first-year ice.

To explain the observed profiles, Untersteiner[64] examined the role of 4 mechanisms which transport brine within the ice. *Brine pocket migration* will occur in the presence of a temperature gradient across the slab. Melting on the warm side of the pocket and ice accretion on the cool side cause the pocket to move opposite to the direction of heat flow. Under normal conditions, this mechanism is too slow to account for a significant amount of brine transport, although if the pocket is large enough to support convective heat transfer, movement can be quite rapid. Another mechanism is *brine expulsion* resulting from the pressure buildup in refreezing pockets. Again brine expulsion does not appear to play a major role in the desalination of sea ice. It is likely to be important only during the early stages of growth when the ice is undergoing rapid cooling. The "salt flowers" and extremely high salinities often found at the surface of young ice are probably the result of this process.

The mechanism primarily responsible for observed salinity changes in first-year ice is almost certainly *gravity drainage*. Because brine pockets typically form in vertical strings, they tend to become interconnected in response to temperature changes and internal stresses, forming long vertical tubes or brine channels which allow the vertical movement of brine through the ice. During most of the year temperature increases with depth and colder denser brine in the upper part of the ice will sink downward to displace the warmer brine. In addition, bottom growth causes the ice sheet to move upward, creating a hydrostatic head that forces brine downward in the channels and out of the ice. Average brine channel diameters near the bottom of the ice are on the order of 0.4 cm. Depending on ice thickness, as many as 50 to 300 channels per m^2 may exist along the bottom of the ice. The cold streamers of brine leaving the ice cause hollow tubes of ice to form around them. These "ice stalactites" have been observed to reach lengths of 6 m in the Antarctic, but lengths of 25 to 50 cm seems to be more representative in the Arctic. Brine movement within the ice causes complex changes within the channels. It has been noted[65] that primary brine channels are often accompanied by smaller tributary channels which provide additional brine and form what is, in effect, a brine drainage system. Stalactite formation and flow in the primary channels have been extensively studied,[48,66,67] but few details are available regarding the formation and role of the tributary channels.

The final mechanism is *flushing* by surface meltwater. As surface temperatures approach the freezing point in spring and early summer, brine volume increases and the ice becomes extremely porous. Melting of the snow cover and the formation of ponds on the surface allow freshwater to percolate into the ice resulting in a further decrease in S_i. Calculations[64] indicate that the greatest salt loss occurs where brine volume is a maximum and that flushing appears to be sufficiently vigorous in the Arctic to explain the observed differences between salinity profiles in first-year and multi-year ice. Flushing does not appear to be a significant factor in the evolution of Antarctic sea ice.

Data from a wide variety of sources and locations were compiled by Cox and Weeks[63] to produce plots of average salinity vs. ice thickness in the Arctic (Figure 11). For cold ice the plots show a roughly linear decrease in $\bar{S}_i$ up to a thickness of about 40 cm, at which point the slope abruptly changes, possibly indicating a shift in the dominant desalination mechanism from brine expulsion to gravity drainage. It should be noted that, in spite of a wide range of growth conditions, $\bar{S}_i(h)$ shows surprisingly little scatter. Measurements of $\bar{S}_i$

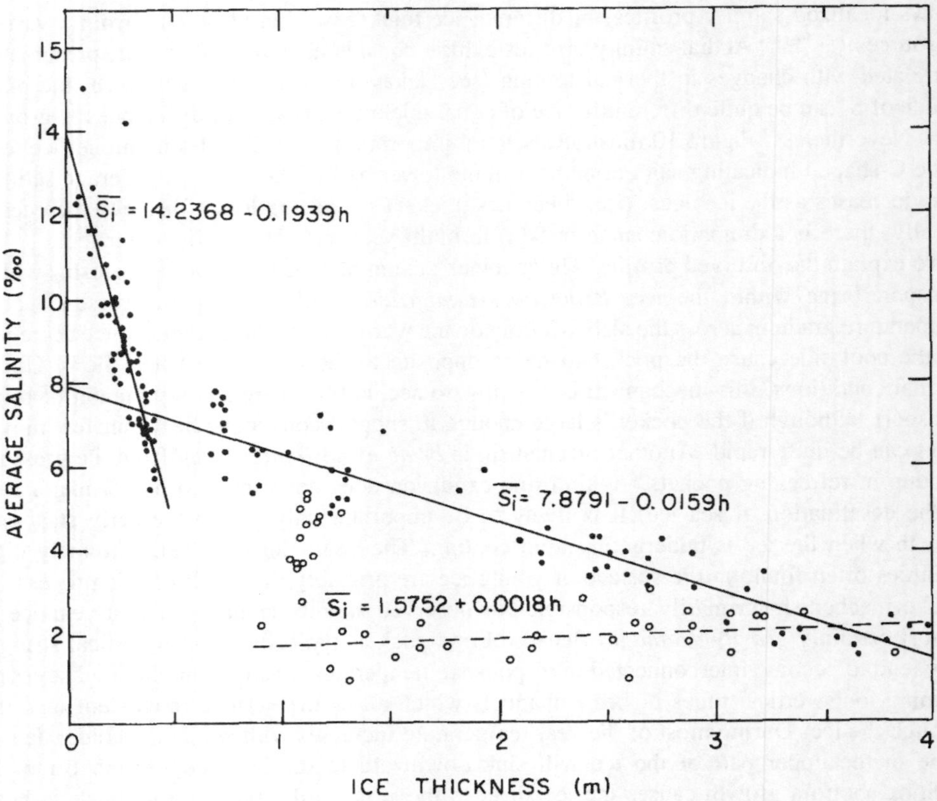

FIGURE 11. Average salinity of Arctic sea ice as a function of ice thickness.[63] Solid lines show $\bar{S}_i(h)$ for cold ice, while dashed line shows $\bar{S}_i(h)$ for warm ice late in the melt season. Data points for the cold ice are given by dots and for the warm ice by open circles.

when the ice is melting are complicated by brine leakage from the cores, but there is at best only a weak dependence on h.

Chemical studies[68] show that sulfate-chlorinity ratios in sea ice are different from the sea-water ratio. On aging, sulfate in the ice is mobilized and removed in a constant ratio to chlorinity, indicating conservative behavior. Comparison of first-year and multi-year ice suggests that fractionation occurs during ice formation, but that further fractionation within the ice does not appear to take place. Because much of the entrapped brine ultimately returns to the ocean, there seems to be no lasting effect on the chemical composition of the underlying water as a result of the freezing process.

The evolution of ice salinity can be summarized as follows. High initial growth rates entrap substantial amounts of brine producing S_i values near the surface of 10 to 20‰. As h increases, f decreases and less brine is entrapped. However, cooling of the upper part of the ice results in brine expulsion that allows S_i in the lower part of the ice to remain high. Substantial quantities of brine are lost to the underlying ocean during this period. By the time h approaches 40 cm, brine channels appear to be sufficiently developed for gravity drainage to control the rate of desalination. The ice continues to lose salt through this process until the onset of snow melting when flushing by freshwater occurs. Beneath elevated, drained areas there is a sharp drop in salinity while beneath low areas close to the freeboard level the decrease is much less pronounced. This pattern persists even in older multi-year ice where horizontal salinity gradients exist which are correlated with surface topography.[63]

D. Annual Layering in Sea Ice

Cores taken in multi-year Arctic sea ice reveal the existence of horizontal layers which are suspected of developing during the annual thaw cycles.[62,69] These layers are of two different types. The first type is a 2 to 5 mm thick layer, milky white in color, with a sharp upper boundary and an irregular lower boundary. The cause of this layer is uncertain, but it appears to form during the summer after the ice has stopped growing. It has been suggested that the milkiness of the ice may be associated with biological activity in the ocean.[14] The second type is much thicker (1 to 10 cm) and appears to result from the freezing of surface meltwater that penetrates beneath the ice. The size of the crystals is much smaller than in the surrounding layers of normal congelation ice and the salinity much lower (1 to 1.5‰). The presence beneath the ice of freshwater from surface melting can be explained by flow through natural or artificial drain holes, or by runoff into nearby leads that subsequently close. If vertical mixing is small, there will be a sharp interface between the freshwater at its freezing point (0°C) and the underlying seawater at its freezing point (~ − 1.8°C). This results in supercooling in the freshwater near the interface and ultimately in the growth of ice crystals throughout the freshwater layer.[70] Since under-ice freshwater does not necessarily occur each summer at a given location, the refrozen meltwater layers are often separated by one or more of the thin milky layers. These layers make it possible to estimate the age at various depths within the ice. Obtaining the total age of the ice cover by this method is difficult because melting and refreezing near the surface tend to obliterate the uppermost layers.

IV. PHYSICAL PROPERTIES OF SEA ICE

As the ice changes temperature, internal melting or freezing within the brine inclusions affect the brine volume of the ice. The fractional volume of brine in the ice (υ_b) can be written as:

$$\upsilon_b = -\frac{\rho_i S_i}{\rho_b S_b} \tag{12}$$

where ρ_i is the density of the ice and ρ_b is the density of the brine. From Equations 7 and 8 we know that S_b is dependent only on T_i, hence υ_b will be a function of S_i and T_i. Application of Equation 12 over a wide temperature range is complicated by the precipitation of solid salts and uncertainties in $\rho_b(T_i)$, however, a table of υ_b for standard sea ice has been tabulated by Assur[58] down to −54°C. In most cases values of sufficient accuracy can be obtained with a simple equation developed by Frankenstein and Garner:[71]

$$\upsilon_b \simeq S_i\left(0.0532 - \frac{4.919}{T_i}\right) \qquad -22.9 \leq T_i \leq -0.5°\text{C} \tag{13}$$

where S_i is in ‰ and υ_b in percent. The brine volume at the bottom of growing sea ice can vary from about 8 to 40%. For multi-year ice υ_b is typically about 10%, except near the surface during the summer when values as high as 40 to 50% have been observed. As will be shown below, the magnitude of υ_b has a major impact on the bulk thermal, mechanical, optical, and electromagnetic properties of sea ice.

In addition to the entrapment of brine, growing sea ice also captures significant numbers of air bubbles. According to Schwerdtfeger,[72] the fractional volume of air bubbles in sea ice (υ_a) is given by:

$$\upsilon_a = 1 - \rho_i\left(\frac{1 - S_i'}{\rho_0} + \frac{4.98 S_i'}{T_i}\right) \qquad T_i > -8.2°\text{C} \tag{14}$$

where ρ_o = 0.9178 mg m^{-3} is the density of pure ice, ρ_i is the bulk density of the sea ice, and S'_i = $S_i/1000$ is the fractional salt content of the ice. Equation 14 can be rewritten and simplified to obtain an approximate relation for $\rho_i(\upsilon_a,T_i,S_i)$:

$$\rho_i \simeq (1 - \upsilon_a)\left(1 - \frac{0.00456\ S_i}{T_i}\right)\rho_o \quad (15)$$

It can be seen from Equation 15 that ρ_i is mainly determined by S_i at warm temperatures and by υ_a at lower temperatures. Since the major changes in thermal and mechanical properties occur when the ice is warm, air bubbles will be neglected in the following discussions.

A. Thermal Properties

If we remove heat from a block of sea ice, T_i decreases causing brine to freeze in the pockets. This freezing releases latent heat which acts to warm the ice. As a result, sea ice cools more slowly than fresh ice. Alternatively, warming produces melting in the brine pockets which slows the rate at which T_i increases. Brine pockets thus behave as thermal buffers which retard the transport of heat in sea ice, affecting both conductivity and specific heat. These changes in brine volume also alter the latent heat of fusion of the ice.

1. Conductivity

The conductive heat flux in the ice (F_c) is given by $F_c = k_i(\partial T_i/\partial z)$, where k_i is the *thermal conductivity* of the ice. The magnitude of k_i and the temperature gradient thus determine the rate of conductive heat transfer. In pure ice the conductivity (k_o) depends on temperature and is approximately k_o = 9.828 exp (−0.0057T), where T is in K and k_o in W m^{-1} K^{-1}.[73] A typical value of k_o for warm ice would be 2.1 W m^{-1} K^{-1} (0.005 cal $cm^{-1}sec^{-1}K^{-1}$). There appears to be a slight difference between k_o measured parallel to the c-axis and that measured normal to it, but this difference is small (~5%) and can usually be neglected. In sea ice, conductivity is substantially lower parallel to the c-axis than normal to it. This is because brine and air bubble layers are regions of lower conductivity and more of these layers are intersected by heat conducted along the c-axis than perpendicular to it. Since horizontal temperature gradients in natural sea ice are usually negligible, only heat flow in the vertical, (i.e., perpendicular to the c-axis) need be considered for practical purposes. Detailed theories, taking into account such factors as brine pocket distribution, air bubbles, and the precipitation of solid salts, have been worked out.[72,74] Tables of $k_i(T_i,S_i,\rho_i)$ are given in Schwerdtfeger.[72] An approximate expression for k_i was proposed by Untersteiner:[41]

$$k_i = k_0 + \frac{\beta S_i}{T_i} \quad (16)$$

where T_i is in °C and $\beta = 3.11 \times 10^{-4}$ cal $cm^{-1}sec^{-1}$ (0.13 W m^{-1}) is a constant. Although somewhat crude, Equation 16 is particularly convenient in applications requiring a simple functional relationship between k_i, T_i, and S_i. Predicted values of $k_i(S_i,T_i)$ are shown in Figure 12. Note that there is little dependence on temperature below about −4°C. Under natural conditions, k_i may vary by 10 to 20% in perennial ice and by up to 50% in young ice.

At first glance it would appear that Equation 16 predicts an infinite value for k_i as $T_i \rightarrow$ 0°C. This turns out not to be the case because T_i cannot approach 0°C unless $S_i \rightarrow 0$. Whereas pure ice has a well defined temperature (0°C) at which phase transitions occur, sea ice does not. Any change in T_i produces internal phase changes. It can be shown that for ice of a given salinity there is a temperature (T_m) above which the ice cannot exist without completely melting away. According to Ono,[75] $T_m \simeq -0.05411\ S_i$. This is consistent with the notion that the temperature of sea ice can reach 0°C only if S_i = 0.

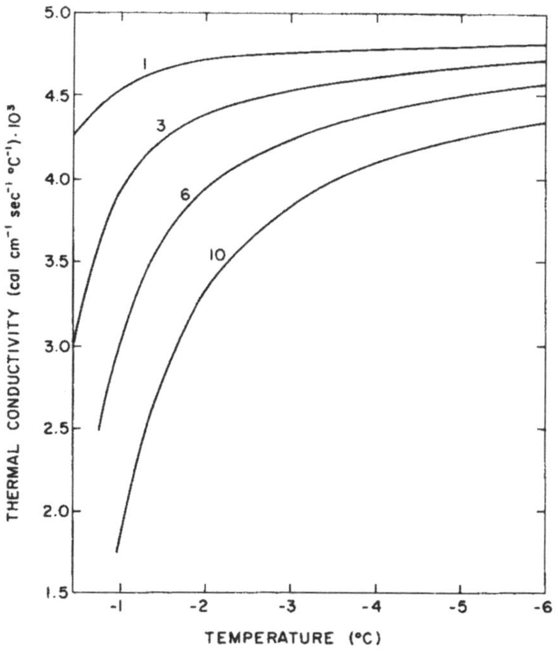

FIGURE 12. Dependence of thermal conductivity in sea ice on ice salinity (‰) and temperature according to Untersteiner.[41]

2. Specific Heat

The *specific heat* of a substance is defined as the amount of heat necessary to raise a unit mass of the material 1°C and is a fundamental property needed in calculating time-dependent temperature changes. The specific heat of sea ice (c_i) is the sum of the heat required to warm the ice and the brine plus that needed for melting in the brine pockets. Ono[75] has derived a theoretical expression for c_i applicable to temperatures above $-8°C$:

$$c_i = c_0 + aT_i + \frac{bS_i}{T_i^2} \qquad (17)$$

where $c_o = 0.505$ cal $g^{-1}°C^{-1}$(2113 J $kg^{-1}°C^{-1}$) is the specific heat of pure ice, $a = 0.0018$ cal $g^{-1}°C^{-2}$(7.53 J $kg^{-1}°C^{-2}$) is a constant, and $b = 4.3115$ cal °C g^{-1} (0.018 MJ °C kg^{-1}) is also a constant. A slightly simpler form which gives similar results is the empirical formula of Untersteiner:[41] $c_i = c_o + \xi S_i/T_i^2$, where $\xi = 4.1$ cal °C g^{-1} (0.0172 MJ °C kg^{-1}). Tables and analytical expressions are also available for colder ice.[72,74] Values of $c_i(T_i,S_i)$ are shown in Figure 13. As with k_i, the largest changes in c_i occur at temperatures above -4 to $-6°C$, but the magnitude of the change is much larger. Under natural conditions c_i can vary by a factor of 20 or more. The Figure demonstrates that it becomes increasingly difficult to change T_i as $T_i \rightarrow T_m$, primarily because mass changes in the brine pockets consume an increasing share of the available energy.

3. Latent Heat of Fusion

The *latent heat of fusion* (L_o) of pure ice is the amount of heat required to melt (or freeze) 1 g of ice at 0°C. Because internal melting and freezing occur continuously over a wide temperature range, the traditional concept of absorption or release of latent heat at a constant temperature does not strictly apply to sea ice. Nevertheless, it is possible to calculate the amount of heat necessary to melt a sample of sea ice with temperature T_i and salinity S_i. It

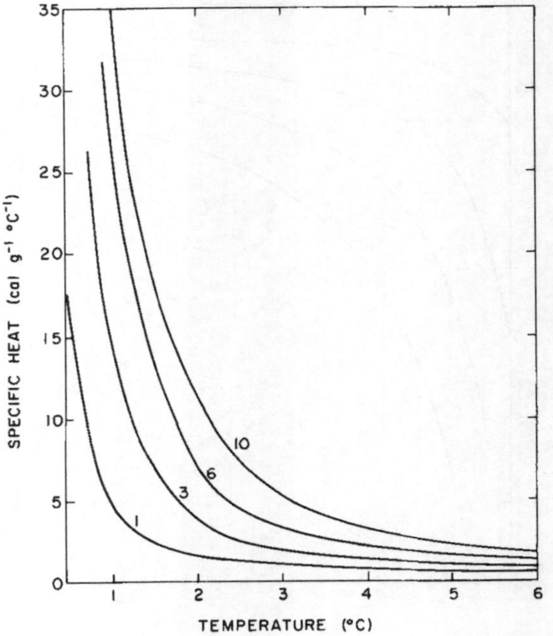

FIGURE 13. Dependence of specific heat in sea ice on ice salinity (‰) and temperature according to data from Ono.[75]

is this quantity that we will take to be the heat of fusion (L_i) of sea ice. A precise formula for L_i was derived by Ono:[75]

$$L_i = L_0 - 0.505T_i - 0.0273S_i + 4.3115\frac{S_i}{T_i} + 0.0008S_iT_i - 0.0009T_i^2 \qquad (18)$$

where $L_o = 79.68$ cal g^{-1}(0.3335 MJ kg^{-1}). Simpler expressions are also possible. Since the amount of ice present in a unit mass of sea ice is $(1 - 0.01\upsilon_b)$, $L_i = (1 - 0.01\upsilon_b)L_o$ and from Equation 13.

$$L_i \simeq \left[1 - S_i\left(0.000532 - \frac{0.04919}{T_i}\right)\right]L_0 \qquad (19)$$

These equations can be used to estimate the heat of formation of sea ice. Suppose, for example, that salinity near the bottom of a growing ice cover is found to be 8‰. Then, if $T_f = -1.8$°C, $\upsilon_b \simeq 22\%$ and $L_i = 61.2$ cal g^{-1}, where L_i represents the amount of heat that must be conducted away from the growing interface to form 1 g of salty ice. As the ice cools below T_f, additional latent heat is released by freezing in the brine pockets, but that heat is accounted for in c_i and is not included in the heat of formation.

4. *Diffusivity and Conductance*

A parameter appearing frequently in the literature is the *thermal diffusivity*, $\kappa_i = k_i/\rho_i c_i$, a simple combination of previously discussed properties. This combination arises naturally from the form of the heat conduction equation (see Section V.B) and can be calculated directly from observations of the rate of change of temperature profiles in the ice.[72] There is a very strong decrease in κ_i with temperature above -4°C. The most complete study of $\kappa_i(T_i,S_i)$ for sea ice was made by Ono.[76]

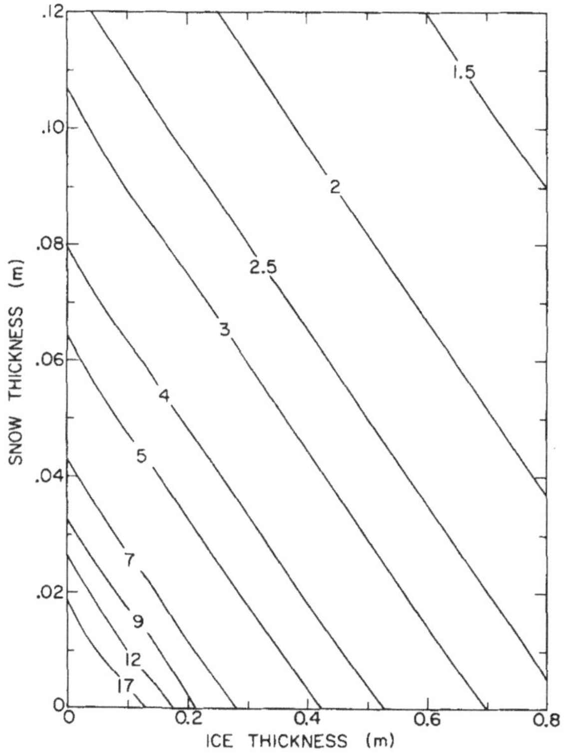

FIGURE 14. Thermal conductance (W m^{-2}°C^{-1}) of Arctic sea ice during winter and spring as a function of ice and snow thickness.[39]

A useful quantity for estimating growth rates and heat conduction in snow covered first-year ice is the *thermal conductance*, $\gamma = k_i k_s/(k_i h_s + k_s h)$. Because $k_s \ll k_i$, small amounts of snow can have a large impact on the amount of heat conducted to the surface (F_c). In situations where h_s undergoes significant changes in time or space, the insulating effect of the snow cannot be ignored. Equation 5 shows that surface temperature and γ are sufficient to obtain F_c when the ice is cold. If F_w is known or assumed to be negligible, the amount of growth can be calculated directly from F_c and L_i. The dependence of γ on h and h_s is shown in Figure 14 for values of k_i and k_s typical of winter conditions in the Arctic.

B. Mechanical Properties

Sea ice is subject to a variety of forces arising from the movement of air and water across its horizontal boundaries and from internal stresses generated by the surrounding ice. The resulting deformations are of two types: elastic strains which are reversible when the stress is removed and inelastic strains which are irreversible. Ice is often described as a visco-elastic material, that is, when deformed rapidly or for brief periods, it behaves elastically while for sustained loadings it tends to creep with a rate depending on time and the amount of stress. The mechanical behavior of nonsaline ice depends on temperature, porosity, grain size, strain rate, and structure. Sea ice is further complicated by the presence of brine inclusions, strong crystal orientation, biological activity, and organized pore geometry. Much of what is known about the rheology and strength of ice has been learned from laboratory measurements. Continual variations in the state of the material make such measurements difficult and much work remains to be done. It is not the intent here to summarize the numerous and sometimes contradictory experimental data. Rather, we shall try and present a more qualitative picture of how sea ice responds to various types of mechanical forcing and how this response depends on other physical properties of the ice. An excellent review of current knowledge regarding the mechanical behavior of sea ice, together with a discussion of the underlying continuum mechanics concepts, has been given by Mellor.[77] Additional information on the small scale physics can be obtained from Doronin and Kheisin.[31]

1. Elastic Constants

The theory of elasticity forms the basis for studies of deformable solids. A perfectly (or linearly) elastic material is one in which strain (ϵ) is directly proportional to σ, i.e., $\sigma = E\epsilon$, where $E \equiv$ *Young's modulus,* the ratio of axial stress to strain. The ratio of the lateral strain to the axial strain is called *Poisson's ratio* (μ). Together, E and μ characterize the elastic behavior of the material. In nonsaline ice, E depends only weakly on T_i, but porosity (air bubble volume) has a significant effect. In sea ice, T_i strongly affects the porosity (brine volume) and hence E. Small scale measurements on first-year ice over a brine volume range of 0 to 8.5% indicate a relationship of the form $E = a + b\upsilon_b$, where a and b are constants.[78] These measurements also suggest that μ ($\simeq 0.3$) is relatively constant in sea ice and independent of T_i and S_i. Theoretical studies of the dependence of E on υ_b have been carried out by Bergdahl[79] assuming that brine pockets and channels within the ice could be modeled as vertical rectangular tubes. For stresses along the tubes he found a relationship with the same form as the empirical studies. However, for stresses normal to the tubes, the relationship was of the form $E = a/(1 - b\upsilon_b)$, where b depends on the assumed geometry of the brine pockets. This result is supported by numerous empirical observations indicating that the direction of applied stress is an important factor in the mechanical response of natural sea ice.

Conditions under which the ice behaves elastically are fairly limited and often are not satisfied in field and laboratory measurements reported in the literature. Nevertheless, elastic analysis is still applied in many problems that involve inelastic processes. Elastic moduli determined under these conditions are called "effective" moduli and are denoted by E′ and μ'. Effective moduli have proved to be useful for many practical purposes and a variety of studies have been made to determine their dependence on T_i, porosity, strain rate, and stress state. In general E′ is substantially less than E, except at very low temperatures and high rates of stress or strain. The rate of decrease in E′ with increasing porosity is much more rapid than for E and there is roughly a 50% increase in E′ as temperature decreases from −5 to −30°C. Values of μ' at very low stress rates approach 0.5 and vary between 0.35 to 0.4 for high stress rates. In contrast to fresh ice, there is a tendency for μ' in saline ice to increase with increasing T_i (and hence υ_b). In natural sea ice, where observations indicate that μ' equals 0 to 0.2 in the vertical and 0.8 to 1.2 in the horizontal, sample orientation is of utmost importance.[80]

2. Strength

The most useful property for engineering purposes is strength, the ratio of the minimum load necessary to fracture the ice to its cross-sectional area. Depending on temperature, structure, and the rate of deformation, either brittle or plastic fracture may occur. Several different parameters are commonly used to characterize ice strength. *Uniaxial compressive strength* (σ_c), the maximum stress that the ice can sustain in a particular direction without fracturing, is usually obtained by compression of ice cylinders at a specific rate of strain. Results show that σ_c in sea ice increases as strain rate ($\dot{\epsilon}$) to approximately the one fourth power. As expected, σ_c decreases more strongly with temperature in sea ice than in either lake or river ice and is much larger when force is applied in the direction of ice growth (vertical) than when applied in the horizontal. The ice is weakest when the stress is 45° to the horizontal plane. *Uniaxial tensile strength* (σ_T) can be measured by applying tension to carefully designed ice specimens, but practical difficulties with this approach have led most investigators to infer σ_T from observations of beam flexure. The behavior of both small ice beams in the laboratory and large ice cantilevers sawn in natural sea ice sheets have been studied. At low strain rates ($\dot{\epsilon} < 10^{-6}s^{-1}$), σ_T, and σ_c are roughly equal, however, for high rates of loading σ_T is much smaller than σ_c. As with σ_c, σ_T increases with decreasing temperature and salinity and is considerably greater for vertically oriented specimens of columnar ice than for horizontal specimens. Other measures of ice strength, (e.g., flexural strength, shear strength, fracture toughness, and bearing capacity) have also been investigated. Results from much of this work have been summarized by Mellor[77] and Doronin and Kheisin.[31] Little work has been published on the mechanical properties of multi-year ice. What evidence is available indicates that its strength is highly variable, with differences of an order of magnitude being observed in ice with similar strain rates. Much more work is needed before it will be possible to draw comparisons with the results from first-year ice.

Brine pockets and channels weaken sea ice and fracturing tends to follow planes which pass through these features. This tendency has been used to simplify the treatment of pore geometry and fracture strength in theoretical sea ice models.[14,81] Brine pockets in these models are assumed to be vertically oriented elliptical cylinders, elongated in the direction of the platelets, which occur in parallel rows. Model results show a relationship between fracture strength and υ_b of the form: $\sigma_f = \sigma_o(1 - a\upsilon_b^k)$, where σ_o is the fracture strength of sea ice containing no brine, and a and k depend on the arrangement of the brine pockets and on how their shape changes with υ_b. Assuming that pockets enlarge in all directions to maintain a constant geometry results in $k = 2/3$, while assuming that expansion of the pockets primarily affects the degree of elongation results in $k = 1$. If, on the other hand, it is assumed that the relative length of the brine columns remain constant and that only the cross-sectional area changes with υ_b to maintain the same geometrical shape, it follows that $k = 1/2$, in agreement with observations.[14]

A subject that has received surprisingly little attention is the effect of biological activity on the strength and breakup of an ice cover. It has long been known that many species of microalgae inhabit sea ice, but the degree to which their presence may alter the mechanical properties of the ice does not appear to have been generally appreciated. A particularly interesting series of observations have been carried out by Buynitskiy[82] during summer cruises in the Antarctic. He found that substantial quantities of algae were present in essentially all the ice between the extreme edge and the shore. Cell densities as high as 33,000/cm^3 were recorded, with the average being about 4,300/cm^3, roughly an order of magnitude larger than in the underlying water column. Late in the summer diatoms occurred throughout the ice cover, while earlier in the year they tended to concentrate in brine channels and in horizontal layers ranging from a few centimeters to tens of centimeters in thickness. The dark color of the algae enhances absorption of shortwave radiation, accelerating the rate of internal melting near the algae, and thereby weakening the ice. Ship travel through such ice was found to be substantially easier than through ice with little algae. In contrast to pure sea ice which splits mainly in directions normal to the plane of freezing, antarctic ice was observed to cleave frequently through horizontal planes populated by the algae. Strength tests were carried out on various ice samples containing algae. It was found that the flexural strength of such ice was reduced by an average of 40% over that of pure sea ice, with the greatest reduction in strength occurring for stresses applied perpendicular to the freezing interface. The bearing capacity of this ice was thus seriously decreased. Compressive tests indicated that σ_c was almost unchanged by the presence of diatoms for vertically applied stresses, but decreased by a factor of 2 for horizontal stresses. Information in the published account is inadequate to identify the exact mechanisms whereby the algae achieve this weakening, but it is probably due more to warmer temperatures and higher brine volumes in the vicinity of the horizontal layers than to any direct effect on the microstructure of the ice. Whether biological activity is high enough anywhere in the Arctic to produce similar effects is uncertain.

3. Large Scale Behavior

While the large scale mechanical behavior of a sea ice cover can be vastly different from that observed in small laboratory samples, the latter provide a basis for understanding and treating large scale behavior. Small scale strength measurements indicate that tensile forces generated by wind and water stresses are much too small to crack a floating sheet of unflawed ice. Fracture analysis similar to that used with the laboratory data shows that leads in natural sea ice must form as a result of the growth of small existing cracks. Such cracks can be produced by thermal stresses or wave action and are very common in natural ice covers. Compressive stresses are likewise inadequate to cause widescale crushing in an ice sheet, although they may produce buckling in areas of very thin ice. This suggests that other processes must be involved in the formation of pressure ridges. A likely factor contributing to ridge building is overthrusting of the ice along a discontinuity (such as an old lead), causing flexural breakage to occur at much lower stress levels than needed for buckling. Overthrusting should take place primarily in thinner ice and does not explain how thicker ice can become involved in ridging. A possible explanation for this can be found in a ridging model developed by Parmerter and Coon.[83] Thin ice in an existing lead is assumed to break up as the lead closes in response to compressional forces. The resulting rubble is piled on and beneath the surrounding thick ice, disturbing its isostatic balance which then leads to flexural breakage adjacent to the lead. Incorporation of fragments from the parent ice into the rubble pile and further closure allow the process to continue until a limit height is attained. Additional activity then acts to widen the ridge, but not to increase its height. Whether this represents a quantitatively accurate simulation of the ridging process is not known, however, predicted limit values appear to be well correlated with ridge height data.[83]

Central to the development of large scale models of ice drift and deformation are a constitutive law relating ice stress to deformation and some measure of the aggregate strength of the ice. Numerous attempts have been made to model the ice as a linear viscous fluid, but results from these models indicate that nonlinear effects are important. Physical considerations suggest that a plastic rheology would provide a more realistic description of large scale ice behavior and both elastic-plastic and viscous-plastic approaches have been used with some success.[84,85] The overall compressive strength of the ice pack (σ_c^*) is determined by the thinnest (weakest) ice present. Convergence causes thinner ice to ridge, increasing its resistance to further compression. A formal theory has been developed for the plastic rheology that relates the rate of work done on the ice through ridging to ice stress and strain, providing a method of calculating σ_c^* from the distribution of thin ice within the pack.[86] Numerical experiments indicate that values of σ_c^* are typically on the order of 10^4 to 10^5 N m^{-1}. A detailed summary of work in this area has been given by Hibler.[87]

C. Optical Properties

The optical properties of a material define how it interacts with solar radiation. Shortwave energy exchange over the polar oceans is complicated by nonuniformities in the ice cover which cause its optical properties to exhibit large variations in both the horizontal and vertical. At the height of the melt season a relatively small area may contain leads, thin ice, surface melt ponds, drained white ice, and snow patches, all having significantly different optical properties. Both the surface and internal structure of the ice are altered by the absorption of shortwave radiation, so that its optical properties undergo temporal variations as well as spatial. While the optical properties of the ice pack do change gradually during the winter months, the largest changes occur during the summer melt cycle with the decay of the snow cover and the formation of melt ponds. In contrast to snow and bare ice, ponds are areas of low reflectivity where large amounts of energy are absorbed and transmitted to the underlying ice and ocean. Accompanying the surface changes are changes in the transparency of the ice caused by internal melting and increasing brine volume. Quantitative knowledge of how light absorption, transmission, and reflection vary in time and space is of fundamental importance in a wide variety of studies involving the ice and upper ocean.

1. Albedo

The *albedo* (α) is defined as the ratio of reflected to incident shortwave radiation and is a measure of net solar input at the surface. Total (integrated over all wavelengths) albedo measurements of polar ice and snow have been made by many investigators (see Grenfell and Maykut[88]). Albedos of cold snow are generally in the 0.80 to 0.85 range, dropping to 0.70 to 0.75 at the onset of melting. Snow free multi-year ice has an albedo of 0.70 to 0.75 when the surface temperature (T_o) is below freezing and 0.55 to 0.70 when $T_o = 0°C$. Thicker (1 to 2 m) first-year ice has α values of 0.5 to 0.6 when cold and 0.3 to 0.5 when melting. The albedo of snow free young ice exhibits a strong dependence on thickness which may be approximated by, $\alpha = 0.44\,h^{.28} + 0.08$, where h is in meters.[89] The higher reflectivity of multi-year ice is due to the presence of a decomposed surface scattering layer (typically 5 to 15 cm in thickness) and to greater vapor bubble density in the underlying ice. Melt pond albedos depend on water depth and the state of the underlying ice. Deep ponds remaining at the end of the melt season can have albedos as low as 0.15, while values of 0.4 are typical of the shallow ponds that form early in the melt season. Contamination by dirt or algae can reduce these values somewhat. Lead albedos vary with wave activity and sun angle, but average about 0.08.

While total albedos are easy to measure and apply, their usefulness is limited. For example, analysis of satellite observations, estimation of cloud effects on α, and calculations of light absorption in the ice require an understanding of how the radiative properties of the ice vary with wavelength (λ). Spectral albedos (α_λ) have been measured over many surface types in the Arctic.[88] Albedos of first-year and multi-year ice are shown in Figure 15. Albedos are relatively constant in the visible region (400 to 700 nm), but begin to decrease in the near infrared at a rate that is related to the liquid water content in the surface layers. Melting first-year ice can be divided into two main types: (1) water saturated blue ice (which includes shallow melt ponds) and (2) white ice consisting of a 5 to 10 cm thick layer of drained ice underlain by blue ice. At longer wavelengths light penetration is small so that α_λ is determined by the properties of the surface layer; at shorter wavelengths backscattering from ice beneath the surface becomes important. This explains the similarity between α_λ in first-year white ice and melting multi-year ice at longer wavelengths, and the difference at shorter wavelengths that results from lower backscattering in the underlying first-year ice. The albedo of melting blue ice is substantially lower than that of first-year white ice at all wavelengths. At shorter wavelengths this is due to the absence of the white scattering layer and at longer wavelengths to increased liquid water in the surface layers. Spectral albedos for snow and melt ponds are plotted in Figure 16. The albedo of dry compact snow is high and has only a weak dependence on λ. Increasing density and liquid water content in the snow decrease α_λ, particularly in the infrared. Pond albedos tend to be high at shorter wavelengths with a sharp decrease between 500 and 800 nm. Because water in the ponds is relatively transparent at short wavelengths, albedos below 500 nm are determined primarily by the scattering properties of the underlying ice. Between 500 and 800 nm α_λ becomes increasingly insensitive to the underlying ice as absorption in the water becomes the dominant factor. Above 800 nm α_λ is determined only by Fresnel reflection from the pond surface. Note that most of the albedo differences between individual ponds occur in the visible and are readily apparent to the naked eye. Ice and snow albedos decrease strongly above 1000 nm, reaching values of 0.1 or less at 1400 nm after which the decrease becomes much more gradual.[90]

The total albedo can be expressed in terms of wavelength dependent quantities as, $\alpha =$

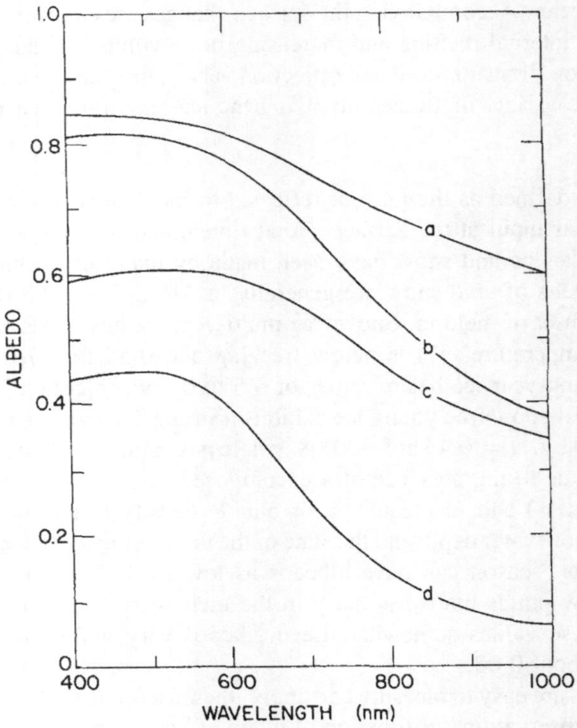

FIGURE 15. Spectral albedos observed over bare sea ice in the Arctic: (a) frozen multi-year ice, (b) melting multi-year ice, (c) melting first-year white ice, and (d) melting first-year blue ice.[88]

$\int_0^\infty \alpha_\lambda F_r(\lambda)d\lambda / \int_0^\infty F_r(\lambda)d\lambda$, where $F_r(\lambda)$ is the downward flux of solar radiation reaching the surface at wavelength λ. The presence of clouds causes measured values of α to increase. The reason is that clouds remove proportionately more energy from $F_r(\lambda)$ at long wavelengths than short ones, so that contributions from the more reflective shorter wavelength region are weighted more heavily in the integral when the sky is overcast. Typical increases observed for the relatively thin clouds found in the Arctic are on the order of 15%.[90] Larger increases can be expected in regions where clouds have greater optical depths.

2. *Extinction Coefficient*

Because ice is a translucent material, light penetrates into the interior where, at any given level, some is absorbed and some transmitted. Light absorbed within the ice produces brine volume changes and provides energy for biological activity. Shortwave radiation transmitted completely through the ice is largely absorbed in the upper 10 to 20 m of the ocean, making an important contribution to the oceanic heat flux at the bottom of the ice. Light attenuation in the ice is characterized by its *spectral extinction coefficient* (κ_λ) defined as:

$$\kappa_\lambda = -\frac{1}{F(\lambda,z)}\frac{dF}{dz} \qquad (20)$$

where $F(\lambda,z)$ is the shortwave flux in the ice at wavelength λ and depth z. Measured values of κ_λ for various types of arctic ice and snow are presented in Figure 17.[88] All curves show relatively constant values of κ_λ in the 400 to 500 nm region, followed by strongly increased attenuation at longer wavelengths. The greatest attenuation was observed in cold snow where

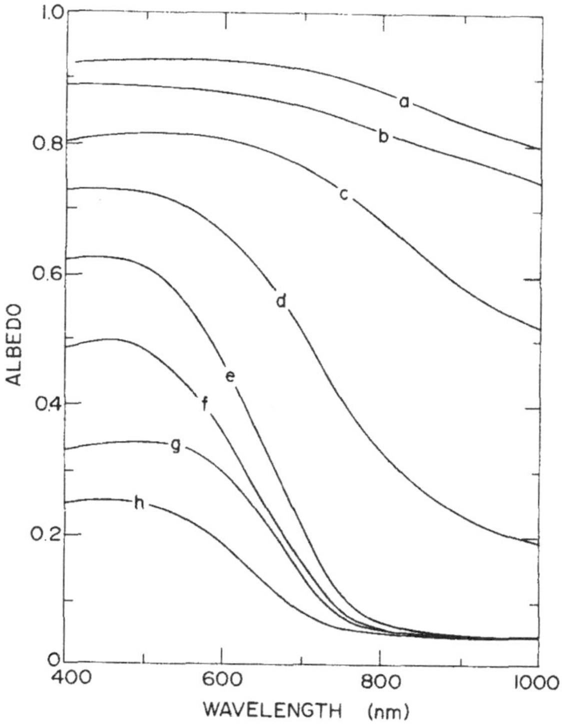

FIGURE 16. Spectral albedos observed over snow and melt ponds in the Arctic: (a) dry snow, (b) wet new snow, (c) melting old snow, (d) frozen melt pond, (e) early season melt pond, (f) mature melt pond on multi-year ice, (g) melt pond on first-year ice, and (h) old melt pond on multi-year ice.[88]

κ_λ's were about 20 times as large as those within the ice. Extinction coefficients in melting snow were about half those in cold snow. The "transition zone" in Figure 17 is a region of rapidly changing κ_λ which occurs within the 5 to 15 cm thick surface scattering layer on multi-year ice as it grades into the more uniform interior ice. Differences between κ_λ's within first-year and multi-year ice result from differences in brine volume and bubble density. Minimum values (curve f) were found beneath mature melt ponds where brine volumes were 20 to 55%.

The strong attenuation at longer wavelengths causes the spectral composition of the light to undergo rapid changes as it passes through the ice. Measurements of downwelling irradiance taken beneath first-year sea ice show energy maxima occurring in the 470 to 480 nm wavelength band, corresponding to the peak in the incident solar spectrum.[91] Red wavelengths are quickly absorbed in the surface layers of the ice so that the amount of energy reaching the ocean is negligible above 700 to 800 nm, even when the ice is very thin. Maximum transmission by the ice occurs in the 450 to 550 nm region, regardless of the surface conditions or ice thickness. The nature of the surface is the dominant factor in determining the amount of energy reaching the ocean. These points are illustrated by Figures 18 and 19 which show the spectral distribution of light beneath different thicknesses of first-year melt pond/blue ice and white ice. Because the spectral composition of the incident radiation field is rarely measured, it is useful to be able to estimate $F(\lambda,h)$ from more commonly available data. Values in Figures 18 and 19 have thus been expressed as a percent of the total incoming shortwave radiation measured under typical Arctic cloud conditions. The presence of snow on the ice drastically reduces the amount of transmitted light as illustrated in Figure 20.

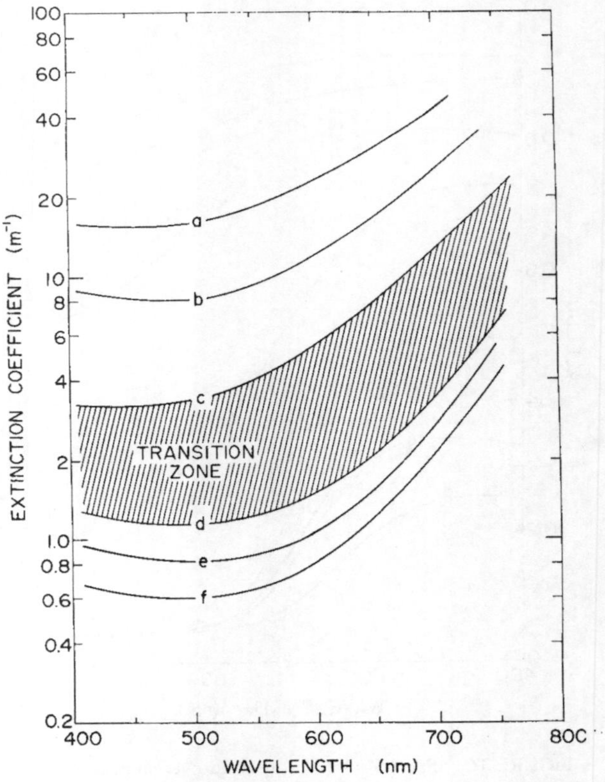

FIGURE 17. Spectral extinction coefficients for various types of arctic ice and snow: (a) dry compact snow, (b) melting snow, (c) the surface scattering layer on melting multi-year ice, (d) interior multi-year ice, (e) interior first-year ice, and (f) ice beneath a late season melt pond.[88]

In many regions a thick layer of algae forms on the underside of the ice in late spring, intercepting most of the transmitted light. With the onset of bottom ablation, this layer tends to slough off, however, algae remaining within the ice are still sufficient to affect the spectral distribution of the transmitted light. The characteristic signature of these algae is a broad peak in the transmission spectra between 450 to 550 nm and a secondary peak or shoulder at about 540 nm.[91] Bottom melting and brine drainage act to reduce the density of interstitial algae and the transmission spectra slowly return to a fairly narrow maximum at 470 to 480 nm later in the melt season. Differences in transmission spectra measured with and without algae indicate that the algae cause maximum absorption at 435 nm, with a secondary peak at about 670 nm.[91]

The spectral distribution of shortwave radiation within the ice can be calculated by integrating equation 20 to obtain:

$$F(\lambda,z) = F(\lambda,0)\, e^{-\kappa_\lambda z} \tag{21}$$

where $F(\lambda,0) = (1 - \alpha_\lambda)F_r(\lambda)$ is the net surface flux of shortwave radiation at wavelength λ. Equation 21 is known as *Beer's law*. In most practical applications it is the total flux, $F_z = \int_0^\infty F(\lambda,z)d\lambda$, which is of interest. By defining a *bulk extinction coefficient* (κ_z) as $\kappa_z = -\int_0^\infty \kappa_\lambda F(\lambda,0)\, e - \kappa_\lambda z d\lambda / \int_0^\infty F(\lambda,0) e - \kappa_\lambda z d\lambda$, it is possible to rewrite Beer's law as:

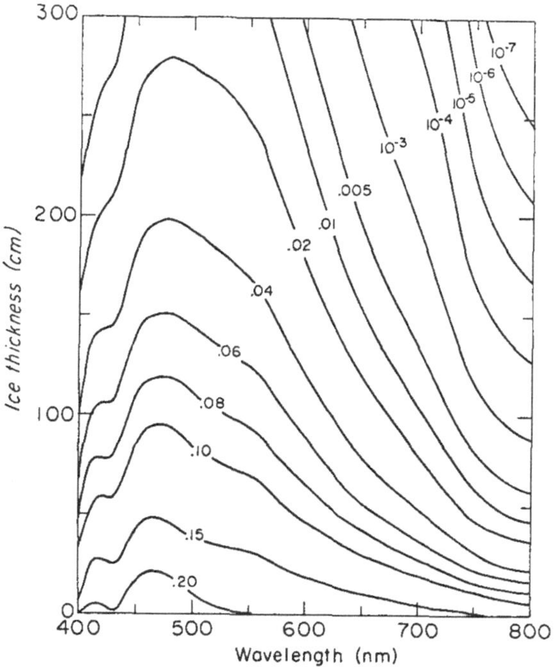

FIGURE 18. Spectral distribution of down-welling irradiance beneath first-year melt pond/blue ice.[91] Isopleths are expressed as a percentage of the total incident irradiance at the surface.

$$F_z = F_0 \exp\left(-\int_0^z \kappa_z \, dz\right) \tag{22}$$

where $F_o = \int_0^\infty F(\lambda,o)d\lambda = \int_0^\infty (1-\alpha_\lambda)F_r(\lambda)d\lambda$ is the net total shortwave radiation at the surface. Grenfell and Maykut[88] have examined the variation of κ_z with depth in both first-year and multi-year ice. They found that κ_z decreases by as much as a factor of 50 in the upper 10 to 20 cm of the ice, after which it exhibits only a weak dependence on z. Very little difference was found between κ_z's in the interior of first-year and multi-year ice. For purposes of estimating F_z, the ice can be thought of as consisting of two layers: an upper layer which filters out nearly all the energy at longer wavelengths and a lower layer where κ_z does not depend strongly on either ice type or cloud conditions. Shortwave energy absorbed in the upper layer is assumed to contribute immediately to temperature or mass changes at the surface, while energy absorbed in the lower layer returns slowly to the surface through conduction. This picture provides a simple method of estimating F_z. If we define i_o to be the fraction of F_o that passes through the upper layer, then:

$$F_z = i_0 F_0 \, e^{-\bar{\kappa}(z-z_0)} \tag{23}$$

where $\bar{\kappa}$ is the average bulk extinction coefficient in the lower layer and z_o is the thickness of the upper layer. Equation 23 provides the best fit to observations when $\bar{\kappa} = 0.015\ cm^{-1}$. The value of i_o depends on both the ice type and the spectral distribution of the incident radiation, (i.e., on cloudiness). Using $z_o = 10$ cm, Grenfell and Maykut[88] calculate that i_o for melting white ice is 0.35 when overcast and 0.18 when clear; for blue ice it is 0.63 when overcast and 0.43 when clear. With Equation 23 it is possible to calculate the amount of energy transmitted to the ocean by setting z = h, while the amount of energy absorbed at a depth $z (z > z_o)$ is simply $\bar{\kappa}F_z$.

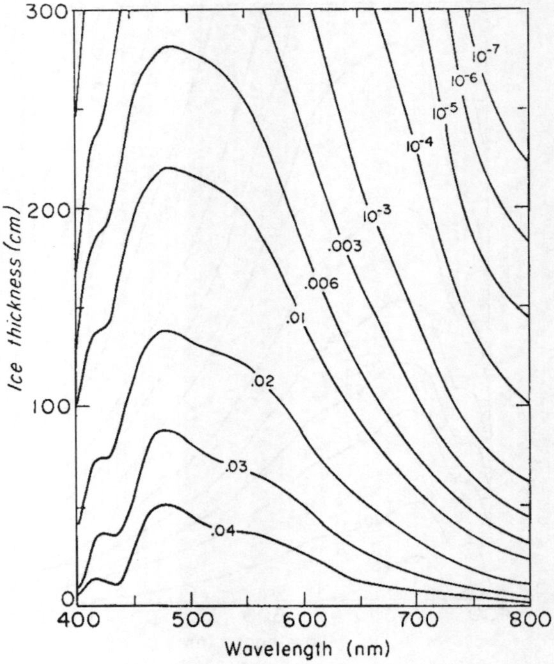

FIGURE 19. Spectral distribution of down-welling irradiance beneath first-year white ice.[91]

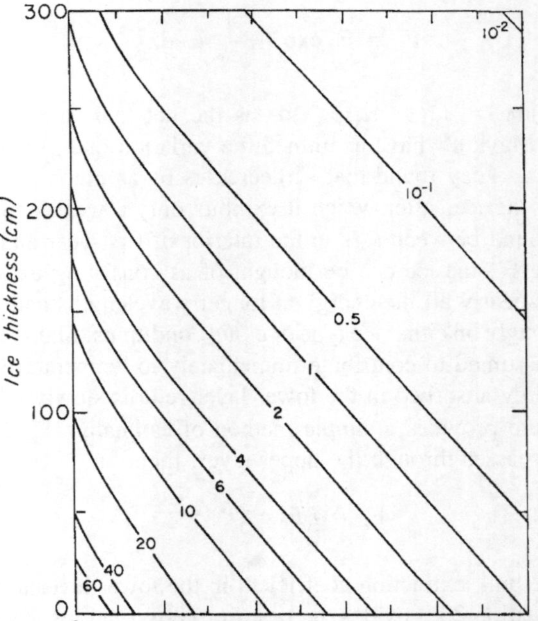

FIGURE 20. Total down-welling irradiance beneath first-year Arctic sea ice as a function of ice thickness and snow depth.[91] Isopleth values are given as a percentage of the total incident irradiance at the surface under overcast conditions.

D. Ice Thickness and Floe Size Distributions

It was pointed out earlier that continual deformation and growth produce an ice cover which is a mixture of many different thicknesses of ice. Because many of the physical properties of the ice cover depend strongly on h, it is useful to have a way to quantify thickness variations within a particular region. To accomplish this, Thorndike et al.[92] have introduced a probability density function g(h) called the *ice thickness distribution*, where $\int_{h_1}^{h_2} g(h)dh$ is the fractional area covered by ice in the thickness range $h_1 \leq h < h_2$. Processes which act to change g(h) are thermodynamic growth, ice advection, ridging, and divergence. A thickness distribution model which includes these processes has been developed and tested using data from the central Arctic.[92] While producing reasonable results, the model parameterizes a number of complex and poorly understood processes related to the mechanical redistribution of h. Considerable work on the mechanical response of the ice to various strain states needs to be carried out before it will be possible to specify these processes with any degree of confidence. Nevertheless, g(h) provides a framework for coupling the dynamics and thermodynamics of the ice pack that has proved to be very useful in a variety of modeling efforts and large scale studies of the ice.

Although the large scale behavior of the ice is directly influenced by the distribution of ice thickness, data on g(h) are difficult to obtain. At present the only practical way to measure g(h) directly is with upward-looking submarine sonar. Figure 21 shows distributions of ice draft derived from submarine transects north of Fram Strait in October.[93] As might be expected, there is a strong peak at about 3 m, close to the thermodynamic equilibrium thickness and a secondary peak at about 50 cm originating from young ice formed earlier in the fall. Level (presumably undeformed) ice makes up about 50% of the ice in the region. The distribution of deformed ice follows a negative exponential at drafts greater than 8 to 10 m. Measurements in the Beaufort Sea[94] show similar characteristics, although the secondary peak in g(h) occurs at about 2 m rather than 50 cm, probably representing second-year ice from the previous season. While a systematic analysis of spatial variations in g(h) has not been carried out, LeShack (see Hibler[87]) has utilized submarine data to calculate the average thickness ($\bar{h}$) at many locations in the Arctic Basin (Figure 22). The dominant feature of the $\bar{h}$ distribution is a sharp increase in thickness north of Greenland and the Canadian Archipelago that is almost certainly the result of greater ridging activity in the region. Comparable data are not available for the Antarctic and little is known about g(h) there.

While providing adequate definition of variations in the amount of thicker ice, submarine sonar data are not sufficiently dense in time to resolve the large and rapid changes that occur in the distribution of young ice. Yet knowledge of temporal changes in $g(h < 1\ m)$ is of fundamental importance because such areas not only make a major contribution to regional ice production and heat exchange, but also control the mechanical behavior of the ice pack. Without routine submarine transects or satellite instruments capable of resolving young ice thicknesses to within 10 to 20 cm, g(h) must be inferred from other data. The most feasible approach is to predict g(h) theoretically by combining models of ice growth and thickness distribution with heat balance and ice velocity data.[92] Thermodynamic forcing for the models is often derived from climatology, modified on the basis of buoy or satellite observations. Information on the velocity field of the ice can be obtained directly from the motion of drifting buoys[6] or calculated from atmospheric pressure data using a dynamic ice model.[84,85]

A geometric property of the ice cover that has received relatively little attention is the *floe size distribution*, N(d), defined to be the number of floes of mean diameter (d) per unit area. Floe size distribution is believed to be a significant factor in problems related to the summer ice pack. Part of the shortwave energy deposited in the upper few meters of a lead goes to lateral melting on floe edges causing a decrease in ice concentration (A_i) and an increase in the amount of open water (A_w). This positive feedback between solar input to the ocean and the rate of change of A_i can have a large effect on the summer decay and

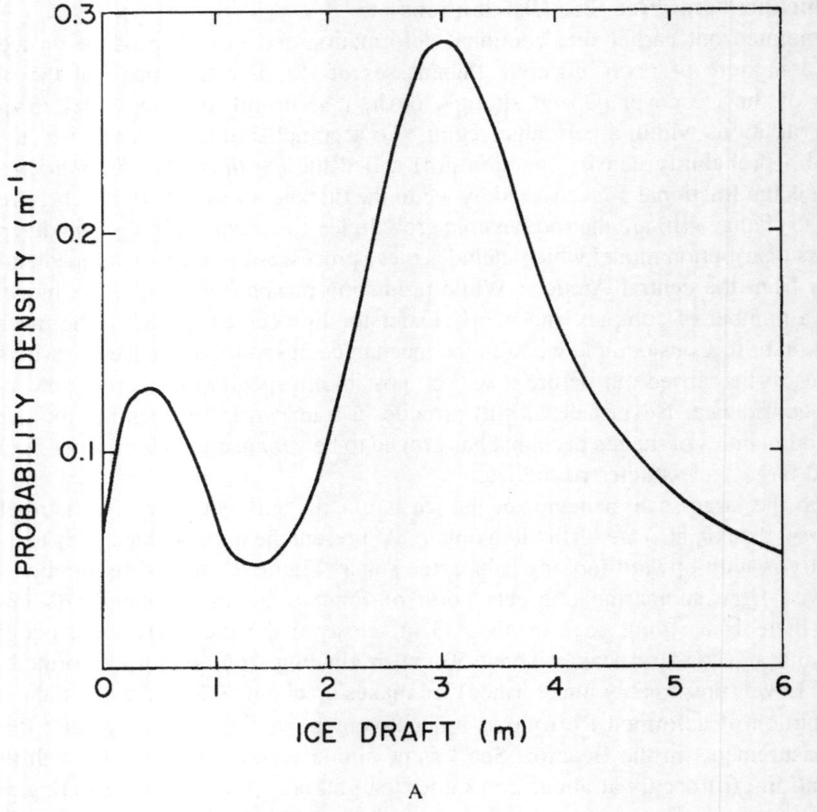

FIGURE 21. October ice thickness distributions observed in the Arctic Ocean north of Fram Strait.[93] (A) the distribution of level ice, and (B) the complete distribution including ridged ice.

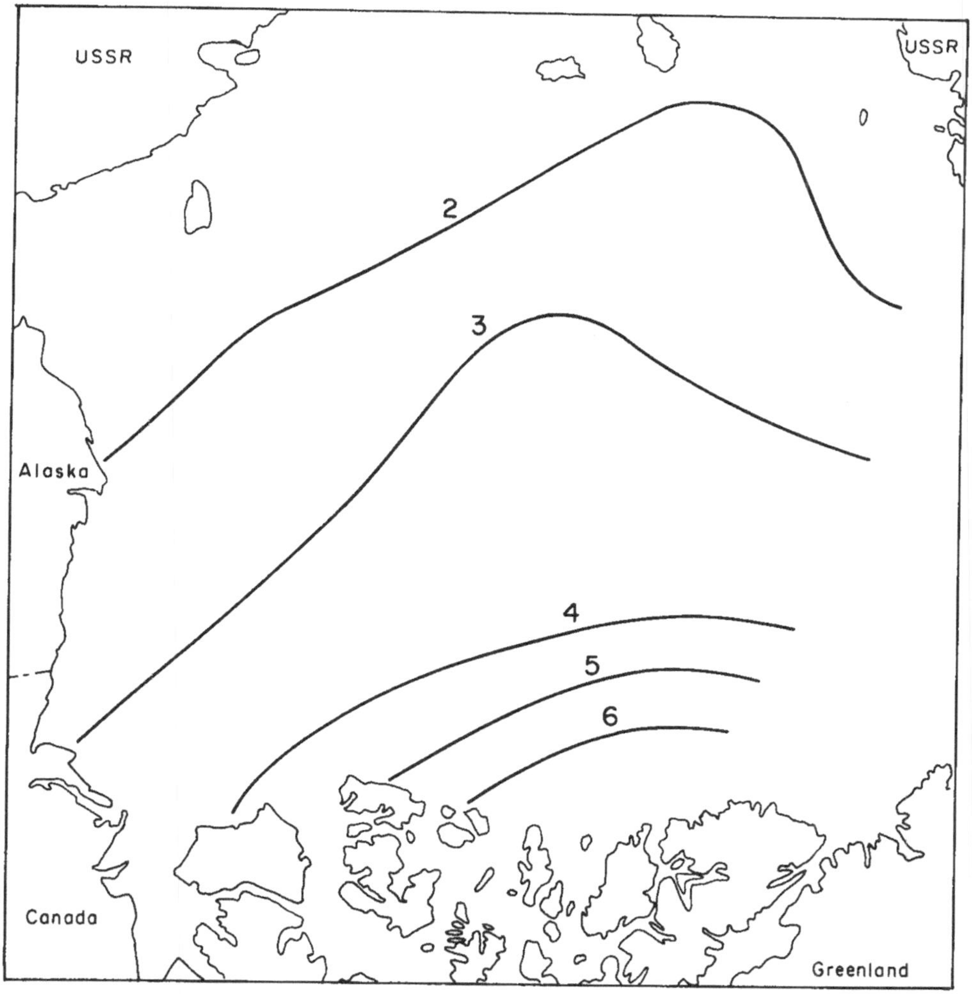

FIGURE 22. Contours (in meters) of mean ice thickness in the Arctic Basin derived from submarine sonar data by LeShack (after Hibler).[87]

retreat of the ice.[51] As the ice cover becomes more broken up, there is a decreasing probability that heat absorbed in the leads will be lost to the atmosphere or underlying ocean and a corresponding increase in the amount of heat transferred to floe edges. Simple model experiments[95] show that for a given ice concentration, the rate of lateral decay depends directly on the number of leads present, i.e., on the total floe perimeter. Ice concentration alone is thus inadequate to characterize the state of the summer ice cover. N(d) provides a measure of total floe perimeter and unlike g(h), is relatively simple to obtain from aerial photography. N(d) appears to be particularly important in modeling the Marginal Ice Zone, affecting not only lateral melting but also drag coefficients, melt pond coverage, and open water distribution. The equation governing N(d) is analogous to that for g(h), containing terms to describe advection, thermodynamics (lateral melting), and mechanical processes (breaking). Except near the ice edge where wave activity fractures the ice in a predictable fashion, processes involved in floe breakup are not well understood. Measurements of floe size distribution have been made in several locations.[96-98] Figure 23, for example, shows mid-August values in the central Beaufort Sea. For this case the decrease in N with d roughly follows a power law between 0.05 and 10 km. In the Marginal Ice Zone N(d) tends to

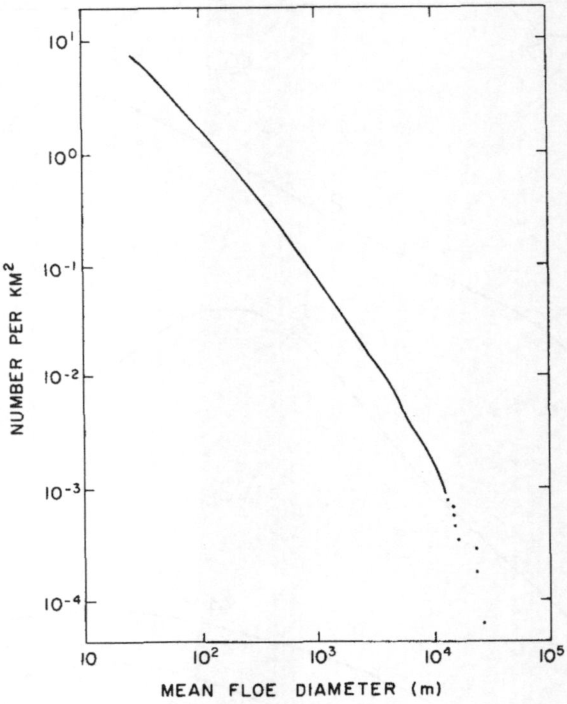

FIGURE 23. Floe size distribution observed in the central Beaufort Sea during mid-August.[98]

assume a negative exponential distribution,[13,97] but N(d) in the interior ice is much more variable and no single analytical form can be used to fit all the data.[96] No systematic studies have yet been carried out to define how N(d) evolves during the summer or how this evolution differs from one region to another.

E. Electromagnetic Properties

In recent years there has been increasing interest in the electrical properties of sea ice in the 0.1 to 100 GHz region. Because of the extensive cloud cover and long periods of darkness, satellite imagery in the visible and infrared are of limited value in monitoring the state of the polar ice pack. Signals in the microwave region are not affected by darkness and are attenuated only slightly by clouds. Microwave imagery thus provides a reliable method of observing the ice throughout the year. Both active and passive microwave measurements have been carried out over the ice. Active methods transmit a radar signal to the ice and measure the resulting reflection. At lower frequencies (0.1 to 0.5 GHz) maximum reflections occur at the top and bottom of the ice, providing in principle a means to obtain thickness information. At higher frequencies (several GHz) radar images primarily show differences in surface roughness. Passive methods measure thermal emission from the ice, generally in the 3 to 100 GHz range. A serious limitation with passive observations is lack of spatial resolution. Whereas satellite-borne radar can resolve surface elements tens of meters across, passive resolution techniques can distinguish several different ice types because of distinct differences in thermal emission. Understanding microwave signatures from various types of ice requires information on the *dielectric constant,* a complex quantity whose real part describes surface reflection and backscattering and whose imaginary part describes attenuation within the ice. Dielectric properties of sea ice depend on the properties of its constituents (ice, brine, air, and solid salts) and various expressions have been derived to express bulk values in terms of temperature, brine volume, and pore geometry.[14,59]

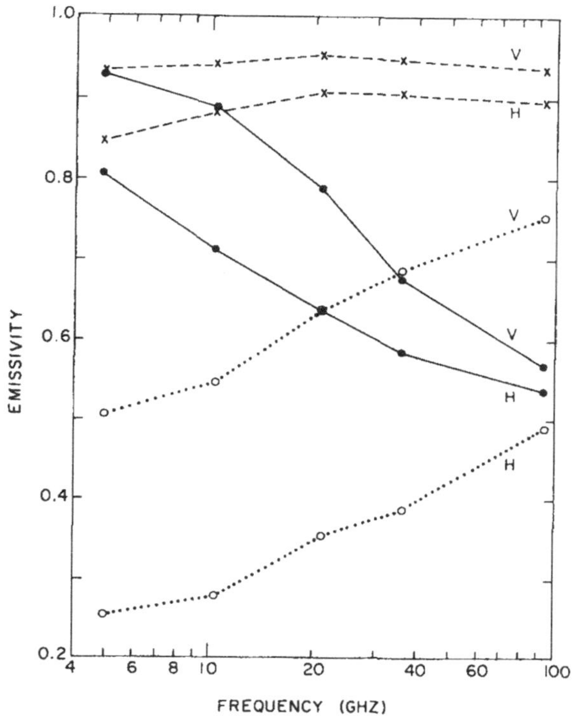

FIGURE 24. Microwave emissivities of multi-year ice (solid lines), first-year ice (dashed lines), and calm seawater (dotted lines) in the Greenland Sea as a function of frequency.[100] The symbols H and V refer to values obtained from horizontally and vertically polarized radiation.

The availability of routine microwave data from polar orbiting satellites has made it possible to procure quantitative information on ice conditions in the polar oceans, (e.g., Zwally el al.[2], Carsey[99]) that cannot be obtained by other techniques. Most of the microwave data have been in the form of passive "brightness temperatures" which, after suitable analysis, can give ice concentration, ice extent, and the relative amounts of first-year and multi-year ice. Unfortunately, the procedure for translating brightness temperature measurements into quantitative estimates of ice state is not straightforward. A single brightness temperature is a spatial integral of energy emitted from a region which is large (30 to 150 km across) compared to the scale of individual features such as leads and floes. Even though such features cannot be resolved by passive microwave imagery, it is possible to estimate their contribution to an observed brightness temperature and so infer their relative abundance. The method makes use of frequency dependent differences in microwave *emissivity,* the ratio of energy actually emitted to that emitted by a blackbody under identical conditions. In the Arctic, multi-year ice, first-year ice, and seawater exhibit markedly different changes in emissivity with frequency (Figure 24).[100] At lower frequencies first-year and multi-year ice appear much the same, while leads and open water have much lower brightness temperatures due to a lower emissivity. At higher frequencies first-year ice appears to be much warmer than multi-year ice. The reason why emissivity decreases with frequency in multi-year ice and not in first-year ice is related to differences in the distribution of brine which controls the depth from which the radiation emanates. Multi-year arctic ice typically has a surface layer containing little brine, but many air pockets. Radiation emitted from brine deeper in the ice is subject to scattering by the air pockets as it moves upward through the ice. The degree of scattering increases with frequency, therefore the observed decrease in emissivity. In first-year ice, brine is present very close to the surface so that the emitted radiation undergoes little scattering and as a result is relatively independent of frequency. It is then possible to use lower frequency data to determine ice concentration and extent, and higher frequency data to calculate first-year/multi-year ratios. The method is of limited value during the summer because melting erases any distinction between ice types and leads cannot be distinguished from surface melt ponds. Ice extent can still be found, regardless of season or the location of the edge.

Uncertainties inherent in the determination of ice concentration are larger than the amount of open water normally present within the arctic ice pack during the growing season, so microwave data are primarily used to map the distribution of first-year and multi-year ice in the Northern Hemisphere. Due to weak surface melting during the summer, differences in microwave emission from first-year and multi-year ice tend to be subdued or totally lacking in the Antarctic, particularly in the Weddell Sea region. Although signatures similar to multi-year ice have been seen in the Ross, Amundsen, and Bellingshausen seas, little is yet known regarding the physical characteristics of this ice. For this reason, microwave analyses of ice conditions in the Southern Hemisphere are generally focused on seasonal and spatial variations in ice concentration.[2]

V. HEAT AND MASS BALANCE

While an ice cover reduces momentum transfer between the air and the water, thermal processes associated with its growth and movement have the greatest impact on the ocean. In particular, solar input to the upper ocean is drastically reduced, limiting biological activity beneath the ice. The cold dense brine rejected by growing ice promotes convective overturning causing any heat contained in the mixed layer to be transported upward where it is lost to the ice. Thus, unlike the open ocean, heat storage in the upper part of the water column is minimal. Heat and mass balance measurements have been carried out from land-based stations in the Arctic and Antarctic, as well as from manned and unmanned drifting stations in the Arctic Basin. Conditions in the coastal regions are influenced by the proximity of the nearby land masses and cannot be considered representative of the ice pack as a whole. As a result our understanding of interactions between the ice, ocean, and atmosphere is based largely on what has been seen in the perennially ice-covered portions of the Arctic Basin. The extent to which such information applies to more dynamically active areas such as the Marginal Ice Zone and the antarctic ice pack is unknown.

A. Energy Fluxes

1. Radiation

Radiation dominates heat exchange in the polar regions throughout the year. All matter radiates energy in proportion to the fourth power of its absolute temperature, i.e., $F^* = \epsilon \sigma T_0^4$, where F^* is the flux of emitted radiation, ϵ is the surface emissivity, and σ is the Stefan-Boltzmann constant $= 5.67 \times 10^{-8}$ W m^{-2} K^{-4} (4.88×10^{-9} cal cm^{-2}s^{-1} K^{-4}). Most of the radiation reaching the earth's surface lies in 2 spectral bands: *shortwave radiation* which originates from the sun and has a narrow intensity maximum near $\lambda = 500$ nm and *longwave radiation* which originates from the earth and atmosphere with a broad intensity maximum near $\lambda = 12000$ nm. Because there is little overlap in the spectral distribution of radiation derived from these two sources, there is a natural separation in their measurement and treatment.

Downward longwave radiation (F_L) emitted from clouds and water vapor in the atmosphere is a major component in the surface heat balance. During most of the winter and fall incoming shortwave radiation (F_r) is negligible and even at the height of the melt season F_r is only

Table 1
MONTHLY AND ANNUAL RADIATION TOTALS (MJ m^{-2}) OVER PERENNIAL ICE IN THE CENTRAL ARCTIC[42,102]

	Jan.	Feb.	Mar.	Apr.	May	Jun.	Jul.	Aug.	Sep.	Oct.	Nov.	Dec.	Year
Incoming longwave radiation (F_L)	435	430	430	495	630	750	800	780	690	580	475	450	6945
Outgoing longwave radiation ($F\uparrow$)	−520	−500	−505	−600	−725	−830	−835	−820	−720	−610	−535	−520	−7720
Net longwave radiation ($F_L + F\uparrow$)	−85	−70	−75	−105	−95	−80	−35	−40	−30	−30	−60	−70	−775
Incoming shortwave radiation (F_r)			100	420	730	785	575	370	155	20			3155
Absorbed shortwave radiation [$(1-\alpha)F_r$]			15	80	145	170	245 (295)	115 (165)	35	5			810 (910)
Net radiation	−85	−70	−60	−25	50	90	210 (255)	75 (120)	5	−25	−60	−70	35 (125)

Note: Numbers in parentheses include the effects of summer melt ponds.

slightly larger than F_L (Table 1). On an annual basis F_L is more than double F_r in the Central Arctic. Despite the magnitude of F_L, the net effect of longwave radiation is to remove heat from the ice. This is because the radiative temperature of the ice usually exceeds that of the atmosphere so that longwave radiation emitted by the ice ($F\uparrow$) is almost always greater than F_L. Measurements of F_L and $F\uparrow$ are difficult in the polar regions because of constant problems with dew and frost formation on the instruments. Fortunately, both can be estimated from other information. Surface temperature data are frequently used to calculate $F\uparrow$ according to the formula, $F\uparrow = \epsilon_L \sigma T_o^4$, where ϵ_L is the longwave emissivity which equals 0.99 for snow and 0.97 for bare ice and melt ponds. Because a large part of the incoming longwave radiation is emitted by clouds and water vapor close to the ground, F_L can be approximated fairly well from surface observations of air temperature (T_a), partial pressure of water vapor (e_a), and fractional cloud cover (C). Longwave parameterizations have the form, $F_L = \epsilon^* \sigma T_a^4$, where ϵ^* is the effective emissivity of the atmosphere. Several empirical expressions have been suggested for ϵ^* in the Arctic, The simplest:

$$\epsilon^* = 0.7829(1 + 0.2232C^{2.75}) \tag{24}$$

obtained by Maykut and Church[101] from observations at Point Barrow, Alaska, gives estimates of F_L accurate to within about 7% under overcast skies and 12% under clear skies. No correlation was found between ϵ^* and e_a. More complex expressions depending on e_a, C, T_a, and a seasonally varying empirical parameter have been obtained by Marshunova[102] and Idso and Jackson.[103]

The sun being continuously above or below the horizon for periods of 2 to 6 months at a time produces seasonal changes in F_r that are much larger than those in F_L. In spite of extensive clouds and low sun angles, monthly totals of F_r in the Arctic can exceed those measured at many midlatitude stations, (e.g., New York, Chicago) during the summer. Spatial variations in F_r pole-ward of 65°N tend to be small during the summer months.[102] Due to the extreme summer cloudiness ($\overline{C}$ = 0.8 to 0.9), most of the shortwave radiation reaching the surface is diffused rather than direct. Clouds over the polar oceans tend to be thin and so typically reduce F_r by 30 to 50% as compared to the 80 to 90% often observed at lower latitudes. The relatively low water vapor content of Arctic clouds allows substantial amounts of near infrared energy to reach the surface. Under clear skies about 55% of F_r lies outside the visible region (400 to 700 nm), while under overcast skies this figure decreases to about 35%. Shortwave parameterizations used in large scale heat balance studies and numerical models are generally written in the form, $F_r = a^*F_{ro}$, where F_{ro} is the incoming shortwave radiation at the surface in the absence of clouds and a* is the fractional attenuation by clouds. Although both F_{ro} and a* can be calculated theoretically, the procedures are too involved for many applications and empirical relationships are frequently employed. A particularly convenient form for F_{ro} was given by Zillman:[104]

$$F_{ro} = \frac{S_0 \cos^2 z}{1.085 \cos z + (2.7 + \cos z)e_a \cdot 10^{-3} + 0.1} \tag{25}$$

where S_o is the solar constant, e_a is in millibars and z is the solar zenith angle, a function of latitude, day, and hour. Limited tests with arctic data[36] indicate that Equation 25 usually underestimates F_{ro}. The reason is probably related to the smaller amount of water vapor in the polar atmosphere and to backscattering of radiation reflected from the ice. This suggests that a high latitude correction may be desirable for formulas based on observations at lower latitudes. On the basis of model calculations, Shine[105] reports that improved performance can be obtained from Equation 25 over ice and snow if the denominator is changed to: $1.2 \cos z + (1.0 + \cos z)\, e_a \times 10^{-3} + 0.046$. Expressions for a* have been obtained by Marshunova[102] ($a^* = 1 - kC$, where k increases from 0.15 in March to 0.5 in July) and Laevastu[106] ($a^* = 1 - 0.6C^3$). A parameterization for the net shortwave flux over ice and snow, claimed to be accurate to within a few percent under cloudy arctic conditions, has also been derived by Shine[105] in terms of solar zenith angle, surface albedo, and the optical thickness of the cloud cover.

As mentioned earlier, the summer melt cycle over perennial ice is driven almost entirely by absorbed shortwave radiation. Although F_r reaches a maximum in June, the largest values of net shortwave radiation occur in July because of the decrease in α following removal of the snow cover. The presence of melt ponds significantly increases the net shortwave radiation and hastens the rate of decay and retreat of seasonal ice. In areas of perennial ice, ponds increase shortwave input to the ice by 20 to 40% (Table 1), but have only a small effect on the annual average ice thickness.[41] Ice concentration is also an important parameter as leads admit large quantities of energy to the ocean. Once the snow cover has vanished, the amount of energy available for primary production in the underlying water is largely determined by the relative area occupied by leads and melt ponds. Table 2 compares how 1 m thick layers of melt pond/blue ice and white ice distribute F_r for different ice concentrations. When A_i is high, transmission through blue ice accounts for most of the energy absorbed beneath the ice. As A_i decreases, leads quickly begin to dominate solar input to the water beneath the ice.

2. *Turbulent Exchange*

Apart from outgoing longwave radiation, the only means of transferring energy from the ice to the atmosphere is through turbulent exchange. The *sensible heat flux* (F_s) results from temperature differences between the ice and atmosphere, while the *latent heat flux* (F_e) is produced by the release of latent heat associated with evaporation, condensation, and sublimation. These fluxes are frequently dismissed as being unimportant because they are typically 1 to 2 orders of magnitude smaller than the radiation fluxes. This argument is misleading, however, because the net radiation balance is small in comparison to its individual components (Table 1). In fact, the magnitude of $F_s + F_e$ in the Central Arctic is on the

Table 2
THE EFFECT OF ICE CONCENTRATION AND SURFACE TYPE ON ABSORPTION AND TRANSMISSION OF INCOMING SHORTWAVE RADIATION IN A REGION OF 1-M THICK BARE ICE

Ice concentration	Surface type	Net input at surface	% Absorbed $0 \leq z \leq 1$ m	% Transmitted below 1 m	Regional input
0.9	Blue ice	61	48	13	70
	White ice	27	25	2	36
	Leads	9	4	5	
0.5	Blue ice	34	27	7	80
	White ice	15	14	1	61
	Leads	46	19	27	

Note: Values are expressed as percentages of F_r, weighted by relative area.

order of 20 to 50% of the total net radiation. Theoretical calculations have demonstrated that the thickness of multi-year ice is very sensitive to changes in F_s and F_e during the summer.[107] Turbulent fluxes near the boundaries of the ice pack tend to be larger and appear to have an even greater impact on summer decay patterns.

The rate of turbulent heat transfer depends on surface roughness, wind speed, boundary layer stability, and the gradients in temperature and water vapor near the surface. Commonly used parameterizations for these fluxes are $F_s = \rho_a c_p C_s u\ (T_a - T_o)$ and $F_e = \rho_a L_v C_e u(q_a - q_o)$, where ρ_a is the density of the air, c_p is the specific heat of the air, u is the wind speed, L_v is the latent heat of vaporization, q_a and q_o are specific humidities in the air and at the surface, and C_s and C_e are bulk transfer coefficients which vary with stability and surface roughness. Eddy correlation measurements over multi-year ice[108] indicate that $\overline{C}_s = 1.2 \times 10^{-3}$ and $\overline{C}_e = 0.55 \times 10^{-3}$, while observations in the unstable boundary layer over a refreezing lead[109] give values of 2.0 to 2.5×10^{-3} for both C_s and C_e. Differences in surface geometry (ridges, hummocks, and floe sizes) can be expected to produce local variations in C_s and C_e of at least a factor of 2.

Studies of turbulent heat exchange over perennial sea ice[110-112] show that, unlike the situation over temperate and tropical oceans, F_e is negligible during much of the polar year (November to April in the Arctic). Even during the summer latent heat losses are only slightly larger than those from sensible heat. During the winter when there is a persistent temperature inversion close to the surface, F_s supplies heat to the ice, compensating for one third to one half the net longwave loss. Although daily totals of F_s during the summer are comparable in magnitude to the winter values, monthly averages are small as a result of frequent alterations in the direction of F_s. All studies in the Central Arctic show a net turbulent heat loss for the summer that acts to reduce the amount of surface melting. Without these turbulent heat losses, surface ablation would be 20 to 100% larger and the equilibrium thickness reduced by up to 50%.[107] At the margins of the ice pack, F_s and F_e are generally downward and contribute substantial amounts of heat to the surface. During May and June the turbulent fluxes can account for as much as one third of the net heat input to coastal ice.[49]

3. Heat Conduction

Except during the melt season, there is an upward conduction of heat in the ice $[F_c = k_i (\partial T/\partial z)]$ which supplies much of the energy needed to balance the longwave loss at the surface. In thin ice $F_c(z = 0) \simeq F_c(z = h)$, greatly simplifying the treatment of young ice growth. In thicker ice internal phase changes and heat storage become important and growth calculations must take into account nonlinearities in the temperature profile. Over an annual cycle the amount of heat conducted to the surface of multi-year ice is the sum of the sensible heat input from the ocean (F_w), the latent heat released by net bottom accretion, and the shortwave energy temporarily stored in brine pockets and surface melt ponds. For perennial Arctic ice $F_c \simeq 6.6$ kcal cm^{-2}year^{-1} (8.8 W m^{-2}), of which 23% is supplied by the ocean, 39% by bottom accretion, 29% by release of stored heat in the brine pockets, and 9% by the refreezing of surface meltwater.[36] F_c in young ice can be more than 2 orders of magnitude larger than in multi-year ice. Over the course of the entire growth season the amount of heat conducted to the surface of seasonal ice is roughly 2 to 3 times the annual total for perennial ice.

4. Oceanic Heat Flux

A variety of observational and theoretical evidence points to a significant input of heat from the ocean to the ice, but direct measurements of this flux (F_w) have not been successful. In the Antarctic where surface ablation is small, the seasonal disappearance of the ice must be caused almost entirely from energy contained in the water. Certainly shortwave radiation absorbed in the upper ocean is a major component of F_w there, however heat from the deeper ocean is also important. Gordon[50] estimates that up to 50% of F_w could be derived from water below the pycnocline, but the uncertainty in this estimate appears to be large. In the Central Arctic F_w is about 1.5 kcal cm^{-2}year^{-1}(2 W m^{-2}). The source of this heat was originally believed to be the layer of relatively warm Atlantic water that occurs across much of the Arctic Basin. More recently it has been suggested that temperature changes in the Atlantic layer are largely due to mixing with colder shelf water[113] and that F_w is determined almost entirely by shortwave input through leads and areas of thin ice.[89] In areas of seasonal ice the upper part of the water column at the onset of freezing can contain 1 to 2 kcal cm^{-2} (40 to 80 MJ m^{-2}) which is slowly lost as the mixed layer deepens in response to salt rejection by the growing ice. The largest values of F_w in the Arctic probably occur in the Fram Strait region where the Atlantic water enters the basin and is still close to the surface.

B. Response of the Ice to Environmental Changes

Because sea ice appears to be an integral part of the global climate system, there are numerous questions that need to be answered regarding its long term response to various environmental factors. For example, how will the polar ice pack be affected by the increasing concentration of carbon dioxide in the atmosphere? Can man artificially alter ice conditions to limit these effects? How does the state of the ice pack change over a glacial cycle? While some information on historical changes in the ice pack can be gained from the fossil record and oxygen isotope analysis of ocean sediments, sensitivity studies with coupled ocean-ice-atmosphere models are needed to fully understand the response of the ice. Such models are still in a relatively crude state of development and do not adequately simulate many system interactions and feedbacks that are likely to affect ice extent and thickness.

Even though prediction of future changes in the state of the ice cannot yet be made with any real degree of confidence, it is possible to determine with some accuracy how specific changes in such factors as air temperature, snowfall, cloudiness, and oceanic heat flux will alter ice thickness. This is the first step needed in addressing the more difficult problem of ice extent. To estimate thickness changes, it is first necessary to formulate energy balance equations at the upper and lower boundaries of the ice. The flux balance at the surface is

$$(1 - \alpha)(1 - i_0)F_r + F_L - F\uparrow + F_s + F_e + k_i\left(\frac{\partial T_i}{\partial z}\right)_0 + F_m = 0 \qquad (26)$$

where $F_m = \rho_i L_i (dh/dt)_o$ is the heat loss due to surface melting. When T_o is below freezing,

$F_m = 0$; when T_o is at the melting point, any surplus of energy will be balanced by melting and a change in h. At the bottom of the ice:

$$\rho_i L_i \left(\frac{dh}{dt}\right)_h - k_i \left(\frac{\partial T_i}{\partial z}\right)_h + F_w = 0 \tag{27}$$

Equations 26 and 27 are coupled via the heat conduction terms, $k_i(\partial T_i/\partial z)_o$ and $k_i(\partial T_i/\partial z)_h$ whose calculation requires knowledge of the vertical temperature distribution within the ice. Changes in $T_i(z)$ can be calculated using a form of the thermal diffusion equation modified to include the effects of internal heating from solar radiation:

$$\rho_i c_i \frac{\partial T_i}{\partial t} = \frac{\partial}{\partial z}\left(k_i \frac{\partial T_i}{\partial z}\right) + \bar{\kappa}_i i_0 (1 - \alpha) F_r e^{-\bar{\kappa}_i (z - z_0)} \tag{28}$$

Simultaneous solution of Equations 26 to 28 then makes it possible to determine the thermal response of the ice to any of the parameters explicitly or implicitly contained in the equations. Inclusion of a snow cover adds an additional boundary condition and heat conduction equation, but does not significantly complicate the solution.[42] Large scale simulations of the ice cover generally retain Equations 26 and 27, but simplify the problem either by eliminating Equation 28 through the assumption of a linear temperature profile in the ice or by resolving $T_i(z)$ only crudely.

Model results[42] indicate that equilibrium thickness (see Section II.B) is especially sensitive to factors which affect the amount of summer melting: air temperature, ice albedo, incoming radiation, and turbulent heat exchange. Complete disappearance of perennial arctic ice would occur with increases of 3 to 5°C in T_a, 25 to 30% in incident solar radiation, or a 15 to 20% decrease in summer albedo. Changes in cloudiness affect both the shortwave and longwave radiation. Decreased cloudiness lowers the amount of incoming longwave radiation while increasing the downward shortwave flux. A decrease in cloudiness thus acts to cool the ice during the winter, but has only a weak effect during the summer because the opposing changes in incident radiation tend to balance. Year to year variations in arctic cloudiness are presently on the order of 25% and do not appear to change h_e significantly.

Theoretical predictions of how variations in snowfall and oceanic heat flux affect h_e in the Central Arctic are shown in Figures 25 and 26. In spite of the large effect that snow depth has on ice temperatures and growth rates, differences in snowfall are found to have little impact on h_e until depths exceed about twice the present day average of 40 cm. With less snow the ice is colder and winter growth greater, but summer ice melting is also greater due to earlier removal of the snow. Changes in surface melting and bottom accretion approximately balance until snow depths become so large that ice ablation is delayed until late in the summer when F_r begins to decrease rapidly with time. Above $h_s \simeq 80$ cm the decrease in ice ablation is greater than the decrease in accretion and h_e increases steeply with increasing h_s. Under present surface conditions the thickness of undeformed perennial ice will not exceed 7 m in the Central Arctic, even in regions where the oceanic heat flux is negligible (Figure 26). However, if h_s is in excess of about 120 cm, summer melting in the Arctic is generally insufficient to completely remove the snow cover and equilibrium ice will grow from above and ablate from below, becoming in effect a "floating glacier". The combination of small F_w and high snowfall can produce ice of almost any thickness and has been used to explain the origin of very thick ice found in Nansen Sound and along the north coast of Ellesmere Island.[114] The calculations also predict that the Arctic ice cover would be seasonal if F_w totaled more than about 5.5 kcal cm^{-2} year^{-1} (7.3 W m^{-2}). It is suspected that average values of F_w in many parts of the Southern Ocean may exceed this value by at least a factor of 2 to 3.[115]

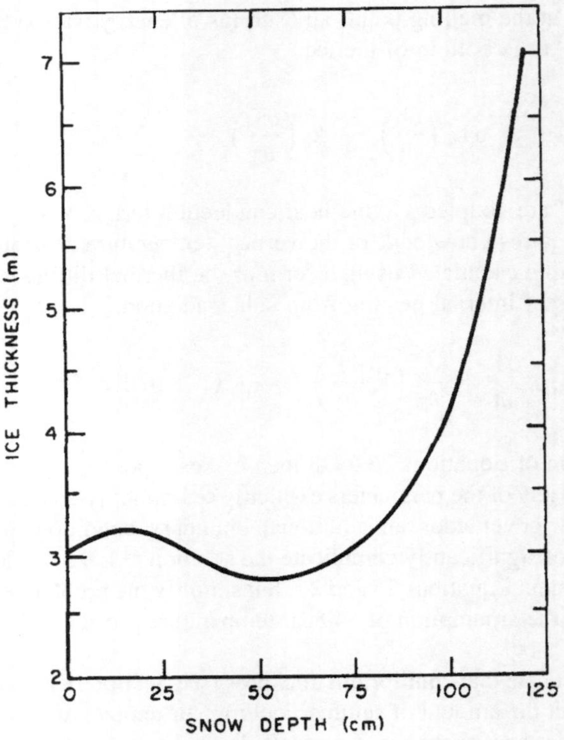

FIGURE 25. Average equilibrium thickness of perennial Arctic sea ice as a function of maximum annual snow depth.[42]

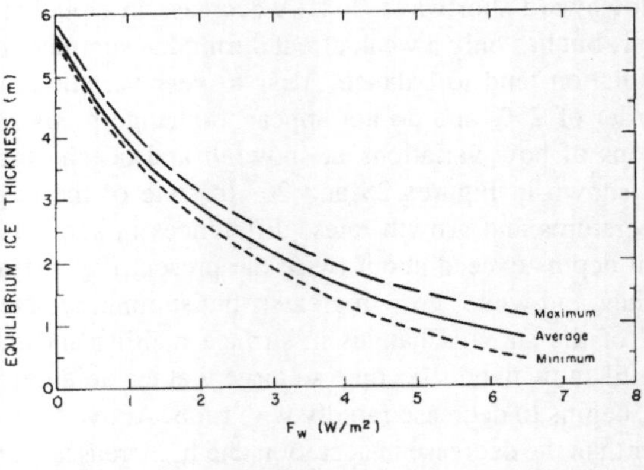

FIGURE 26. Equilibrium thickness of Arctic sea ice as a function of the average annual oceanic heat flux. Average annual thickness, as well as the absolute annual maximum and minimum are shown.[42]

C. Effects of Ice Thickness on Surface Heat Exchange

It was shown in Figure 5 that for constant thermal forcing, winter growth rates in young ice are strongly dependent on thickness. The same holds true for the surface heat balance. Surface temperatures in thin ice decrease rapidly with increasing thickness (Figure 27) affecting the conductive heat flux, the turbulent exchange, and the longwave radiation loss. In addition, the increase of albedo with thickness produces a decrease in shortwave absorption as the ice grows. Heat exchange over thin ice tends to be much more vigorous than over

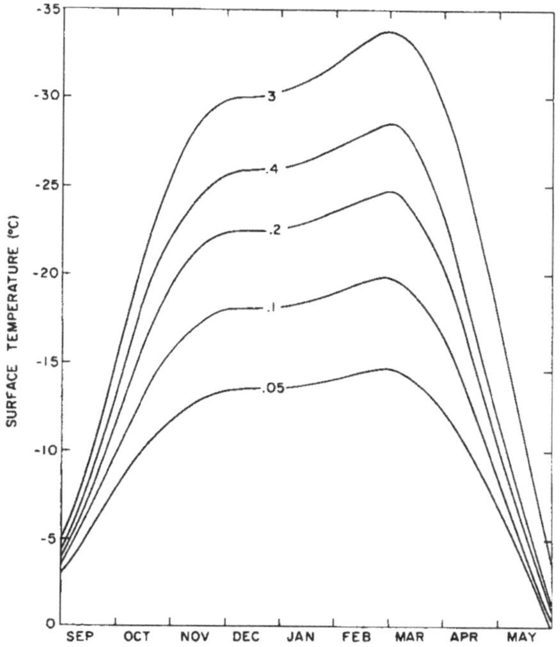

FIGURE 27. Seasonal variations in surface temperature for various ice thicknesses (in meters) in the Central Arctic.[39]

thicker ice. Turbulent heat losses over a refreezing lead can be more than two orders of magnitude larger than those over nearby thick ice. Thus, relatively small areas of thin ice and open water can have a significant impact on regional heat, mass, and salt balances.

Theoretical calculations have been carried out to quantify how thickness affects the surface heat balance in regions of perennial Arctic ice.[39] Results are summarized in Table 3. When the ice is thin, heat input at the surface is controlled by F_c and temperatures are relatively insensitive to changes in atmospheric forcing. Heat exchange with the atmosphere over young ice is dominated by the sensible heat flux. In March, for example, F_s ranges from a small heat gain over thick ice to a loss over open leads that is about 3.5 times the total of the incident radiation fluxes. While F_e over open leads is comparable to the radiation fluxes during most of the winter, its importance decreases rapidly with h due to a steep decline in saturation vapor pressure below 0°C. Over thicker first-year ice F_s is comparable to the net longwave radiation. Energy needed to support the large energy fluxes over young ice is derived largely from latent heat released by the growth of ice and not by heat transport in the ocean. In all cases the relative differences in heat exchange tend toward zero in the spring as the ice approaches the melting point. The calculations also demonstrate that the surface heat balance becomes insensitive to thickness once h exceeds 80 to 100 cm.

D. Regional Fluxes

Large changes in the amount of thin ice and open water can occur on time scales of hours to days, producing corresponding changes in the regional heat and mass balance which are independent of the incident energy fluxes. For obvious reasons most field data have been gathered from thick, relatively stable sea ice. However, because they do not take into account contributions made by areas of thin ice and open water, such data are not necessarily representative of the region as a whole. The large spatial variability in local values makes direct determination of regional fluxes difficult. An alternate approach is to utilize theoretical models to infer large scale values from routine local observations. The method works as

Table 3
SURFACE HEAT BALANCE COMPONENTS IN THE CENTRAL ARCTIC AS A FUNCTION OF ICE THICKNESS[39]

	Sept.	Oct.	Nov.	Dec.	Jan.	Feb.	March	April	May	June
					Open Water					
Net shortwave radiation	89	24					7	83	209	281
Net longwave radiation	−28	−69	−114	−138	−142	−147	−149	−140	−98	−43
Sensible heat flux	−68	−259	−458	−540	−543	−575	−615	−520	−276	−33
Latent heat flux	−34	−108	−139	−145	−145	−147	−150	−144	−112	−31
Oceanic heat flux	130	436	711	823	830	869	914	804	486	107
					0.5 m					
Available shortwave energy[a]	60	16					5	56	141	189
Net longwave radiation	−22	−42	−70	−87	−91	−93	−93	−93	−76	−51
Sensible heat flux	−44	−142	−261	−311	−312	−334	−362	−313	−179	−74
Latent heat flux	−31	−50	−49	−47	−46	−45	−44	−51	−62	−42
Conductive heat flux	37	218	380	445	449	472	494	401	176	−17
					0.1 m					
Available shortwave energy	56	15					4	52	131	175
Net longwave radiation	−20	−33	−55	−69	−73	−75	−73	−77	−68	−50
Sensible heat flux	−36	−101	−186	−222	−222	−238	−262	−232	−143	−73
Latent heat flux	−27	−34	−30	−27	−26	−25	−24	−31	−47	−47
Conductive heat flux	27	153	271	318	321	338	355	288	127	−5

0.2 m										
Available shortwave energy	53	14					4	49	124	166
Net longwave radiation	−19	−25	−41	−53	−57	−58	−55	−62	−61	−49
Sensible heat flux	−29	−64	−115	−137	−136	−147	−165	−154	−109	−68
Latent heat flux	−23	−22	−16	−14	−14	−13	−12	−17	−34	−46
Conductive heat flux	18	97	172	204	207	218	228	184	80	−3
0.4 m										
Available shortwave energy	48	13					4	46	114	153
Net longwave radiation	−17	−19	−31	−41	−45	−45	−42	−51	−55	−47
Sensible heat flux	−22	−36	−62	−72	−70	−77	−91	−93	−82	−60
Latent heat flux	−19	−14	−9	−6	−6	−6	−5	−10	−26	−44
Conductive heat flux	10	56	102	119	121	128	134	108	49	−2
0.8 m										
Available shortwave energy	45	12					3	42	104	140
Net longwave radiation	−16	−16	−24	−33	−37	−37	−34	−43	−51	−46
Sensible heat flux	−17	−18	−27	−30	−27	−31	−42	−53	−61	−54
Latent heat flux	−16	−9	−4	−3	−2	−2	−2	−5	−19	−39
Conductive heat flux	4	31	55	66	66	70	75	59	27	−1
3.0 m										
Available shortwave energy	16	4					1	17	42	59
Net longwave radiation	−14	−12	−18	−26	−31	−29	−25	−32	−38	−35
Sensible heat flux	−5	0	6	11	17	16	12	9	−2	−8
Latent heat flux	−9	−4	0	0	0	0	0	0	−4	−10
Conductive heat flux	12	12	12	15	14	13	12	6	2	−6

Note: Energy fluxes are given in W m^{-2} and apply to conditions on the first day of the month.

[a] $(1 - i_o)(1 - \alpha) F_r$.

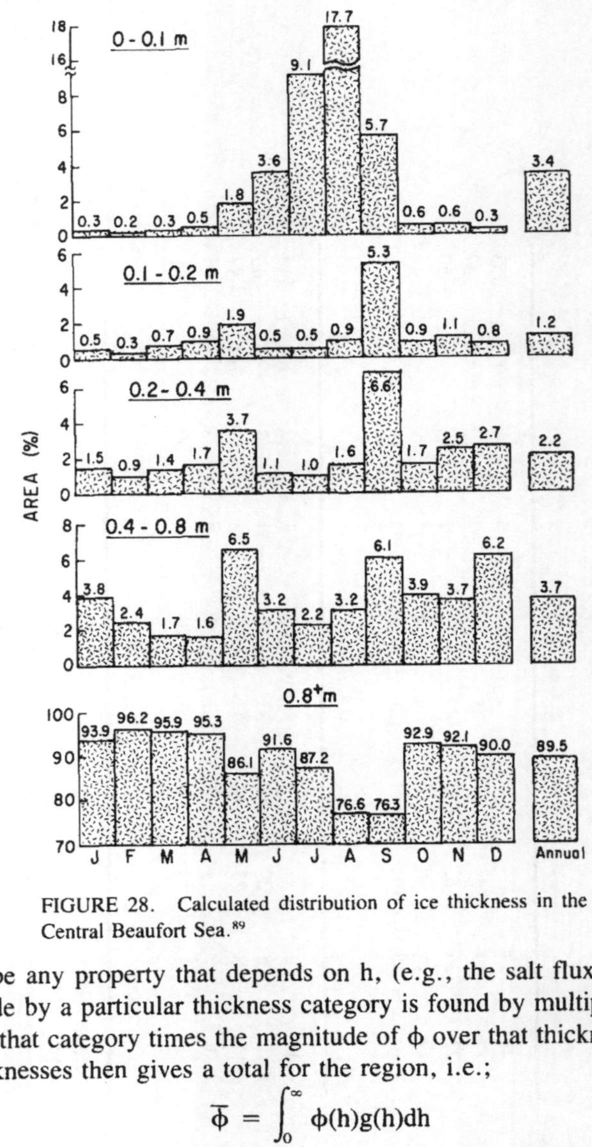

FIGURE 28. Calculated distribution of ice thickness in the Central Beaufort Sea.[89]

follows. Let ϕ be any property that depends on h, (e.g., the salt flux to the ocean). The contribution made by a particular thickness category is found by multiplying the fractional area covered by that category times the magnitude of ϕ over that thickness. Summing over all possible thicknesses then gives a total for the region, i.e.;

$$\overline{\phi} = \int_0^\infty \phi(h)g(h)dh \tag{29}$$

$\overline{\phi}$ is defined to be the *regional* or *large scale* average. In practice it is usually necessary to work with discrete thickness categories. If g(h) is partitioned into n categories, Equation 29 can be approximated by:

$$\overline{\phi} = \sum_{i=1}^{n} \overline{\phi}_i \Delta G_i \tag{30}$$

where $\overline{\phi}_i$ is the average value of ϕ in the ith category, $\Delta G_i = \overline{g}_i \Delta h_i$, $\Delta h_i = h_i - h_{i-1}$, and $\overline{g}_i$ is the average value of g in the ith category. Since ϕ and g are also usually functions of time, we use the notation $<\phi> = \int_0^t \overline{\phi}(t)dt$ to denote spatial averages integrated over time. Application of Equation 29 or 30 requires general heat balance and ice movement information. A thermodynamic ice model is used with heat balance data to obtain growth rates and energy fluxes over all thicknesses of ice. Growth rates and ice velocities are then used to drive an ice thickness distribution model and calculate g(h) for the region, thereby allowing the determination of $\overline{\phi}$ from Equation 30.

A study of the effects of thickness variations on regional fluxes has been carried out using data from the central Beaufort Sea.[89] Results are summarized in Table 4 and Figures 28 and

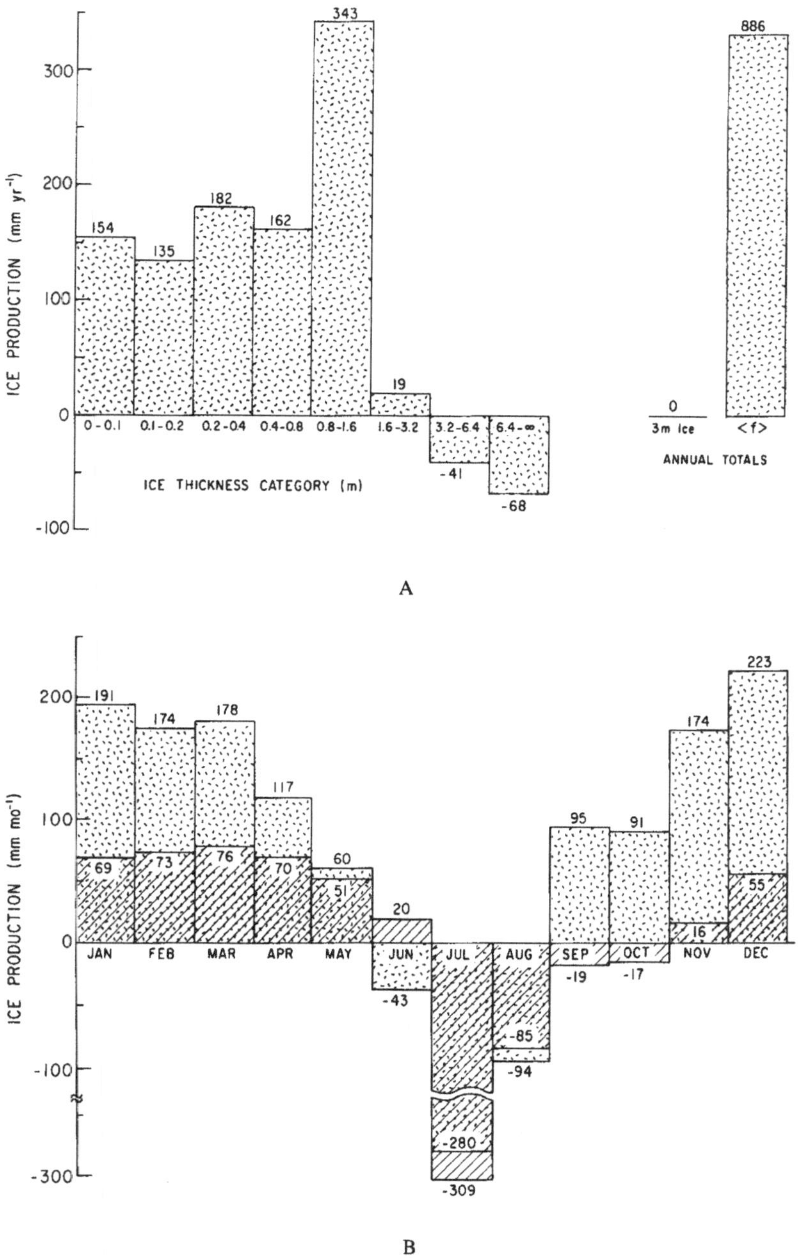

FIGURE 29. Regional ice production in the Central Beaufort Sea: (A) annual totals in each thickness category, (B) monthly totals taking into account ice thickness variations (stippled) and assuming a uniform 3m thick ice cover (cross-hatched).[89]

29. The calculations clearly demonstrate that relatively small amounts of thin ice have a major impact on the overall heat and mass balance. Between October and April refreezing leads generally covered no more than 0.5% of the area, while thin ice ($10 < h < 40$ cm) covered only 2 to 3%. Greater divergence and melting of thin ice caused a large increase in the amount of open water during the spring and summer. Freezing of the open water in late August produced a September peak in the distribution of young ice, but ridging and continued growth combined to move most of this ice into the 80^+ cm category by October.

Table 4
AREA WEIGHTED FLUXES OF HEAT (MJ m^{-2} $month^{-1}$) AND SALT (kg m^{-2} $month^{-1}$) IN THE CENTRAL BEAUFORT SEA[89]

	Ice thickness category	Jan.	Feb.	Mar.	Apr.	May.	Jun.	Jul.	Aug.	Sep.	Oct.	Nov.	Dec.	Annual
Net radiation	0—0.1	−0.6	−0.4	−0.6	0.6	7.4	21.6	44.4	46.6	5.5	−0.7	−1.2	−0.8	121.8
	0.1—0.2	−0.8	−0.5	−0.8	1.4	6.7	2.1	1.5	1.7	3.5	−0.7	−1.6	−1.3	11.2
	0.2—0.4	−2.0	−1.1	−1.1	2.7	11.7	4.4	3.1	2.8	2.9	−0.9	−2.7	−3.3	16.5
	0.4—0.8	−4.1	−2.4	−0.9	1.9	18.0	10.8	5.9	5.0	2.1	−1.3	−3.3	−5.9	25.8
	0.8—∞	−74.2	−65.2	−50.2	−10.9	48.5	85.6	181.8	86.5	−1.7	−30.8	−55.4	−60.9	53.1
	Total	−81.7	−69.6	−53.6	−4.3	92.3	124.5	236.7	142.6	12.3	−34.4	−64.2	−72.2	228.4
	3 m	−79.0	−67.8	−52.4	−10.8	56.5	94.3	207.9	111.2	−1.8	−33.1	−60.2	−67.6	97.2
Sensible heat flux $\langle F_s \rangle$	0—0.1	−2.1	−1.4	−2.9	−2.9	−4.7	−1.0	−1.2	−0.7	−8.1	−2.9	−4.4	−2.7	−35.0
	0.1—0.2	−2.3	−1.6	−3.4	−3.2	−3.5	−0.2	−0.1	−0.2	−5.0	−2.6	−4.7	−3.8	−30.6
	0.2—0.4	−3.9	−2.6	−4.4	−4.0	−5.7	−0.6	−0.1	−0.3	−5.0	−2.7	−6.3	−7.4	−43.0
	0.4—0.8	−4.7	−3.6	−2.8	−2.6	−9.3	−1.9	−0.3	−0.6	−3.1	−3.0	−4.7	−8.2	−44.8
	0.8—∞	43.7	32.8	27.8	11.1	−13.4	−15.2	−11.8	−12.0	−5.7	4.6	20.5	30.9	113.3
	Total	30.7	23.6	14.3	−1.6	−36.6	−18.9	−13.5	−13.8	−26.9	−6.6	0.4	8.8	−40.1
	3 m	46.5	34.1	29.0	11.3	−15.6	−16.7	−13.6	−15.7	−7.5	4.9	22.3	34.3	113.3
Latent heat flux $\langle F_e \rangle$	0—0.1	−0.3	−0.2	−0.4	−0.7	−2.2	−1.3	−2.5	−3.1	−4.8	−0.8	−0.8	−0.4	−17.5
	0.1—0.2	−0.3	−0.2	−0.3	−0.7	−1.5	−0.2	−0.1	−0.3	−2.9	−0.6	−0.6	−0.4	−8.1
	0.2—0.4	−0.4	−0.2	−0.4	−0.8	−2.5	−0.6	−0.3	−0.5	−2.7	−0.6	−0.7	−0.8	−10.5
	0.4—0.8	−0.5	−0.3	−0.2	−0.5	−4.3	−1.8	−0.6	−0.9	−1.8	−0.9	−0.5	−0.8	−13.1
	0.8—∞	0.0	0.0	0.0	−3.2	−16.2	−26.1	−24.0	−20.5	−12.7	−7.3	−0.1	0.0	−110.1
	Total	−1.5	−0.9	−1.3	−5.9	−26.7	−30.0	−27.5	−25.3	−24.9	−10.2	−2.7	−2.4	−159.3
	3 m	0.0	0.0	0.0	−3.5	−18.8	−28.4	−27.5	−26.7	−16.8	−7.9	−0.1	0.0	−129.7

Conductive heat flux $<F_c>$	0—0.1	3.1	2.0	4.1	4.1	6.1	0.1	0.0	3.3	13.7	4.6	6.4	3.9	51.4
	0.1—0.2	3.4	2.3	4.7	4.2	3.6	0.0	0.0	0.1	8.1	3.9	6.7	5.5	42.5
	0.2—0.4	6.3	3.9	6.2	4.6	4.1	0.0	0.0	0.0	7.8	4.4	9.8	11.4	58.5
	0.4—0.8	9.3	6.2	4.2	2.3	3.6	0.0	0.0	0.0	4.4	5.5	8.5	14.9	58.9
	0.8—∞	30.6	32.4	22.4	3.1	−19.0	−4.5	0.0	0.0	20.1	33.5	35.0	29.9	183.5
	Total	52.7	46.8	41.6	18.3	−1.6	−4.4	0.0	3.4	54.1	51.9	66.4	65.6	394.8
	3 m	32.5	33.7	23.4	3.0	−22.1	−5.0	0.0	0.0	26.1	36.1	38.0	33.3	199.0
Salt flux $< F_{salt}>$	0—0.1	0.14	0.09	0.18	0.21	0.27	−0.05	0.00	−0.06	0.83	0.26	0.30	0.18	2.35
	0.1—0.2	0.20	0.14	0.28	0.25	0.14	−0.05	−0.06	−0.04	0.51	0.26	0.42	0.34	2.39
	0.2—0.4	0.42	0.26	0.40	0.27	0.13	−0.13	−0.13	−0.08	0.51	0.31	0.66	0.76	3.38
	0.4—0.8	0.66	0.44	0.29	0.15	0.07	−0.35	−0.29	−0.18	0.29	0.40	0.61	1.06	3.15
	0.8—∞	2.43	2.65	2.38	1.59	0.71	−3.96	−6.20	−1.91	−0.20	0.57	1.33	2.06	1.45
	Total	3.85	3.58	3.53	2.47	1.32	−4.54	−6.68	−2.27	1.94	1.80	3.32	4.40	12.72
	3 m	1.57	1.77	1.84	1.71	1.25	−3.60	−7.30	−2.09	−0.45	−0.41	0.21	1.24	−4.26

The annual net radiation balance for the region was more than double what would have been expected on the basis of measurements over thicker ($h > 1$ m) ice. Most of this additional energy input came from solar radiation absorbed in summer leads and ultimately went to lateral melting on floe edges and F_w. The amount of solar radiation absorbed beneath the ice during the summer was about 105 MJ m^{-2}, more than a factor of 50 greater than the amount that would be absorbed beneath a continuous 3 m thick ice cover. Conduction of heat to the surface increases by a factor of 2 when thickness variations were taken into account. Most of this heat was lost to the atmosphere during the fall and winter via larger turbulent fluxes. Turbulent heat input to the atmosphere totaled about 200 MJ m^{-2} for the year, due in large part to greater sensible heat losses associated with young ice formation in the fall. It is notable that the open water category (0 to 10 cm) did not dominate the balance, rather it was the intermediate thicknesses of young ice (10 to 80 cm) that were responsible for over half the total loss.

Total ice production within the region $<f>$ was also dramatically larger when contributions from young ice were considered. Figure 29 shows annual ice production in each of the thickness categories. The mass produced in the 0 to 10 cm category was equivalent to a 15.4 cm thick layer covering the entire region. For the year $<f>$ totaled about 89 cm, in contrast to the zero mass balance expected for 3 m ice. Growth in areas of first-year ice accounted for essentially all of the net ice production. Despite the extremely large growth rates in new leads, only about 15% of the first-year ice production occurred in the 0 to 10 category, reflecting the very small area occupied by open water during the winter. As expected, the greatest production took place following the fall freezeup. Seasonal variations in the flux of salt to the ocean paralleled the ice production. Net salt input began in mid-September, correlating with the observed time of minimum salt content in the mixed layer. Without young ice, the salt content of the mixed layer would not begin to increase significantly until early December.

Numerical experiments were also conducted to determine how uncertainties in specified strains might affect regional heat and mass fluxes. It was found that the average divergence of the ice was poorly correlated with monthly or annual heat flux totals, but that $<f>$ was related to the time average of the strain components. Ice typically alternates between diverging and converging motions over periods of days, producing continual changes in the amount of open water, ice production, and ridging activity. It appears that these short term variations in strain are more important to the thermodynamics than the longer term averages. The tendency for $<f>$ to increase with increased divergence suggests that variations in the export of ice from the Arctic Basin may not have a great impact on total ice volume because of the compensating changes in ice production.

These calculations have shown that thin ice and open water resulting from differential ice movement cause interactions between the ocean and atmosphere to be much more vigorous than would be the case for a uniform, static ice cover. Further, it is evident that care must be exercised when interpreting local measurements made in areas of nonuniform ice. Failure to consider thickness variations and how they affect area averages can lead to substantial errors in understanding large scale interactions between the ice, ocean, and atmosphere.

ACKNOWLEDGMENTS

I am indebted to Steve Ackley, Don Perovich, and Ed Waddington for technically reviewing the material in this chapter. Special thanks go to Rita Horner for her patience and many helpful suggestions during preparation of the manuscript. Finally, I would like to acknowledge the contribution made by the Office of Naval Research whose long term commitment to sea ice research produced much of the information summarized above and whose support under Contract N00014-84-C-0111 made possible this review.

REFERENCES

1. Central Intelligence Agency, Polar Regions Atlas, GC 78-10040, U.S. Government Printing Office, Washington, D. C., 1978, 12.
2. **Zwally, H. J., Comiso, J. C., Parkinson, C. L., Campbell, W. J., Carsey, F. D., and Gloersen, P.,** Antarctic Sea Ice, 1973-1976: Satellite Passive-Microwave Observations, NASA SP-459, National Aeronautics and Space Administration, Washington, D. C., 1983, chap. 5.
3. **Walsh, J. E. and Johnson, C. M.,** An analysis of arctic sea ice fluctuations, 1953-77, *J. Phys. Oceanogr.*, 9, 580, 1979.
4. **Ackley, S. F.,** Mass-balance aspects of Weddell Sea pack ice, *J. Glaciol.*, 24, 391, 1979.
5. **Gordienko, P.,** Arctic ice drift, in *Proc. Conference on Arctic Sea Ice*, Publ. 598, National Academy of Sciences, Washington, D. C., 1958, 210.
6. **Colony, R. and Thorndike, A. S.,** An estimate of the mean field of arctic sea ice motion, in *Proc. Eighth Conference on Probability and Statistics in Atmospheric Sciences*, American Meteorological Society, Boston, 1983, 10.
7. **Muench, R. D. and Ahlnas, K.,** Ice movement and distribution in the Bering Sea from March to June 1974, *J. Geophys. Res.*, 81, 4467, 1976.
8. **Cavalieri, D. J., Martin, S., and Gloersen, P.,** Nimbus 7 SSMR observations of the Bering Sea ice cover during March 1979, *J. Geophys. Res.*, 88, 2743, 1983.
9. **Budd, W. F.,** Antarctic sea ice variation from satellite sensing in relation to climate, *J. Glaciol.*, 15, 417, 1975.
10. **Ackley, S. F. and Keliher, T.,** Antarctic sea ice dynamics and its possible climatic effects, *AIDJEX Bull.*, 33, 53, 1976.
11. **Herbert, W.,** The first surface crossing of the Arctic Ocean, *Geogr. J.*, 136, 511, 1970.
12. **Tucker, W. B. and Govoni, J. W.,** Morphological investigations of first-year sea ice pressure ridge sails, *Cold Regions Sci. Tech.*, 5, 1, 1981.
13. **Wadhams, P.,** Ice characteristics in the seasonal sea ice zone, *Cold Regions Sci. Tech.*, 2, 37, 1980.
14. **Weeks, W. F. and Ackley, S. F.,** The Growth, Structure, and Properties of Sea Ice, Monograph 82-1, Cold Regions Research and Engineering Laboratory, Hanover, 1982.
15. **Paige, R. A.,** Crystallographic Studies of Sea Ice in McMurdo Sound, Antarctica, U.S. Naval Civil Engineering Laboratory Technical Report R494, Washington, D.C., 1966.
16. **Andreas, E. L. and Ackley, S. F.,** On the differences in ablation seasons of the Arctic and Antarctic sea ice, *J. Atmos. Sci.*, 39, 440, 1982.
17. **Gill, A. E.,** Circulation and bottom water production in the Weddell Sea, *Deep-Sea Res.*, 20, 111, 1973.
18. **Carsey, F. D.,** Microwave observations of the Weddell Sea polynya, *Mon. Weather Rev.*, 108, 2032, 1980.
19. **Streten, N. A.,** Satellite observations of the summer decay of the Antarctic sea ice, *Arch. Meteorol. Geophys. Bioklimatol. Ser. A*, 22, 119, 1973.
20. **McPhee, M. G.,** Physical oceanography of the seasonal sea ice zone, *Cold Regions Sci. Tech.*, 2, 93, 1980.
21. **Kovacs, A. and Mellor, M.,** Sea ice morphology and ice as a geological agent in the Southern Beaufort Sea, in *The Coast and Shelf of the Beaufort Sea*, Reed, J. C. and Sater, J. E., Eds., The Arctic Institute of North America, Arlington, Va. 1974, 113.
22. **Tucker, W. B., Weeks, W. F., Kovacs, A., and Gow, A. J.,** Nearshore ice motion at Prudhoe Bay, Alaska, in *Sea Ice Processes and Models*, Pritchard, R. S., Ed., University of Washington Press, Seattle, 1980, 261.
23. **Newbury, T. K.,** Under landfast ice, *Arctic*, 36, 328, 1983.
24. **Tucker, W. B., Weeks, W. F., and Frank, M.,** Sea ice ridging over the Alaskan continental shelf, *J. Geophys. Res.*, 84, 4885, 1979.
25. **Campbell, W. J., Gloersen, P., Webster, W. J., Wilheit, T. T., and Ramseier, R. O.,** Beaufort Sea ice zones as delineated by microwave imagery, *J. Geophys. Res.*, 81, 1103, 1976.
26. **Bauer, J. and Martin, S.,** Field observations of the Bering Sea ice edge properties during March 1979, *Mon. Weather Rev.*, 108, 2045, 1980.
27. **Alexander, V.,** Interrelationships between the seasonal sea ice and biological regimes, *Cold Regions Sci. Tech.*, 2, 157, 1980.
28. **McPhee, M.,** Greenland Sea ice/ocean margin, *EOS*, 64, 82, 1983.
29. MIZEX West Study Group, MIZEX West: Bering Sea marginal ice zone expriment, *EOS*, 64, 578, 1983.
30. **Neumann, G. and Pierson, W. J.,** *Principles of Physical Oceanography*, Prentice - Hall, Englewood Cliffs, N.J., 1966, 43.
31. **Doronin, Yu. P. and Kheisin, D. E.,** *Sea Ice*, Gidrometeoizdat Publ., Leningrad, 1975. (Translation by National Science Foundation, TT 75-52088, 1977.)

32. **Martin, S. and Kauffman, P.**, A field and laboratory study of wave damping by grease ice, *J. Glaciol.*, 27, 283, 1981.
33. **Anderson, D. L.**, Growth rate of sea ice, *J. Glaciol.*, 3, 1170, 1961.
34. **Armstrong, T., Roberts, B., and Swithinbank, C.**, *Illustrated Glossary of Snow and Ice*, Scott Polar Research Institute, Cambridge, 1973, 40.
35. **Bilello, M. A.**, Formation, growth and decay of sea ice in the Canadian Arctic Archipelago, *Arctic*, 14, 3, 1961.
36. **Maykut, G. A.**, The surface heat and mass balance, in *The Geophysics of Sea Ice*, Untersteiner, N., Ed., Plenum Press, New York, in press, chap. 5.
37. **Nakawo, M. and Sinha, N. K.**, Growth rate and salinity profile of first-year sea ice in the high Arctic, *J. Glaciol.*, 27, 315, 1981.
38. **Allison, I.**, Antarctic sea ice growth and oceanic heat flux, in *Sea Level Ice and Climatic Change*, Allison, I., Ed., International Association Hydrological Sciences, Publ., 131, Washington, D.C., 1981, 161.
39. **Maykut, G. A.**, Energy exchange over young sea ice in the Central Arctic, *J. Geophys. Res.*, 83, 3646, 1978.
40. **Zubov, N. N.**, *L'dy Arktiki (Arctic Ice)*, Izdatel'stvo Glavsevmorputi, Moscow, 1945. (Translated from Russian by U.S. Naval Oceanographic Office and American Meteorological Society, 1965.)
41. **Untersteiner, N.**, On the mass and heat balance of arctic sea ice, *Arch. Meteorol. Geophys. Bioklimatol. Ser. A*, 12, 151, 1961.
42. **Maykut, G. A. and Untersteiner, N.**, Some results from a time dependent, thermodynamic model of sea ice, *J. Geophys. Res.*, 76, 1550, 1971.
43. **Semtner, A. J.**, A model for the thermodynamic growth of sea ice in numerical investigations of climate, *J. Phys. Oceanogr.*, 6, 379, 1976.
44. **Dayton, P. K., Robillard, G. A., and DeVries, A. L.**, Anchor ice formation in McMurdo Sound, Antarctica, and its biological effects, *Science*, 163, 273, 1969.
45. **Morecki, V. N.**, Underwater sea ice (in Russian), *Probl. Arkt. Antarkt.*, 19, 32, 1965. (English translation, Def. Res. Board Can., Translation T497 R, 1968.)
46. **Larsen, L.**, Sediment-laden sea ice: concepts, problems, and approaches, NOAA, *Arctic Proj. Bull.*, 29, 59, 1980.
47. **Martin, S.**, Frazil ice in rivers and oceans, *Annu. Rev. Fluid Mech.*, 13, 379, 1981.
48. **Martin, S.**, Ice stalactites: comparison of a laminar flow theory with experiment, *J. Fluid Mech.*, 63, 51, 1974.
49. **Langleben, M. P.**, On the factors affecting the rate of ablation of sea ice, *Can. J. Earth Sci.*, 3, 431, 1966.
50. **Gordon, A. L.**, Seasonality of southern ocean sea ice, *J. Geophys. Res.*, 86, 4193, 1981.
51. **Langleben, M. P.**, The decay of an annual ice cover, *J. Glaciol.*, 11, 337, 1972.
52. **Hall, R. T.**, Seasonal photo mosaic of the AIDJEX triangle, *AIDJEX Bull.*, 39, 79, 1978.
53. **Hallett, J.**, Crystal growth and the formation of spikes in the surface of supercooled water, *J. Glaciol.*, 3, 698, 1960.
54. **Weeks, W. F. and Hamilton, W. L.**, Petrographic characteristics of young sea ice, Point Barrow, Alaska, *Am. Mineral.*, 47, 945, 1962.
55. **Cherepanov, N. V.**, Spatial alignment of sea ice crystal structure (in Russian), *Probl. Arkt. Antarkt.*, 38, 176, 1971.
56. **Weeks, W. F. and Gow, A. J.**, Preferred crystal orientations along the margins of the Arctic Ocean, *J. Geophys. Res.*, 84, 5105, 1978.
57. **Harrison, J. D. and Tiller, W. A.**, Controlled freezing of water, in *Ice and Snow*, Kingery, W. D., Ed., MIT Press, Cambridge, 1963, 215.
58. **Assur, A.**, Composition of sea ice and its tensile strength, in *Arctic Sea Ice*, National Academy of Sciences, National Research Council Publ. 598, Washington, D.C., 1958, 106.
59. **Morey, R. M., Kovacs, A., and Cox, G. F. N.**, Electromagnetic Properties of Sea Ice, Report 84-2, Cold Regions Research and Engineering Laboratory, Hanover, 1984.
60. **Cox, G. F. N. and Weeks, W. F.**, Brine Drainage and Initial Salt Entrapment in Sodium Chloride Ice, Research Report 345, Cold Regions Research and Engineering Laboratory, Hanover, 1975.
61. **Weeks, W. F. and Lee, O. S.**, The salinity distribution in young sea ice, *Arctic*, 15, 93, 1958.
62. **Schwarzacher, W.**, Pack-ice studies in the Arctic Ocean, *J. Geophys. Res.*, 64, 2357, 1959.
63. **Cox, G. F. N. and Weeks, W. F.**, Salinity variations in sea ice, *J.. Glaciol*, 13, 109, 1974.
64. **Untersteiner, N.**, Natural desalination and equilibrium salinity profile of perennial sea ice, *J. Geophys. Res.*, 73, 1251, 1968.
65. **Lake, R. A. and Lewis, E. L.**, Salt rejection by sea ice during growth, *J. Geophys. Res.*, 75, 583, 1970.
66. **Eide, L. and Martin, S.**, The formation of brine drainage features in young sea ice, *J. Glaciol.*, 14, 137, 1975.

67. **Niedrauer, T. M. and Martin, S.,** An experimental study of brine drainage and convection in young sea ice, *J. Geophys. Res.*, 84, 1176, 1979.
68. **Reeburgh, W. S. and Springer-Young, M.,** New measurements of sulfate and chlorinity in natural sea ice, *J. Geophys. Res.*, 88, 2959, 1983.
69. **Cherepanov, N. V.,** Using the method of crystal optics for determining the age of drift ice (in Russian), *Probl.*, 2, 179, 1957.
70. **Martin, S. and Kauffman, P.,** The evolution of under-ice melt ponds, or double diffusion at the freezing point, *J. Fluid Mech.*, 64, 507, 1974.
71. **Frankenstein, G. and Garner, R.,** Equations for determining the brine volume of sea ice from $-0.5°$ to $-22.9°C$, *J. Glaciol.*, 6, 943, 1967.
72. **Schwerdtfeger, P.,** The thermal properties of sea ice, *J. Glaciol.*, 4, 789, 1963.
73. **Yen, Y. C.,** Review of Thermal Properties of Snow, Ice, and Sea Ice, Research Report 81-10, Cold Regions Research and Engineering Laboratory, Hanover, 1981.
74. **Anderson, D. L.,** A model for determining sea ice properties, in *Arctic Sea Ice*, National Academy of Sciences, National Research Council Publ. 598, Washington, D.C., 1958, 148.
75. **Ono, N.,** Specific heat and heat of fusion of sea ice, in *Physics of Snow and Ice*, Oura, H., Ed., Hokkaido University, Hokkaido, Japan, 1967, 599.
76. **Ono, N.,** Thermal properties of sea ice IV: thermal constants of sea ice (in Japanese), *Low Temp. Sci.*, A26, 329, 1968.
77. **Mellor, M.,** Mechanical Behavior of Sea Ice, Monograph 83-1, Cold Regions Research and Engineering Laboratory, Hanover, 1983.
78. **Langleben, M. P. and Pounder, E. R.,** Elastic parameters of sea ice, in *Ice and Snow*, Kingery, W. D., Ed., MIT Press, Cambridge, 1963, 69.
79. **Bergdahl, L.,** Physics of Ice and Snow as Affects Thermal Pressure, Chalmers University, Department of Hydraulics, Report Series A:1, Göteborg, Sweden, 1977.
80. **Wang, Y. S.,** Uniaxial comparison testing of Arctic Sea ice, *International Conference on Port & Ocean Engineering under Arctic Conditions, POAC*, Vol. 1, Quebec, 346, 1981.
81. **Weeks, W. F. and Assur, A.,** Fracture of Lake and Sea Ice, Research Report 269, Cold Regions Research and Engineering Laboratory, Hanover, 1969.
82. **Buynitskiy, V. Kh.,** The influence of microalgae on the structure and strength of antarctic sea ice, *Oceanology*, 8, 771, 1968.
83. **Parmerter, R. R. and Coon, M. D.,** Model of pressure ridge formation in sea ice, *J. Geophys. Res.*, 77, 6565, 1972.
84. **Coon, M. D., Maykut, G. A., Prichard, R. S., Rothrock, D. A., and Thorndike, A. S.,** Modeling the pack ice as an elastic-plastic material, *AIDJEX Bull.*, 24, 1, 1974.
85. **Hibler, W. D.,** A dynamic thermodynamic sea ice model, *J. Phys. Oceanogr.*, 9, 815, 1979.
86. **Rothrock, D. A.,** The energetics of the plastic deformation of pack ice by ridging, *J. Geophys. Res.*, 80, 4514, 1975.
87. **Hibler, W. D.,** Sea ice growth, drift, and decay, in *Dynamics of Snow and Ice Masses*, Colbeck, S. C., Ed., Academic Press, New York, 1980, 141.
88. **Grenfell, T. C. and Maykut, G. A.,** The optical properties of ice and snow in the Arctic Basin, *J. Glaciol.*, 18, 445, 1977.
89. **Maykut, G. A.,** Large-scale heat exchange and ice production in the Central Arctic, *J. Geophys. Res.*, 87, 7971, 1982.
90. **Grenfell, T. C. and Perovich, D. K.,** Spectral albedos of sea ice and incident solar irradiance in the southern Beaufort. Sea, *J. Geophys. Res.*, 89, 3573, 1984.
91. **Maykut, G. A. and Grenfell, T. C.,** The spectral distribution of light beneath first-year sea ice in the Arctic Ocean, *Limnol. Oceanogr.*, 20, 554, 1975.
92. **Thorndike, A. S., Rothrock, D. A., Maykut, G. A., and Colony, R.,** The thickness distribution of sea ice, *J. Geophys. Res.*, 80, 4501, 1975.
93. **Wadhams, P.,** Sea-ice topography of the Arctic Ocean in the Region 70°W to 25°E, *Philos, Trans. R. Soc. London Ser. A*, 302, 45, 1981.
94. **Wadhams, P. and Horne, R. J.,** An analysis of ice profiles obtained by submarine sonar in the Beaufort Sea, *J. Glaciol.*, 25, 401, 1980.
95. **Perovich, D. K.,** On the Summer Decay of a Sea Ice Cover, Ph.D. thesis, University of Washington, Seattle, 1983.
96. **Wadhams, P.,** The Effect of a Sea Ice Cover on Ocean Surface Waves, Ph.D. thesis, University of Cambridge, England, 1973.
97. **Vinje, T. E.,** Sea Ice Studies in the Spitzbergen-Greenland Sea Area, Landsat Report E77-10206, U.S. Department of Commerce, Washington, D. C., 1977.
98. **Rothrock, D. A. and Thorndike, A. S.,** Sea ice floe size distribution, *J. Geophys. Res.*, 89, 6477, 1984.

99. **Carsey, F. D.,** Arctic sea ice distribution at the end of summer 1973-76 from satellite microwave data, *J. Geophys. Res.*, 87, 5809, 1982.
100. **Svendsen, E., Kloster, K., Farrelly, B., Johannessen, O. M., Johannessen, J. A., Campbell, W. J., Gloersen, P., Cavalieri, D., and Mätzler, C.,** Norwegian remote sensing experiment: evaluation of the Nimbus 7 scanning multichannel microwave radiometer for sea ice research, *J. Geophys. Res.*, 88, 2781, 1983.
101. **Maykut, G. A. and Church, P. E.,** Radiation climate of Barrow, Alaska, 1962-66, *J. Appl. Meteorol.*, 12, 620, 1973.
102. **Marshunova, M. S.,** Principal characteristics of the radiation balance of the underlying surface and of the atmosphere in the Arctic (in Russian), *Tr. Arkt. Antarkt. Nauchno-Issled. Inst.*, 229, 5, 1961. (Translation in *Soviet Data on the Arctic Heat Budget and Its Climatic Influence,* Fletcher, J. O., Keller, B., and Olenicoff, S. M., Eds., Rand Corporation Memorandum RM-5003-PR, Santa Monica, 1966, 51.)
103. **Idso, S. B. and Jackson, R. D.,** Thermal radiation from the atmosphere, *J. Geophys. Res.*, 74, 5397, 1969.
104. **Zillman, J. W.,** A study of some aspects of the radiation and heat budgets of the southern hemisphere oceans, Meteorological Studies, 26, Bureau of Meteorology, Department of Interior, Canberra, Australia, 1972.
105. **Shine, K. P.,** Parameterization of shortwave flux over high albedo surfaces as a function of cloud thickness and surface albedo, *Q. J. R. Meteorol. Soc.*, 110, 747, 1984.
106. **Laevastu, T.,** Factors affecting the temperature of the surface layer of the sea, *Commentat. Phys. Math. Soc. Sci. Fenn.*, 25, 8, 1960.
107. **Maykut, G. A. and Untersteiner, N.,** Numerical Prediction of the Thermodynamic Response of Arctic Sea Ice to Environmental Changes, Rand Corporation Memorandum RM-6093-PR, Santa Monica, 1969.
108. **Thorpe, M. R., Banke, E. G., and Smith, S. D.,** Eddy correlation measurements of evaporation and sensible heat flux over arctic sea ice, *J. Geophys. Res.*, 78, 3573, 1973.
109. **Lindsay, R. W.,** Wind and Temperature Profiles Taken During the Arctic Lead Experiment, Master's thesis, Department of Atmospheric Sciences, University of Washington, Seattle, 1976.
110. **Doronin, Yu, P.,** On the heat balance of the Central Arctic (in Russian), *Tr. Arkt. Antarkt. Nauchno-Issled. Inst.*, 253, 178, 1963. (Translation in *Soviet Data on the Arctic Heat Budget and Its Climatic Influence,* Fletcher, J. O., Keller, B., and Olenicoff, S. M., Eds., Rand Corporation Memorandum RM-5003-PR, Santa Monica, 1966, 193.)
111. **Badgley, F. I.,** Heat balance at the surface of the Arctic Ocean, in *Proc. Symp. Arctic Heat Budget and Atmospheric Circulation,* Fletcher, J. O., Ed., Rand Corporation Memorandum RM-5233-NSF, Santa Monica, 1966, 267.
112. **Leavitt, E., Albright, M., and Carsey, F.,** Report on the AIDJEX meteorological experiment, *AIDJEX Bull.*, 39, 121, 1978.
113. **Aagaard, K., Coachman, L. K., and Carmack E.,** On the halocline of the Arctic Ocean, *Deep-Sea Res.*, 28, 529, 1981.
114. **Walker, E. R. and Wadhams, P.,** Thick sea-ice floes, *Arctic*, 32, 140, 1979.
115. **Gordon, A. L., Chen, C. T. A., and Metcalf, W. G.,** Winter mixed layer entrainment of Weddell deep water, *J. Geophys. Res.*, 89, 637, 1984.

Chapter 3

ECOLOGY OF SEA ICE MICROALGAE

Rita Horner

TABLE OF CONTENTS

I. INTRODUCTION

Sea ice is an important component of the physical environment in polar regions and strongly affects all organisms living in these areas. Ice effectively reduces the amount of light reaching the water column and reduces heat and gas exchange and vertical mixing. At the same time, the upper surface provides a special habitat for a number of birds and mammals, while the lower surface acts as a substrate for microalgae, meiofauna, epifauna, and some fish.

Microalgae that live in close association with sea ice, either attached to ice crystals or in the interstitial water found between ice crystals, are here collectively called ice algae. Algal cells may be found scattered throughout the ice thickness, as in the Arctic in winter, or more often, are concentrated in distinct layers in the ice, including the snow-ice interface, the interior of the ice, and the ice-water interface.

In addition to the ice itself, microalgae may be found in the seawater at the outer margin of the ice, in the so called ice edge or marginal ice zone. This is not an edge at all, but a relatively wide band of broken ice. The physical features of this region are discussed in more detail in Chapter 2. The organisms found in this region may include cells that originally grew in the ice and were released into the water as the ice melted or they may be strictly planktonic forms that normally grow in the water column. The planktonic cells may be washed onto ice floes by wave action or they may become attached to the bottom and sides of floes as they both drift in the water column. Planktonic algal cells may thus be mistaken for cells that actually originated in the ice, i.e., true ice algae. This ice edge habitat provides a whole new set of conditions for organisms living in it and is considered here in comparison with the true ice habitat because the two are often difficult to separate.

II. TERMINOLOGY

A number of terms have been used to describe ice algal communities. Zubov[1] used the terms *ice plankton* and *cryoplankton* for organisms peculiar to sea ice that develop during the summer in and around ice and form communities. Other early terminology included *colored ice* and *brown ice*.[2] Matsuda[3] was apparently the first person to use the name *plankton ice* and other Japanese investigators have also used this term,[2,4,5] sometimes qualifying it by adding bottom or surface depending on whether the community in question occurred at the ice-water interface or at the snow-ice interface. Plankton ice is not a good name because this implies that the organisms are planktonic in origin and this may not always be true. In fact, the ice community has been referred to as a reverse or upside down benthos.[6,7]

Bunt and Wood[8] referred to the bottom ice community as the *epontic* community (epi = on, ontos = being) and said it consisted of both attached and nonattached species, but epontic has also been used to describe all attached forms regardless of whether they are on plants, animals, rocks, debris, or at the surface of muds.[9,10] As Bunt and Wood[8] pointed out, ice algae become part of the plankton when the ice melts, but their development takes place under conditions quite unlike those that typical planktonic organisms experience and therefore they should not be called plankton. *Meroplankton* and *ice edge* forms are likewise misleading and unacceptable terms because they do not indicate the origin of the cells.

A number of somewhat more specific words have been used, some of which indicate the relationship of the organisms with ice. *Psychrophilic*, used by Bunt et al.,[11] refers to organisms living at low temperatures, but does not specifically indicate a relationship with ice. A better word is *cryophilic*, used by Usachev,[12] which does imply association with ice. Round[13] proposed the terms *epicryotic* for those cells living attached to ice and *endocryotic* for cells found actually within ice. Still another term, *cryophyton*, has been used for algae of snowfields and ice floes.[14] Kol[15] used *cryobiont* for organisms inhabiting snow and ice, while Bursa[16] used this term for organisms initially found on the ice surface. Whitaker,[17]

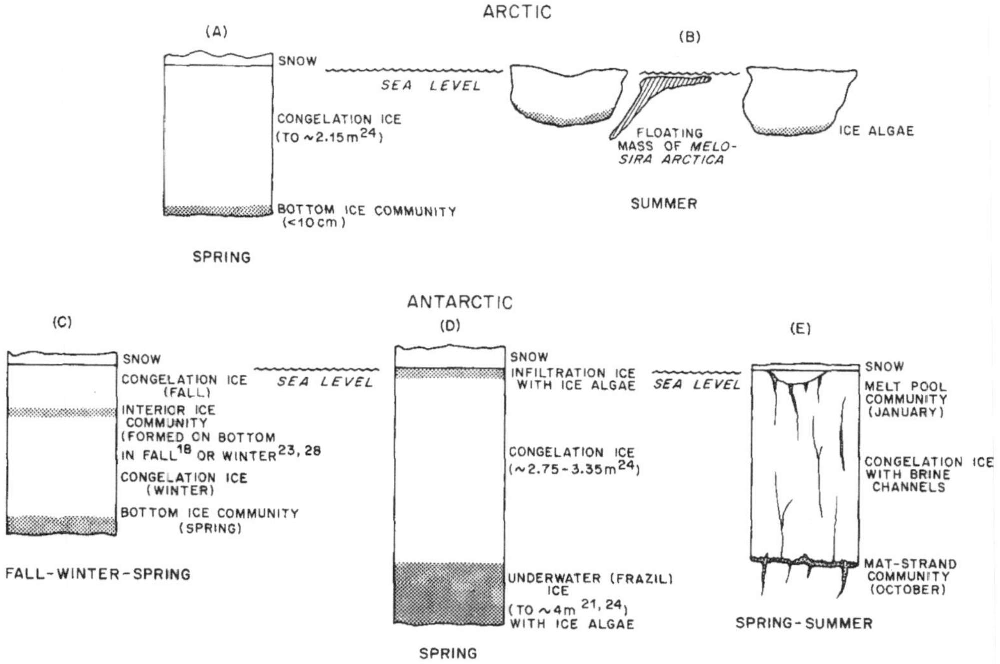

FIGURE 1. Schematic representation of the kinds of algal communities found in sea ice. The thickness of the ice and snow layers varies annually and with location. (E is redrawn from McConville and Wetherbee.[23])

saying that epontic meant "out of the sea", proposed the term *sympagic,* meaning "with ice", instead. Even though it is not a very accurate descriptive work, epontic has been used most often in the literature to indicate the community found on the under surface of the ice and here called the bottom ice community.

Ackley et al.[18] referred to *bottom, surface,* and *interior* ice algal communities depending on the location of the community in the ice. These names are more easily understood than most of the other terms and are more correct than plankton ice. They indicate not only the origin of the cells, but also where in the ice the community is found. However, as will be discussed later, there are at least three kinds of bottom ice communities depending on where and how the ice is formed, and there is also some variability with regard to the interior ice community. Perhaps the best terminology is also the simplest, i.e., bottom, surface, and interior, with qualifiers added to indicate variations depending on geographic location or origin, e.g., Arctic bottom ice community.

III. FORMATION AND DISINTEGRATION OF ICE COMMUNITIES

There are three main types of ice communities and their formation and disintegration have been generally described, but a number of variations occur depending on location (Figure 1). This seems to be more generally true in the Antarctic than in the Arctic, although as more locations are studied in the Arctic, more kinds of communities may be found there as well.

A. Bottom Ice

In the Arctic, cells are incorporated into the ice as it forms in the fall. The cells are present in brine pockets and channels throughout the ice thickness in winter. The ice increases in thickness until late spring, reaching a maximum thickness near 2 m in late May. As the air temperature increases in spring, brine drainage occurs and cells are carried toward the bottom few centimeters of the ice. The bottom ice containing the algae consists of unconsolidated ice crystals, is relatively soft when compared with the harder congelation ice above, and is easily broken or dislodged. When light levels at the bottom of the ice reach some threshold level, between 2 to 10 $\mu E\ m^{-2}\ sec^{-1}$* at Barrow and Narwhal Island, Alaska,[19,20] the cells begin to grow and divide and by early April form a visible, golden brown layer about 3 to 5 cm thick. Productivity of the ice algae continues until late May and is also highest at that time. As the air temperature continues to increase, the snow melts and the ice begins to decay. It is likely that increased radiation contributes to the disintegration of the algal layer because of selective absorption of solar energy by the cells.[5] Water currents under the ice also hasten the disappearance of the ice algal community by washing away the soft ice. The ice algae are released from the ice over a relatively short time period ranging from a few to perhaps 10 days.

The bottom ice community in the Antarctic is often found in a layer of unconsolidated ice crystals that may be 0.5 to 4 cm thick.[21] This layer of frazil ice is composed of aggregations of ice platelets. Cells are trapped between these unconsolidated ice crystals that are assumed to form in the water column and float up to the ice-water interface. The ice crystals are held in place by their positive buoyance, although fusion occurs at the frazil-congelation ice interface.[22] As in the Arctic, the algae grow when adequate light reaches the cells.[18] Major algal growth in the ice occurs from early November to early December at McMurdo Sound.[22] The annual cycle of bottom ice algal growth has not been completely described in the Antarctic with little or no information available on the end of the ice algal bloom. This bottom, frazil ice community probably occurs only in areas where there is little water movement under the ice, otherwise the bottom ice layer would be destroyed. Presumably the cells in this bottom community are also released over a relatively short time span.

The two additional kinds of bottom ice communities, the mat-strand community and a congelation ice community similar to the Arctic bottom ice community, have been reported from East Antarctica,[23] but their formation was not described.

B. Surface Ice

Two surface ice communities occur in the Antarctic. The first forms when the weight of the snow layer is sufficient to depress the ice and seawater containing algal cells floods the snow-ice interface.[2] This is also called infiltration ice. It is porous with an irregular, granular structure and random crystal layering.[24] This layer also disintegrates when warming from selective absorption of solar radiation by the algal cells weakens the ice. The layer separates from the rest of the ice and the cells are rapidly released into the water column.

The second surface community develops on fast ice when there is not enough snow to depress the ice. The melt pool community develops within the ice rather than in the overlying snow layer. Algal cells are transported up brine channels to melt pools and flourish to produce this surface community.[23]

The only report of surface ice communities in the Arctic is from Davis Strait[25] where samples of surface ice taken before and after a floe was washed over by the surrounding sea water showed that planktonic algae had been frozen into the previously barren ice surface. There was no indication of the growth or survival of these algae. Patches of a mat comprised primarily of *Melosira arctica* (Ehrenberg) Dickie and also containing *Coscinodiscus* sp. and *Gymnodinium* sp., were found on the surface of a honeycombed block of ice. There was no indication of the source of this mat, but large, mucilaginous masses of *M. arctica* have been reported from a number of Arctic areas.[12,26] The surface communities in Davis Strait

* All values for light measurements were converted to $\mu E\ m^{-2}\ sec^{-1}$ using 1 ft c ≅ 0.21 $\mu E\ m^{-2}\ sec^{-1}$; 1 lux ≅ 0.0195 $\mu E\ m^{-2}\ sec^{-1}$; 1 ly h^{-1} = 54.24 $\mu E\ m^{-2}\ sec^{-1}$.

are thought to be rare and insignificant because there is no insulating cover of snow or ice to protect the cells from low air temperatures and high light intensity.[25]

C. Interior Ice

The interior ice community is also apparently found only in the Antarctic, although there is at least one report of an interior ice algal community from the Bering Sea near Nome, Alaska.[27]

This community forms on the bottom of sea ice through a process similar to the growth of the spring-summer community. The interior community is presumably the result of the spring-summer bottom community being trapped as the ice increases in thickness below it and is therefore a remnant population.[18] The interior community described by Hoshiai[28] from Syowa Station and by McConville and Wetherbee[23] from Casey Station was present in winter (May-September), while Ackley et al.[18] found a similar community in the Weddell Sea in February-March. This community depends on air temperatures at or slightly below the freezing point to initiate brine drainage which also enriches the interior of the ice with nutrients, thus promoting increased algal growth. The interior community in the Weddell Sea releases cells to the water column over a relatively long period of time as the ice floes are slowly eroded through mechanical abrasion by wind and waves.[18]

IV. GEOGRAPHIC DISTRIBUTION

Ice algae are present in annual sea ice in all areas where sea ice is a regular feature of the environment. In higher latitudes, the ice community is well-developed and consists primarily of a special group of pennate diatoms that, in the northern hemisphere, are widely distributed throughout the area. Organisms other than diatoms are also present in the ice, but these are not so well-known and have been identified and studied in relatively few areas. Some of the organisms occur only in ice and are not found in the water column or sediments at other times of the year, while some taxa, especially centric diatoms, are nearly always found in the water column when ice is present. Whether they depend on the ice during some part of their life cycle is not known, although spores of some centric diatoms are regularly found in the ice. In lower latitudes of the northern hemisphere, such as in the Gulf of St. Lawrence, centric diatoms form a major part of the ice community.[29] In other areas, such as the Bering Sea, the ice edge community in the water column has been studied extensively,[30,31] but the ice community itself has not, although a few measurements have been made of primary production[32,33] and chlorophyll,[30] and a few cores have been analyzed for species present.[31]

The distribution of ice algae in old sea ice, i.e., ice older than one year, is not known. Gran[34] said ice algae were not found during the second year of the Norwegian North Polar Expedition because the ship was too far into the permanent pack ice and away from shore. Little additional information can be added to this statement after nearly a century. Schell[35] has found ice algae 160 km from shore off Prudhoe Bay, Alaska, but sampling was always done on relatively smooth, thin, annual ice in refrozen leads where it was easier to core. I have examined the bottom of blocks of sea ice approximately 7 m thick from an area north of Greenland in early April.[36] The bottom few centimeters of the ice consisted of relatively large platelets. The only organisms present were spores of *Thalassiosira* sp., probably *T. gravida* Cleve or *T. antarctica* Comber, and skeletons of the silicoflagellate *Distephanus speculum* (Ehrenberg) Haeckel, but perhaps this was too early in the spring for algal growth to occur. In the Antarctic, most of the sampling has been done in annual ice close to shore. A few samples of drifting ice have been collected offshore, but the age and previous history of the ice were not known. Most of the offshore samples have been of seawater collected near drifting ice.[37]

V. ORIGIN OF ICE ALGAE

The origin of the ice algae is still a matter for speculation. Early investigators[38] thought the cells may have been carried to the ice by wind or rivers, with the marine forms washed up during stormy weather or frozen in the ice when it formed. Ross[39] suggested that diatoms form spores in the fall and these are incorporated in the ice as it forms. With the onset of winter, the diatoms are embedded in the ice. While all the species can survive, most individuals succumb, but the few survivors provide the new crop of cells for the next spring growing seasons.

In the Arctic, cells are apparently frozen into the ice when it forms in the fall. Pennate diatoms from the benthos may be brought up into the water column during the fall storms and then trapped when the water freezes. There is, however, no significant species correlation between the benthic and ice diatom floras, and microflagellates are often the most abundant cells in the the ice, although fall blooms have been reported from the Eskimo Lakes, Northwest Territories,[40] and Narwhal Island, Alaska.[35] The origin of these fall blooms is also not known. They may be remnant populations having remained in the ice since the preceding spring. At Narwhal Island, the same species complement was present as in the spring ice algal bloom. In the Arctic it is not known how long the fall blooms last, what happens to the cells, or if they are a regular event.

Relatively few cells are present in the ice in winter and they are distributed throughout the ice thickness. Perhaps more cells are present, but are relatively well dispersed so that infrequent winter sampling gives a false impression of their abundance.

In the central Arctic Ocean, Mel'nikov[41] reported that as ice freezes, it concentrates various biogenous and organic forms present in the water. As the ice grows, these elements migrate from the brine and become redistributed in the ice mass. The concentration of organic matter in the ice is thus several times greater than in the water from which the ice formed.

In the Bering Sea, a patchy, subsurface chlorophyll maximum has been reported[30] and it has been suggested that ice diatoms sink during the spring phytoplankton bloom to form this subsurface layer.[42] Whether the cells in this layer remain viable and are brought back into the photic zone in the fall is not known. Perhaps the cells come into contact with frazil crystals in the water column where they are concentrated and frozen into the ice again.

Another possibility is the formation of a thick, soft ice layer on the underside of the ice. This phenomenon is widespread in the Beaufort, Chukchi, and Bering seas and varies from season to season, but it is most likely associated with fall freezing events.[43] It is not known how this layer forms, but it may be similar to the frazil layer reported from the Antarctic. This soft ice layer may act as a reservoir in concentrating organic material, bacteria, and other organisms into the ice canopy, thus acting as a mechanism for the transport and supply of nutrients and biota. Just how the organisms become trapped in the ice is not explained.

Bunt[44] observed benthic debris trapped in ice and also reported ice crystals rising from the bottom of McMurdo Sound. Perhaps many of the cells reach the ice layer by becoming associated with ice crystals when they form in the water column. Watanabe[45] also reported cells from the water column settling on ice crystals and growing to form a colored layer. He cited as evidence two species of centric diatoms, *Porosira pseudodenticulata* (Hustedt) Jousé and *Coscinodiscus furcatus* Karsten, that are found in the ice and in the water column.

The bottom ice community in the Antarctic was thought to be formed when phytoplankton cells were trapped between unconsolidated ice crystals that form in the water column and float up under the pack ice.[18] Cells are accumulated in the ice by a combination of scavenging and nucleation.[46] Scavenging occurs when ice crystals moving up in the water column collide with and collect algal cells from the water, while nucleation occurs when a frazil ice crystal nucleates on an algal cell in water at its freezing point.

Garrison et al.[47] proposed a physical mechanism for establishing algal populations in frazil ice whereby frazil ice crystals rising to the surface to form new ice harvest cells from the water column. This mechanism may explain the origin of cells in ice in some areas, particularly in the Weddell Sea, where frazil ice is continuously formed and where the same species are found in the ice and in the water column, but in shallow areas, such as the near shore Chukchi and Beaufort seas where few cells are found in the water column during ice formation, harvesting may not be the only way cells are concentrated in the ice.

Anchor ice also occurs in the Antarctic. This is ice that forms on lines or on the bottom in water less than 33 m deep.[21] It is composed of aggregations of ice platelets 10 to 15 cm in diameter and 0.2 to 0.3 cm thick that have frozen together. The platelets begin to form in July and disappear in mid-December when the water warms from -1.9 to -1.8k°C.[21] The platelets collect cells as they move upward in the water column.

Anchor ice is important because it provides a means by which benthic organisms can be transported to the under surface of the ice and because sediment transported by it and subsequently incorporated into the ice cover increases turbidity in the ice and decreases light penetration that could affect algal growth.[48] Little is known about anchor ice, including factors that govern its formation such as water depth (but see Dayton et al.[21]), circulation patterns, salinity, or bottom substrates.[48]

Surface ice communities are formed when seawater containing algal cells floods the snow-ice interface[2] and when cells are transported upward through brine channels in the ice to melt pools located on the ice surface.[23]

Freshwater algae also occur on sea ice, often on the surface and in melt ponds of freshwater that form on the ice surface. Cleve[49] assumed that freshwater forms had been transported from land by winds. Usachev[50] found freshwater diatoms, green, and bluegreen algae in freshwater ice formed in the Yenisey River, but he did not say how the cells actually got into the ice. Bursa[16] found freshwater green algae and flagellates in meltwater ponds on the ice and said these algae were first found on the ice surface as conspicuous brown-green patches, but he also did not say how the cells go to on the ice. These freshwater forms were said to mix with the seawater and plankton when the ice broke up,[16] but there was no mention of their survival in seawater.

There are also organisms that, while apparently not important members of the ice community, are nearly always found associated with the ice. They may be found in the water column immediately under the ice and are found only when ice is present or they may be found in the ice. The origin of these organisms is not known, nor is their whereabouts when ice is not present.[51,52]

VI. ANNUAL CYCLE

The annual cycle of the ice algal community is relatively well-known only from Barrow, Alaska.[19] Studies from other areas have been mostly in spring or have missed some part of the productive season (Table 1). From the Barrow study, it is known that cells are present in the ice from the time it forms. Presumably cells in the water column are incorporated into the ice during freezing (Figure 2). Microflagellates are usually the most abundant organisms in the new ice, but pennate diatoms are also present. Growth of the diatoms and flagellates probably continues into November when light becomes limiting. A fall bloom in the ice has not been found at Barrow, but has been reported in November from the Eskimo Lakes[40] and from near Narwhal Island.[35] The Eskimo Lakes population, consisting primarily of pennate diatoms, contained 3.9×10^3 cells m^{-2}, while the Narwhal Island bloom, also comprised of pennate diatoms, contained concentrations of 1×10^8 cells m^{-2}, comparable to levels found during the earlier spring ice algal bloom in the same area.[20] A fall bloom is considered to be a regular occurrence at Syowa Station[28] and has also been found at Casey Station[23] in the Antarctic.

Table 1
LOCATIONS AND SEASONS FOR PRODUCTIVITY AND CHLOROPHYLL STUDIES

Location	Season	Ref.
Arctic		
Chukchi Sea	July	5
Barrow, Alaska	All year	19
Stefansson Sound, Alaska	May	20
	Spring	72
Narwhal Island, Alaska	April—June	20
Eskimo Lakes, N.W.T.	Fall, spring	40
Eclipse Sound	Spring	40
Frobisher Bay	Fall, spring	40, 106
Robeson Channel	May	29
Barrow Strait	May	29
Hudson Bay	March	29
Manitounuk Sound	February—May	76, 84
Gulf of St. Lawrence	February—March	29
Pond Inlet	May—June	83
Davis Strait	April—May	25
Godhavn, West Greenland	March—April	107
Jones Sound, Devon Island	April—June	108
Bering Sea	February—April	32, 33
	March—May	30, 31
Lake Saroma, Japan	February	109
Antarctic		
Antarctic Ocean	January	2
McMurdo Sound	December—January	8, 67
	November—January	55
	June—December	22
	November—December	110
Lützow—Holm Bay	December—February	111
Syowa Station	March—December	28, 112
Mirny	April—June	113
Weddell Sea	February—March	18, 114
Antarctic Peninsula	February	65
Signy Island	May—December	17
Casey Station	November—January	23

Cells remain scattered throughout the ice in winter,[40,53] with chlorophyll *a* concentrations being low and variable throughout the ice from January to mid-March.[19] By mid-March, the algae become concentrated in the bottom few centimeters of the ice and by early April, a brown layer is visible. The concentration mechanism is not known, but is probably the result of a combination of brine drainage and active migration of cells through brine channels in the ice.

Growth begins in spring in response to increasing light levels. At Barrow, the threshold light levels initiating growth were 2.3 and 9.3 $\mu E\ m^{-2}\ sec^{-1}$ in successive years.[19] Growth continues throughout the spring although adverse weather conditions, such as storms or periods of fog that decrease the light, may interrupt or slow growth during this time. Highest production occurs in mid to late May when the ice is thickest, the snow cover is gone, and light levels beneath the ice are 10 to 65 $\mu E\ m^{-2}\ sec^{-1}$. At the same time, the bottom of the ice containing the ice algae becomes softer and begins to deteriorate. The ice algae probably contribute to the disintegration by selective absorption of solar energy.[5,54] The decaying ice may remain associated with the harder ice above for a short time, but weak water movements gradually wash the algal layer away and by early June, the brown layer is no longer visible.

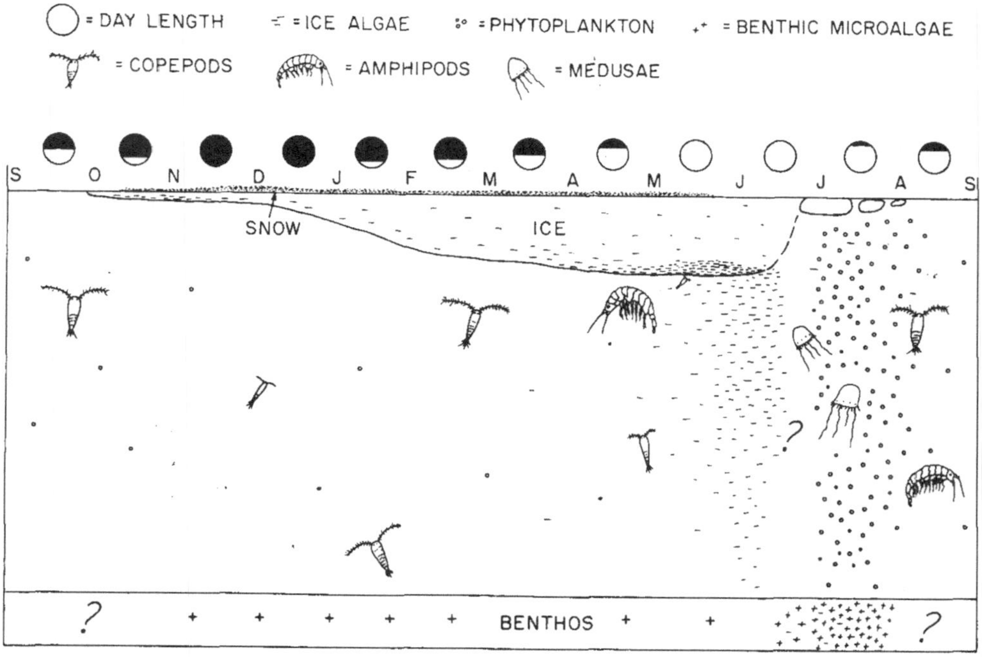

FIGURE 2. Schematic representation of the annual cycles of ice algae, phytoplankton, and benthic microalgae in the nearshore area of the Beaufort Sea. (From Horner, R. and Schrader, G. G., *Arctic*, 35, 485, 1982. With permission.)

Pennate diatoms were the dominant organisms in the ice community at Barrow, but a few centric diatoms, unidentified pigmented and nonpigmented microflagellates, dinoflagellates, cryptomonads, and euglenoids may also be present. At Barrow, the dominant species are generally *Nitzschia frigida* Grunow and *Navicula marina* Ralfs. Other abundant species include *Navicula transitans* Cleve, *N. trigonocephala* Cleve, and *N. sibirica* (Grunow) Cleve, *Nitzschia grunowii* Hasle, *N. delicatissima* Cleve, *Cylindrotheca closterium* (Ehrenberg) Reimann and Lewin, *Entomoneis* spp., and *Gomphonema exiguum* Kützing.[19]

Changes in community structure occurred during the ice algal bloom at Narwhal Island.[20] *Nitzschia cylindrus* (Grunow) Hasle was dominant during April and early May, but by mid-May, this species became less healthy and empty frustules were common in the ice and water column. As *N. cylindrus* declined, diversity increased and other species, including *Gomphonema exiguum* var. *arctica* Grunow, *Navicula directa* (W. Smith) Ralfs, *N. transitans, Nitzschia frigida,* and *Pinnularia quadratarea* (A. Schmidt) Cleve, became more abundant; flagellates also increased after mid-May.

At McMurdo Sound, Bunt and Lee[22] found increased numbers of algae in the ice from mid-July to early December. They counted cells in both interstitial water and melted frazil ice crystals. Cell numbers increased even during the winter dark period, remained relatively constant from late August through October, and then increased sharply by early December. The winter increase may have been due to cells in the water column being carried up to the ice by ice crystals, while the summer increase was the result of growth and cell division.[22] Chlorophyll *a* concentrations followed the same pattern. Primary productivity was measured only on interstitial water and experiments were run in an incubator under several light conditions. Photosynthesis was saturated at the lowest light intensity, approximately 42 $\mu E\ m^{-2}\ sec^{-1}$, used in their experiments. There was no discussion of species present. Bunt[55] reported peak development of the ice algae in December or early January. The ice algae disappeared when the bottom layer of the ice was lost. This was brought about by slightly higher water temperature and perhaps also by water currents.

At Syowa Station two brown layers were found: the upper layer was formed during the austral autumn; and the bottom layer during the austral spring and summer.[28] A layer of ice algae formed on the bottom of ice about 30 cm thick in late March and was visible until October. Chlorophyll *a* in this layer was extremely high, being approximately 10^3 mg m^{-3} in April. The ice continued to grow downward during the winter and a second algal layer formed on the bottom of the ice in November when the ice was 150 cm thick. By mid-December, the chlorophyll *a* concentration in the new bottom ice layer was $> 10^3$ mg m^{-3}. There was also a seasonal change in the species composition of the two ice communities with *Nitzschia* spp., a dinoflagellate, and *Fragilariopsis* sp. being present in the bottom layer in April. These species, plus *Nitzschia frigida,* were still present in this layer in mid-August when the layer was in the middle part of the total ice thickness. A *Peridinium* sp. was present in the bottom of the ice. In October when no brown layer was visible, *Nitzschia* spp. and *Fragilariopsis* sp. were present throughout the ice and *Amphiprora* sp. was also present in the bottom of the ice. By mid-December when the spring-summer bottom ice community was present, *Navicula* sp., *Stephanopyxis* sp., and *Amphiprora* sp. were present in the bottom layer; *Nitzschia* spp., *Fragilariopsis* sp., and a different dinoflagellate were present throughout the rest of the ice except at the surface where *Biddulphia* sp. was present. The horizontal extent of the brown ice layer was patchy. Cells were present in the ice at other sampling sites near Syowa Station, but no brown layer was present, perhaps indicating uneven growth rates for the ice algae.[28]

How cells actually survive the long winter dark period is not known (see Chapter 5). Some organisms are known to produce resting spores or cysts under adverse conditions. This was suggested for choanoflagellates in the Antarctic where cyst production occurred at the end of the austral summer in areas that would soon be ice-covered[56] and for diatoms found in Arctic sea ice.[5] Morphologically distinct resting cells have been reported for only a few pennate diatoms and there are few, if any, endemic pennate diatom species in polar waters that produce resting spores,[57] although *Nitzschia grunowii,* an important ice species, has been pictured (as *Fragilaria oceanica* Cleve) with resting spores.[58] Heterotrophy has also been suggested as a possible mechanism for winter survival.[59,60] Horner and Alexander[61] found *in situ* heterotrophic activity to be negligible, while Bunt and Lee[62] reported that organic supplements did not affect dark survival. Palmisano and Sullivan,[63] however, found that three ice diatoms could take up 0.003 to 0.3% of their total carbon requirement by heterotrophy. This might be adequate for maintenance metabolism when photosynthesis does not occur and maintenance is considered to be more important than growth when grazing pressure is minimal.[63] It appears that slow respiration of stored food reserves at low temperatures during winter may be the most important factor for dark survival.

It is difficult to compare rates of production between areas in the Arctic or Antarctic and even more difficult between the two polar regions because few measurements have been made, methods have often been different, and ice conditions and communities are different. Often production has been calculated from chlorophyll *a* measurements (Table 2) which can give misleading values unless carbon:chlorophyll ratios have been carefully determined.[64] Production has also often been given on a volume rather than an areal basis which adds more problems.[30] Except for the surface snow community near the Antarctic Peninsula studied by Burkholder and Mandelli,[65] annual production for most ice algal communities is <5 g C m^{-2} or about 2% of the total annual primary production in most areas. The 5 g C m^{-2} $year^{-1}$ calculated for Barrow, Alaska, however, has been estimated to be as much as 25 to 30% of the total annual production in that area.[66]

Table 2
ANNUAL ICE ALGAL PRODUCTIVITY DETERMINED FROM ^{14}C UPTAKE EXPERIMENTS AND CHLOROPHYLL *a* CONCENTRATIONS

Location	Annual productivity ($g\ C\ m^{-2}$)	Ref.
Arctic		
Davis Strait	0.15	107
Barrow, Alaska	5.0	66
Stefansson Sound, Alaska	0.9	20
	1.4[a]	72
Simpson Lagoon, Alaska	0.18[a]	72
Offshore Beaufort Sea	1.7[a]	72
Narwhal Island, Alaska	0.7	20
Bering Sea	0.22—0.5	32
Antarctic		
Antarctic Peninsula, snow community	19.0	65
McMurdo Sound	4.1[a]	110
Interstitial water	0.5—1.1[a]	22
Casey Station, fast ice	0.15—4.37[b]	23
Pack ice	2.11[b]	23
Signy Island	1—2	17

[a] Calculated from chlorophyll *a* concentrations.
[b] Calculated from $\mu g\ C\ \ell^{-1}$, 3 cm depth of ice algal layer, 18-hr day length, 100-day growing season.

VII. ENVIRONMENTAL FACTORS

Environmental factors that influence the development of the ice algal community include light, nutrient concentrations, salinity, and temperature, with light usually considered to be the most important.

A. Light

Unfortunately, relatively few light measurements have been made at the same time that photosynthesis has been measured in the field (Table 3). It is apparent from these few light measurements that the ice algae are adapted to live and grow successfully at light levels near 0.1% of the total surface irradiance, although the 1% light level is usually used to indicate the bottom of the euphotic zone in the ocean. At Devon Island, when snow was cleared from the ice, the chlorophyll *a* level was about one fifth that of measurements made when snow was present.[54] About 90% of the total radiation was reflected from the snow surface, so that only about 11 $\mu E\ m^{-2}\ sec^{-1}$ penetrated 23 cm of snow. About twice that amount was needed to initiate photosynthesis and cells were therefore adapted to very low light intensities.

Extreme shade adaptation was also reported at McMurdo Sound where ice algae were living at 2 $\mu E\ m^{-2}\ sec^{-1}$ and photosynthesizing at 20% of the net photosynthetic capacity.[67] Ice algae were reported to be light saturated at 22 $\mu E\ m^{-2}\ sec^{-1}$ with photosynthesis clearly inhibited at 240 $\mu E\ m^{-2}\ sec^{-1}$.[68] The compensation point was about 1 $\mu E\ m^{-2}\ sec^{-1}$ and perhaps as low as 0.5 $\mu E\ m^{-2}\ sec^{-1}$.[68]

It is not known what changes must occur in the photosynthetic apparatus that enable ice algae to be photosynthetically efficient at the low light levels found in their natural habitat.

Table 3
AMOUNT OF LIGHT ($\mu E\ m^{-2}\ s^{-1}$) NECESSARY TO INITIATE PHOTOSYNTHESIS AND FOR MAXIMUM PHOTOSYNTHESIS

Compensation point	Maximum photosynthesis	Ref.
Arctic		
~21		55
1.3		73
2.3—9.3	16—65	19
1—2		75
7.7	10—25	20
	1.8	25
Antarctic		
~1	~21	68

It is known, however, that the effects of high light intensities on ice algae can be either physical, with cells selectively absorbing heat and thus causing disintegration of their ice habitat, or biological, with photosynthesis being inhibited.[54]

Meguro et al.[5] did not measure light, but speculated that cells living in the ice were living at the highest light levels available to them in the spring and furthermore, the ice algae could use this available light more effectively than cells in the water column beneath the ice. Once the ice algal layer formed, light intensity under the ice was further reduced and cells in the water column received even less light.[5] Diatoms in the Arctic were thought to utilize sunlight more efficiently if they lived in sea ice from April to July and in seawater in August and September.[5]

Light may be severely limited by snow on the ice surface or by sediments trapped in the ice. The light-attenuating property of snow is very high when compared with that of clear sea ice.[69] Only about 2% of surface light reached the algal layer through 1.8 m of sea ice when no snow was present, but 3 to 4 cm of snow reduced the light level to 1% and 20 cm to <0.1% of the surface level.[20] An inverse relation between light and snow depth was found when light was measured under the ice along transects with varying snow depths.[19,20]

In the Antarctic, Sullivan et al.[70] found light levels of 0.20 to 3.12 $\mu E\ m^{-2}\ sec^{-1}$ under 148 to 180 cm of ice and 4 cm of snow. They reported chlorophyll *a* values under these conditions increased from 3.3 to 100 mg m^{-2} from late October to mid-December. Irradiance for the ice algae was <1% of the surface value. In an adjacent area, that had been artificially covered with 70 cm of snow, they found biomass and metabolic rates to be significantly lower.

In Stefansson Sound, near Prudhoe Bay, Alaska, light attenuation by turbid ice is thought to be the primary factor controlling the development and distribution of the ice algal community.[71,72] Horner and Schrader[20] also reported finding one or more layers of sediment and no algal layer present in Stefansson Sound. Sediment layers in the ice have apparently not been a problem in other areas where ice algae have been studied.

B. Nutrient Concentrations and Salinity

Nutrient concentrations and salinity in the ice are difficult to measure and techniques are needed for the *in situ* determination of these parameters in the brine channels. Generally, ice cores are cut into sections, the sections are melted, and then nutrient and salinity concentrations are measured. This does not give a true picture of the environment in which the algae are living because melted ice crystals dilute the hypersaline interstitial water that is formed by a combination of salt rejection during ice formation and brine drainage in the spring as the ice responds to increasing air temperature. Alexander et al.,[19] trying to reduce dilution, collected bottom ice cores with their combination corer-incubation chamber,[73] allowed the core to drain for a short time, and determined nutrient concentrations and salinity on the so-called interstitial water. Salinity, ranging from 20.4 to 34.1‰, was lower than expected and was near that of the underlying seawater, which ranged from 27.7 to 32.9‰. Salinity determined on melted ice cores ranged from 4.4 to 8.0‰ in surface ice to 5.2 to 12.5‰ in bottom ice.

Nutrient concentrations in the interstitial water were variable with ammonia either the same or an order of magnitude higher in the ice than in the water column. Nitrate and nitrite were about twice as high in the ice as in the seawater, while phosphate and silicate had similar concentrations in both habitats.[19]

At Narwhal Island, nutrient concentrations and salinity were also determined on cores collected with a combination corer-incubation chamber,[74] but the ice was allowed to melt completely before measurements were made. These nutrient and salinity values for the ice should, therefore, be considered minimal.[20] Salinity in the water column was near 34 to 35‰ until June when the ice started to melt and the seawater salinity dropped to 24‰. Salinity of the bottom ice was generally 1 to 2‰ less than that of the underlying sea water, ranging from 30 to 34‰ until June, when the lowest bottom ice salinity was 16.4‰. Measured nutrient concentrations in the ice and water column were generally similar except for high nitrate and ammonia peaks in the ice late in the season, perhaps caused by bacterial recycling (nitrate) and heavy grazing pressure (ammonia). Ammonia was low to barely detectable in both habitats; nitrate was near 10 μg - at ℓ^{-1} and silicate ranged from 20 to 25 μg - at ℓ^{-1}. Phosphate was the most variable with unexplained high concentrations in the water column in early May.

Grainger[75] reported nitrate and phosphate accumulation in sea ice during the winter with very low levels in the upper part of the ice and high levels, >14 μg - at ℓ^{-1} nitrate and <4 μg - at ℓ^{-1} phosphate, at the bottom. Measurements were made using sections of melted ice cores that were kept frozen and dark until analyses began. Salt loss after 1 week of ice growth was estimated to be 70% of the amount in the original water with a continued loss of 3 to 4% of the original water content per month. At the end of one growing season, salinity of the melted cores was about 10% of that in the original seawater.[75]

Highest salinity occurred at the top and bottom of ice cores from Robeson Channel, Barrow Strait, and Hudson Bay, and at the bottom of cores from the Gulf of St. Lawrence.[29] High salinity at the ice surface was thought to be caused by salt sprayed over the top of the ice from open water. Nutrient concentrations also tended to be higher at the top and bottom of the ice cores. In Davis Strait,[25] salinity values of 7 to 5‰ were found in top, middle, and bottom ice sections. Nutrient concentrations showed no vertical gradients, but were higher and more variable in the bottom 6 cm.[25] Nutrient concentrations, however, could have been 10 to 15 times higher than reported because determinations were made on melted ice sections.[25]

In southeastern Hudson Bay, salinity of the underlying surface waters was found to be a major environmental factor controlling the distribution and abundance of species.[76] At this site, a freshwater river plume established a salinity gradient in surface water under the ice leading to a parallel gradient in ice dynamics toward the upper end of Manitounuk Sound. The surface water salinity gradient from the mouth of the Great Whale River to the upper section of Manitounuk Sound influences the rate of ice formation and its thickness, as well as its structure. It generates a change in the ice microenvironment which becomes markedly more favorable for the settlement and growth of ice algae toward the upper end of the Sound, while the decline of salinity toward the river mouth causes a gradual disappearance of the ice algae. Thus, there are two distinct gradients in the ice related to the surface water salinity gradient: the ice thickness and structure control the microalgal biomass, while the ice salinity is responsible for the composition of the bottom ice community.[76]

In the Antarctic, gradients between the frazil ice and water column with regard to salinity, nitrate, and phosphate were reported with low values in the ice, higher values in the interstitial water, and highest concentrations in the water column 5 m beneath the ice.[22] Salinity values were not reported from Signy Island, but salinity was that at which freezing would be expected to occur.[17] When prolonged melt reduced the salinity to < 5‰, the diatoms changed from their normal yellow-brown color to a pale bluish-green,[17] indicating moribund or dead cells.

In the Weddell Sea, salinity was lower in the upper layers than in the middle layers of ice indicating that only partial warming and brine drainage had occurred.[18] Also in these cores, chlorophyll *a* maxima generally occurred at the same depth in the ice as the salinity maxima.

Under laboratory conditions, McMurdo Sound ice algae photosynthesized over a fairly wide salinity range, up to 36‰, but maximum carbon uptake occurred over a relatively narrow range, 13.6 to 18.1‰.[68] Growth of diatoms isolated from ice collected in the Chukchi and Beaufort seas occurred over a salinity range of 5 to 60‰, averaging 0.4 to 0.8 divisions day^{-1} over the range of 10 to 50‰.[77] Greater mortality and longer lag periods occurred at the low salinities indicating greater cell stress at lower salinities than at higher values. This is supported by the presence of dead or unhealthy cells in low-salinity seawater under melting ice in spring.[17,20]

C. Temperature

Few measurements have been made of the effect of temperature on the growth of ice algae. Net carbon fixation by ice algae at 14 μE m^{-2} sec^{-1} was markedly temperature dependent in the range −1.5 to +5°C.[68] Maximum carbon uptake occurred between 10 to 15°C with a rapid decline above 15°C. *Fragilaria sublinearis* Van Heurck was found to be an obligate psychrophile,[64] but maximum growth occurred at a temperature higher than that of the sea ice, regardless of light intensity. Conditions in the ice at McMurdo Sound must be close to the limits of growth, with growth at 0°C and low light levels being extremely slow.[64]

In a study of motility, the prasinophyte *Pyramimonas gelidicola* McFadden, Moestrup, and Wetherbee from ice in hypersaline Antarctic lakes, but also found in sea ice at Prydz Bay, Antarctica, stopped swimming at −10°C, but its flagella continued to beat at −15°C.[78] The characteristic lobed, pyramidal shape of this organism is maintained by microtubules, which, in most organisms depolymerize near 4°C, but even at −14°C, *P. gelidicola* maintained its shape, indicating that microtubules remain intact inside the cell as well as in the flagella. It is not known how *P. gelidicola* is able to modify its internal environment to maintain the polymerized tubulin that comprises the skeletal microtubules.[78]

VIII. FATE OF THE ICE ALGAE

The literature contains a number of conflicting views concerning the fate of the ice algae. Perhaps the most widespread hypothesis is that the ice algae produce the spring phytoplankton bloom in the water column. At Barrow, Alaska, however, it has been shown for the nearshore area, that the ice algae and phytoplankton blooms are clearly separated by time and by the species present.[19,61,73,79] The ice algal bloom consists primarily of pennate diatoms that are usually not found in the water column during the phytoplankton bloom. Moreover, the phytoplankton bloom occurs after the ice algal bloom is over and consists primarily of centric diatoms.[80] This is presumably also the situation in the nearshore Beaufort Sea.[20] No data are available for the spring phytoplankton bloom, but phytoplankton production was low and few live cells were present in the water column in early June when the ice algal bloom ended.[20]

Offshore in the Chukchi Sea, the high chlorophyll *a* concentration in the water column diluted with ice meltwater was thought to be a result of ice diatoms being liberated into the water column.[4,5] Large numbers of ice algae were released into the water column when the ice melted in the Bering Sea and, at the same time, the spring phytoplankton bloom occurred in the stable surface waters.[42] The ice diatoms sink during the phytoplankton bloom leaving a surface community of spring species. During the summer the spring species sink to the bottom and disintegrate and are replaced in the surface layer by summer species, while the ice algae remain as a patchy subsurface layer.[42]

In the Bering Sea, high primary productivity of cells from brash ice, high photosynthetic rates of ice algae incubated at surface light intensities, and the presence of similar species in slush ice, ice cores, and sea water led Alexander and co-workers[30,31] to conclude that ice algae do play a significant role in the ice edge bloom. Perhaps the ice algal contribution here is more important early in spring when fewer planktonic species are present in the water column. The ice edge phytoplankton community may be quite different from a phytoplankton community found under the ice well away from the ice edge, but no information is available.

Alexander and Chapman[30] have also reported a layer of photosynthetically active cells close to the bottom at 30 to 40-m depth near the ice edge and over a wide area of the northeastern Bering Sea. This layer might be a remnant of the ice and ice edge blooms, but a species list is necessary to confirm this. High chlorophyll *a* concentrations have also been found near the bottom at the ice edge in the eastern Chukchi Sea,[81] but this layer consisted mostly of centric diatoms that are spring water column species.

In the Eskimo Lakes region of Canada, the ice algae were released from the melting ice more or less at the start of the phytoplankton bloom, but this did not imply that the ice algae initiated the phytoplankton bloom.[82] In Frobisher Bay, there is often a time lag between the release of the ice algae and the beginning of the phytoplankton bloom, again indicating that the ice algae do not directly initiate the phytoplankton bloom.[75]

At Pond Inlet, sheets of algae partially detached from the ice and persisted as strands and amorphous masses.[83] Large numbers of meio- and macrofauna were associated with these masses so much of the ice algal production must be consumed at or near the undersurface of the ice. In Hudson Bay, increased solar energy simultaneously initiated melting of the ice cover, release of the ice algae into the water column, and increased phytoplankton production in the water column.[84]

In the Antarctic, Hasle[37] thought the phytoplankton bloom in the pack ice region was due to the release of spores and vegetative cells grown in the ice. Walsh[85] also believed that ice algae were an important inoculum for summer blooms and said they could account for 99% of the standing crop in the water column in the Ross Sea. In the Weddell Sea, there is a constant dispersion of cells from an interior ice community as the ice floes disintegrate.[18] According to Palmisano and Sullivan,[63] the phytoplankton bloom in the Antarctic is dominated by the prymnesiophyte *Phaeocystis* and is distinct from the ice algal bloom.

At Arthur Harbor (Palmer Station), *Navicula glaciei* Van Heurck was numerically dominant in sea ice.[86] It was also abundant in the water column in October to November following its release from the ice, but it was nearly absent in the water column after the disappearance of the ice in late November and after the spring bloom. Diatoms from sea ice were considered to be an important inoculum for the phytoplankton bloom in this region because of their huge standing crops and high species diversity.[86]

There is no quantitative information and very little qualitative information on the survival of the ice algal cells in the water column, so even if large numbers of cells are released from the ice, we do not know how long they may remain alive or if they are able to grow in the water column. Survival is probably dependent on a number of factors, but salinity of the water column may be among the most important considering the number of unhealthy and dead ice algal cells often found in the water column.[17,19,20,77]

Grazing is another possible fate for the ice algae and again there are conflicting views. Meguro et al.,[5] Bunt and Lee,[22] and Krebs[86] thought little grazing occurred in the ice and the net production was released into the water column. The abundant meiofauna reported in the ice[61,83,87-90] and the large numbers of amphipods and fish found on the undersurface of the ice or in cracks and brine channels[6,20,53,83,88-95] all indicate that grazing does occur and ice algae are an important source of food. This is especially true in the early spring when little or no primary production occurs in the water column. The ice algae-based food web must therefore, be an important part of polar ecosystems.

In the Chukchi and Beaufort seas where the ice algae are released in a single pulse, much of the ice algal production may be rapidly utilized leaving few cells in the water column. Cells that are not grazed may sink to the bottom. In shallow, nearshore areas, however, few ice algae cells are found in the sediments,[20,79] although Sancetta[96,97] has reported an assemblage of diatoms associated with melting ice in the surface sediments of the Bering Sea and Carey[98] believes the benthic-oriented ice community sinks rapidly, by-passing the pelagic community, but being available to the benthos. The amount of material reaching the benthos may also depend on the structure of the grazing community, such that during cold years when the ice edge in the Bering Sea is over oceanic waters, a diverse assemblage of large copepods is present to ingest a large portion of the ice algae when they are released into the water column. During warm years, the numerically dominant grazers are small copepods that are not as efficient and more organic material from the ice settles to the bottom.[99] In the Weddell Sea where large ice floes are present and cells are gradually released into the water column, ice algae may be present in the water column for longer periods, which increases the potential for grazing by zooplankton.

The fate of the ice algae appears to be different in different geographic areas and may depend to some extent on the ice habitat and on the regional environment. Before the fate of the ice algae can be determined more specifically, additional information is needed on the quantitative distribution of algal species present in the ice and in the water column; the length of time ice algal cells remain in the water column; the number and kinds of animals present in the ice and at the ice-water interface; quantitative utilization of ice algae by meiofauna and epifauna in the ice community; and carbon fluxes between the ice and the benthos.

IX. USE OF ICE ALGAE AS ENVIRONMENTAL INDICATORS

Recently, a number of papers[96,97,100-105] have suggested the use of ice algae as possible environmental indicators. Zhuse et al.[100] found seven distinct diatom complexes, each corresponding to a particular climatic zone, in upper layer sediments from the Pacific Ocean: In their Antarctic complex, most species were considered to be endemic, neritic ice algae.

Distinct diatom assemblages that appear to be related to environmental regimes have also been reported from the Bering and Okhotsk seas.[96,97,101] The sea ice assemblage, composed primarily of *Thalassiosira nordenskioeldii* Cleve, *Nitzschia grunowii*, and *N. cylindrus*, is present on the northern Bering Sea shelf and in the southern Chukchi Sea where annual sea ice is present for 6 months of the year and where productivity is high very early in spring at the edge of melting ice. While these assemblages are not necessarily representative of the living plankton, they appear to be consistent with hydrographic and productivity regimes, which lend support to the hypothesis that diatoms in sediments can be useful in determining past hydrographic conditions. In particular, the ice assemblage may be used to estimate the duration and extent of sea ice during glacial intervals.[96]

Also in the Bering Sea, Baldauf[102] found the modern diatom flora from surface sediment in the Navarin Basin to be dominated by *Denticulopsis seminae* (Simonsen and Kanaya)

Simonsen in deeper waters of the slope-basin area and by *Nitzschia oceanica* (Cleve) Hasle in shallow shelf waters where it is associated with ice cover. Within sediment cores these species alternate in dominance in late Quaternary sediments suggesting responses to climatic fluctuations. The interval where *D. seminae* is replaced down core by *N. oceanica* as the dominant species represents the transition from glacial to interglacial conditions and the Holocene-Pleistocene boundary.

Significant numbers of archaeomonads, thought to be fossil marine chrysophyte cysts that cannot be assigned to modern forms, have recently been reported from sea ice in the Bering, Ross, and Weddell seas.[103,104] Although the cysts have also been found in small numbers in areas where ice is not present like the California Current, their modern distribution implies association with sea ice and it may be that marine chrysophytes only produce large numbers of cysts under extreme environmental conditions. If so, then the distribution of archaeomonads in modern sediments suggests they may be useful fossil indicators of sea ice environments.[103,104]

A number of dinoflagellates may also be members of the sea ice algal community and it is well-known that many dinoflagellates produce cysts under adverse conditions. There is now some evidence from Quaternary glacial sediments and recent equivalents suggesting that smooth-walled *Protoperidinium* cysts (*P. conicum* [Paulsen] Balech) may be abundant near permanent or semipermanent ice.[105]

X. CONCLUSIONS

The ecology of the ice algae is complex. Even providing adequate terminology to describe the ice communities is difficult. The organisms are able to live and grow in a variety of ice environments, usually at times when other plant life in the ocean is sparse or nonexistent. The origin of the cells in the ice is often not known because few cells are present in the water column when the ice forms, but it is likely that these cells are collected and concentrated by frazil ice crystals forming in the water column. Similarly, the fate of the ice algae is still not well-known. The presence of these communities early in spring before the phytoplankton start to grow suggests that they provide a valuable source of food for planktonic and benthic grazers, thereby lengthening the growing season and moderating the harsh seasonal variability found in polar regions. In some areas, the ice algae may also seed the water column and initiate, or help initiate, the spring phytoplankton bloom.

Ice algae are common in annual sea ice in both polar regions and even in some more temperate areas such as the Baltic Sea, but their distribution in sea ice more than 1-year-old is not known. The annual cycle of ice algae is well documented at only one site, but factors affecting the growth of the ice algae have been studied at a number of localities. Light is apparently the most important factor regulating the growth and distribution of ice algae, but temperature and salinity may also be significant, either alone or in combination with light. It is not known how cells are able to survive the winter dark period, although some progress is being made toward answering this question. Rates of production are difficult to determine both *in situ* and under laboratory conditions because of the difficulties in adequately sampling the ice communities and providing natural conditions for incubations. These difficulties are caused in part by the structure of the ice itself, but also by the limitations imposed on humans working in polar environments. Some of these problems are gradually being overcome using new sampling techniques and better logistic support. Different kinds of ice algal communities are being described as more geographical areas are investigated and this adds to the difficulties in comparing communities. The recent use of ice algae as indicators of past climatological conditions adds a new facet to the already fascinating study of ice algae.

REFERENCES

1. **Zubov, N. N.,** *L'dy Arktiki (Arctic Ice),* Izdatel'stvo Glavsevmorputi, Moscow, 1945, 1. (Translated from Russian by U.S. Navy Oceanographic Office and The American Meteorological Society).
2. **Meguro, H.,** Plankton ice in the Antarctic Ocean, *Nankyoku Shiryo,* 14, 1192, 1962.
3. **Matsuda, T.,** Biological investigation during Antarctic Expedition voyage and at East and West Ongul Islands in Antarctica, *Nankyoku Shiryo,* 11, 159, 1961.
4. **Meguro, H., Ito, K., and Fukushima, H.,** Diatoms and the ecological conditions of their growth in sea ice in the Arctic Ocean, *Science,* 152, 1089, 1966.
5. **Meguro, H., Ito, K., and Fukushima, H.,** Ice flora (bottom type): a mechanism of primary production in polar seas and the growth of diatoms in sea ice, *Arctic,* 20, 114, 1967.
6. **Mohr, J. L.,** Marine biological work, in *Scientific Studies at Fletcher's Ice Island T-3 (1952-1955),* Bushnell, V., Ed., *Geophys. Res. Pap.,* 63, 82, 1959.
7. **Mohr, J. L. and Tibbs, J.,** Ecology of ice substrates, in *Proc. Arctic Basin Symposium,* Tidewater Publ., Centreville, 1963, 245.
8. **Bunt, J. S. and Wood, E. J. F.,** Microalgae and Antarctic sea-ice, *Nature (London),* 199, 1254, 1963.
9. **Crosby, L. H. and Wood, E. J. F.,** Studies on Australian and New Zealand diatoms. II. Normally epontic and benthic genera, *Trans. R. Soc. N.Z.,* 86, 1, 1959.
10. **Wood, E. J. F.,** Studies in microbial ecology of the Australian region. V. Microbiology of some Australian estuaries, *Nova Hedwigia,* 8, 461, 1964.
11. **Bunt, J. S., Owens, O. van H., and Hoch, G.,** Exploratory studies on the physiology and ecology of a psychrophilic marine diatom, *J. Phycol.,* 2, 96, 1966.
12. **Usachev, P. I.,** The microflora of polar ice, *Tr. Inst. Okeanol., Akad. Nauk SSSR,* 3, 216, 1949 (In Russian).
13. **Round, F. E.,** Benthic marine diatoms, *Oceanogr. Mar. Biol. Ann. Rev.,* 9, 83, 1971.
14. **Round, F. E.,** *The Ecology of Algae,* Cambridge University Press, New York, 1981, 42.
15. **Kol, E.,** The snow and ice algae of Alaska, *Smithson. Misc. Collect.,* 101(16), 1, 1942.
16. **Bursa, A. S.,** Phytoplankton in coastal waters of the Arctic Ocean at Point Barrow, Alaska, *Arctic,* 16, 239, 1963.
17. **Whitaker, T. M.,** Sea ice habitats of Signy Island (South Orkneys) and their primary productivity, in *Adaptations within Antarctic Ecosystems,* Llano, G. A., Ed., Gulf Publ., Houston, 1977, 75.
18. **Ackley, S. F., Buck, K. R., and Taguchi, S.,** Standing crop of algae in the sea ice of the Weddell Sea region, *Deep-Sea Res.,* 26A, 269, 1979.
19. **Alexander, V., Horner, R., and Clasby, R. C.,** Metabolism of Arctic sea ice organisms, Rep. R74-4, Institute of Marine Science, University of Alaska, Fairbanks, 1974.
20. **Horner, R. and Schrader, G. C.,** Relative contributions of ice algae, phytoplankton, and benthic microalgae to primary production in nearshore regions of the Beaufort Sea, *Arctic,* 35, 485, 1982.
21. **Dayton, P. K., Robilliard, G. A., and DeVries, A. L.,** Anchor ice formation in McMurdo Sound, Antarctica, and its biological effects, *Science,* 163, 273, 1969.
22. **Bunt, J. S. and Lee, C. C.,** Seasonal primary production in Antarctic sea ice at McMurdo Sound in 1967, *J. Mar. Res.,* 28, 304, 1970.
23. **McConville, M. J. and Wetherbee, R.,** The bottom-ice microalgal community from annual ice in the inshore waters of east Antarctica, *J. Phycol.,* 19, 431, 1983.
24. **Lewis, E. L. and Weeks, W. F.,** Sea ice: some polar contrasts, in *Symposium on Antarctic Ice and Water Masses,* Deacon, Sir G., Ed., W. Heffer & Sons Ltd., Cambridge, 1971, 23.
25. **Booth, J. A.,** The epontic algal community of the ice edge zone and its significance to the Davis Strait ecosystem, *Arctic,* 37, 234, 1984.
26. **Dickie, G.,** Notes on the algae, in *Journal of a Voyage in Baffin's Bay and Barrow Straits, in the Years 1850-1951, Performed by H. M. Ships "Lady Franklin" and "Sophia," Under the Command of Mr. William Penny, in Search of the Missing Crews of H. M. Ships "Erebus" and "Terror,"* Vol. 2, Sutherland, P. C., Longman, Brown, Green, and Longmans, London, 1852, cxci.
27. **Burns, J. J.,** personal communication, 1970.
28. **Hoshiai, T.,** Seasonal change of ice communities in the sea ice near Syowa Station Antarctica, in *Polar Oceans,* Dunbar, M. J., Ed., The Arctic Institute of North America, Calgary, 1977, 307.
29. **Dunbar, M. J. and Acreman, J. C.,** Standing crops and species composition of diatoms in sea ice from Robeson Channel to the Gulf of St. Lawrence, *Ophelia,* 19, 61, 1980.
30. **Alexander, V. and Chapman, T.,** The role of epontic algal communities in Bering Sea ice, in *The Eastern Bering Sea Shelf: Oceanography and Resources,* Vol. 2, Hood, D. W. and Calder, J. A., Eds., University of Washington Press, Seattle, 1981, 773.
31. **Schandelmeier, L. and Alexander, V.,** An analysis of the influence of ice on spring phytoplankton population structure in the southeast Bering Sea, *Limnol. Oceanogr.,* 26, 935, 1981.

32. **McRoy, C. P. and Goering, J. J.,** The influence of ice on the primary productivity of the Bering Sea, in *Oceanography of the Bering Sea with Emphasis on Renewable Resources,* Hood, D. W. and Kelley, E. J., Eds., University of Alaska, Fairbanks, 1974, 403.
33. **McRoy, C. P., Goering, J. J., and Shiels, W. E.,** Studies of primary production in the eastern Bering Sea, in *Biological Oceanography of the Northern North Pacific Ocean,* Takenouti, A. Y., Chief Ed., Idemitsou Shoten, Tokyo, 1972, 199.
34. **Gran, H. H.,** Diatomaceae from the ice-floes and plankton of the Arctic Ocean, *Sci. Res. Norw. N. Polar Exped.,* 4(11), 3, 1904.
35. **Schell, D. M.,** personal communication, 1980.
36. **Horner, R.,** unpublished data, 1972.
37. **Hasle, G. R.,** An analysis of the phytoplankton of the Pacific Southern Ocean, *Hvalradets Skr.,* 52, 1, 1969.
38. **Østrup, E.,** Marine diatomeer fra Østgrønland, *Medd. Groenl.,* 18, 395, 1895.
39. **Ross, R.,** Algae: planktonic, in *Cryptogamic Flora of the Arctic, Bot. Rev.,* 20, 400, 1954.
40. **Hsiao, S. I. C.,** Quantitative composition, distribution, community structure and standing stock of sea ice microalgae in the Canadian Arctic, *Arctic,* 33, 768, 1980.
41. **Mel'nikov, I. A.,** Hydrobiological investigations in the central Arctic Ocean, *Oceanology,* 16, 314, 1976.
42. **Saito, K. and Taniguchi, A.,** Phytoplankton communities in the Bering Sea and adjacent seas. II. Spring and summer communities in seasonally ice-covered areas, *Astarte,* 11, 27, 1978.
43. **Larsen, L.,** Sediment-laden sea ice: concepts, problems, and approaches, *OLSEAP NOAA, Actic Proj. Bull.,* 29, 59, 1980.
44. **Bunt, J. S.,** Microalgae of the Antarctic pack ice zone, in *Symposium on Antarctic Oceanography,* Currie, R. I., Ed., W. Heffer & Sons Ltd., Cambridge, 1968, 198.
45. **Watanabe, K.,** Centric diatom communities found in the Antarctic sea ice, *Nankyoku Shiryo,* 74, 119, 1982.
46. **Ackley, S. F.,** Ice scavenging and nucleation: two mechanisms for incorporation of algae into newly-formed sea ice, *Eos,* 63, 54, 1982.
47. **Garrison, D. L., Ackley, S. F., and Buck, K. R.,** A physical mechanism for establishing algal populations in frazil ice, *Nature (London),* 306, 363, 1983.
48. **Schell, D. M.,** Biological interactions, *Cold Regions Sci. Tech.,* 2, 336, 1981.
49. **Cleve, P. T.,** Microscopical examination of dust from drift-ice north of Jan Mayen, *Oefv. K. Vetenskapsakad, Foerh.,* 1900(4), 393, 1900.
50. **Usachev, P. I.,** Biological analysis of ice-floes, *C. R. (Dokl.) Acad. Sci. URSS,* 19, 645, 1938.
51. **Horner, R.,** *Discoasteromonas calciferus* Bursa, an unusual organism occurring in Arctic sea ice, *J. Phycol.,* Suppl. 8, 7, 1972.
52. **Horner, R.,** unpublished data, 1973.
53. **Horner, R.,** Sea ice organisms, *Oceanogr. Mar. Biol. Ann. Rev.,* 14, 167, 1976.
54. **Apollonio, S.,** The chlorophyll content of Arctic sea-ice, *Arctic,* 14, 197, 1961.
55. **Bunt, J. S.,** Primary productivity under sea ice in Antarctic waters. I. Concentrations and photosynthetic activities of microalgae in the waters of McMurdo Sound, Antarctica, *Antarct. Res. Serv.,* 1, 13, 1964.
56. **Silver, M. W., Mitchell, J. G., and Ringo, D. L.,** Siliceous nanoplankton. II. Newly discovered cysts and abundant choanoflagellates from the Weddell Sea, Antarctica, *Mar. Biol.,* 58, 211, 1980.
57. **Hargraves, P. E. and French, F. W.,** Diatom resting spores: significance and strategies, in *Survival Strategies of the Algae,* Fryxell, G. A., Ed., Cambridge University Press, New York, 1983, 49.
58. **Hustedt, F.,** Die Kieselalgen, in L. Rabenhorst's *Kryptogamen-Flora von Deutschland, Österreich und der Schweiz,* VII(2), 148, 1959.
59. **Rodhe, W.,** Can plankton production proceed during winter darkness in subarctic lakes?, *Verh. Int. Verein. Limnol.,* 12, 117, 1955.
60. **Allen, M. B.,** High latitude phytoplankton, *Ann. Rev. Ecol. Syst.,* 2, 261, 1971.
61. **Horner, R. and Alexander, V.,** Algal populations in Arctic sea ice: an investigation of heterotrophy, *Limnol. Oceanogr.,* 17, 454, 1972.
62. **Bunt, J. S. and Lee, C. C.,** Data on the composition and dark survival of four sea-ice microalgae, *Limnol. Oceanogr.,* 17, 458, 1972.
63. **Palmisano, A. C. and Sullivan, C. W.,** Physiology of sea ice diatoms. I. Response of three polar diatoms to a simulated summer-winter transition, *J. Phycol.,* 18, 489, 1982.
64. **Bunt, J. S.,** Some characteristics of microalgae isolated from Antarctic sea ice, *Antarct. Res. Ser.,* 11, 1, 1968.
65. **Burkholder, P. R. and Mandelli, E. F.,** Productivity of microalgae in Antarctic sea ice, *Science,* 149, 872, 1965.
66. **Alexander, V.,** Primary productivity regimes of the nearshore Beaufort Sea, with reference to potential roles of ice biota, in *The Coast and Shelf of the Beaufort Sea,* Reed, J. C. and Sater, J. E. Eds., The Arctic Institute of North America, Arlington, 1974, 609.

67. **Bunt, J. S.,** Diatoms of Antarctic sea-ice as agents of primary production, *Nature (London)*, 199, 1255, 1963.
68. **Bunt, J. S.,** Primary productivity under sea ice in Antarctic waters. II. Influence of light and other factors on photosynthetic activities of Antarctic marine microalgae, *Antarct. Res. Ser.*, 1, 27, 1964.
69. **Thomas, C. W.,** On the transfer of visible radiation through sea ice and snow, *J. Glaciol.*, 4, 481, 1963.
70. **Sullivan, C. W., Palmisano, A. C., Kottmeier, S., and Moe, R.,** Development of the sea ice microbial community in McMurdo Sound, *Antarct. J. US*, 17, 155, 1982.
71. **Schell, D. M.,** Chemical overview of biological-physical process interactions, *OLSEAP NOAA, Actic Proj. Bull.*, 29, 25, 1980.
72. **Schell, D. M.,** Foodweb and nutrient dynamics studies in nearshore Alaskan Beaufort Sea waters, *NOAA*, 2, 467, 1980.
73. **Clasby, R. C., Horner, R., and Alexander, V.,** An in situ method for measuring primary productivity of Arctic sea ice algae, *J. Fish. Res. Board Can.*, 30, 835, 1973.
74. **Schrader, G. C., Horner, R., and Smith, G. F.,** An improved chamber for in situ measurement of primary productivity by sea ice algae. *Can. J. Fish. Aquat. Sci.*, 39, 522, 1982.
75. **Grainger, E. H.,** The annual nutrient cycle in sea-ice, in *Polar Oceans*, Dunbar, M. J., Ed., The Arctic Institute of North America, Calgary, 1977, 285.
76. **Poulin, M., Cardinal, A., and Legendre, L.,** Réponse d'une communauté de diatomées de glace á un gradient de salinité (baie d'Hudson), *Mar. Biol.*, 76, 191, 1983.
77. **Grant, W. S. and Horner, R.,** Growth responses to salinity variation in four Arctic ice diatoms, *J. Phycol.*, 12, 180, 1976.
78. **Burch, M. D. and Marchant, H. J.,** Motility and microtubule stability of Antarctic algae at sub-zero temperatures, *Protoplasma*, 115, 240, 1983.
79. **Matheke, G. E. M. and Horner, R.,** Primary productivity of the benthic microalgae in the Chukchi Sea near Barrow, Alaska, *J. Fish. Res. Board Can.*, 31, 1779, 1974.
80. **Horner, R.,** Do ice algae produce the spring phytoplankton bloom in seasonally ice-covered waters? A review of recent literature, in *Proc. Seventh Symposium on Living and Fossil Diatoms*, Mann, D., Ed., J. Cramer, Braunschweig, 1984, 401.
81. **Horner, R.,** unpublished data, 1974.
82. **Grainger, E. H. and Evans, M. S.,** Seasonal variations in chlorophyll and nutrients in a Canadian Arctic estuary, *Estuaries*, 5, 294, 1982.
83. **Cross, W. E.,** Under-ice biota at the Pond Inlet ice edge and in adjacent fast ice areas during spring, *Arctic*, 35, 13, 1982.
84. **Legendre, L., Ingram, R. C., and Poulin, M.,** Physical control of phytoplankton production under sea ice (Manitounuk Sound, Hudson Bay), *Can. J. Fish. Aquat. Sci.*, 38, 1385, 1981.
85. **Walsh, J. J.,** Vertical distribution of antarctic phytoplankton. II. A comparison of phytoplankton standing crops in the Southern Ocean with that of the Strait of Florida, *Limnol. Oceanogr.*, 14, 86, 1969.
86. **Krebs, W. N.,** Ecology of neritic marine diatoms, Arthur Harbor, Antarctica, *Micropaleontology*, 29, 267, 1983.
87. **Nansen, F.,** Protozoa from the ice-floes of the North Polar Sea, *Sci. Res. Norw. N. Polar Exped.*, 5(16), 1, 1906.
88. **Andriashev, A. P.,** The problem of the life community associated with the Antarctic fast ice, in *Symposium on Antarctic Oceanography*, Currie, R. I., Ed., W. Heffer & Sons Ltd., Cambridge, 1968, 147.
89. **Carey, A. G., Jr. and Montagna, P. A.,** Arctic sea ice faunal assemblages: first approach to description and source of underice meiofauna, *Mar. Ecol. Prog. Ser.*, 8, 1, 1982.
90. **Kern, J. C. and Carey, A. G., Jr.,** The faunal assemblage inhabiting seasonal sea ice in the nearshore Arctic Ocean with emphasis on copepods, *Mar. Ecol. Prog. Ser.*, 10, 159, 1983.
91. **Holmquist, C.,** An observation on young specimens of *Ammodytes dubius*, *Medd. Groenl.*, 159(2), 9, 1958.
92. **English, T. S.,** Some biological oceanographic observations in the central North Polar Sea, Drift Station Alpha, 1957-1958, Sci. Rep. No. 15, The Arctic Institute of North America, Washington, D. C., 1961.
93. **McAllister, D. E.,** Ecology of the marine fishes of arctic Canada, in *Proc. Circumpolar Conference on Northern Ecology*, Vol. 2, National Research Council of Canada, Ottawa, 1975, 49.
94. **Green, J. M. and Steele, D. H.,** Observations on marine life beneath sea ice, Resolute Bay, N.W.T., in *Proc. Circumpolar Conference on Northern Ecology*, Vol. 2, National Research Council of Canada, Ottawa, 1975, 77.
95. **Gruzov, Ye. N.,** Seasonal alterations in coastal communities in the Davis Sea, in *Adaptations within Antarctic Ecosystems*, Llano, G. A., Ed., Gulf Publ., Houston, 1977, 263.
96. **Sancetta, C.,** Diatoms as hydrographic tracers: example from Bering Sea sediments, *Science*, 211, 279, 1981.
97. **Sancetta, C.,** Oceanographic and ecologic significance of diatoms in surface sediments of the Bering and Okhotsk seas, *Deep-Sea Res.*, 28A, 789, 1981.

98. **Carey, A. G., Jr.,** personal communication, 1980.
99. **Neibauer, H. J., Alexander, V., and Cooney, R. T.,** Primary production at the eastern Bering Sea ice edge: the physical and biological regimes, in *The Eastern Bering Sea Shelf: Oceanography and Resources,* Vol. 2, Hood, D. W. and Calder, J. A., Eds., University of Washington Press, Seattle, 1981, 763.
100. **Zhuse, A. P., Kozlova, O. G., and Mukhina, V. V.,** Vidovoi sostav i zonal'noe rasprelelenie diatomei v poverkhnostnom sloe osadkov Tikhogo Okeana, *Dokl. Akad. Nauk SSSR,* 172, 1183, 1967, (In Russian, cited from Hasle.[37])
101. **Sancetta, C.,** Distribution of diatom species in surface sediments of the Bering and Okhotsk seas, *Micropaleontology,* 28, 221, 1982.
102. **Baldauf, J. G.,** Identification of the Holocene-Pleistocene boundary in the Bering Sea by diatoms, *Boreas,* 11, 113, 1982.
103. **Mitchell, J. G.,** Sea ice environments are indicated by archaeomonads, *Eos,* 63, 47, 1982.
104. **Mitchell, J. G. and Silver, M. W.,** Modern archaeomonads indicate sea-ice environments, *Nature (London),* 96, 437, 1982.
105. **Dale, B.,** Dinoflagellate resting cysts: "benthic plankton," in *Survival Strategies of the Algae,* Fryxell, G. A., Ed., Cambridge University Press, New York, 1983, 69.
106. **Grainger, E. H. and Hsiao, S. I. C.,** A study of the ice biota of Frobisher Bay, Baffin Island 1979-1981, *Can. Manuscr. Rep. Fish. Aquat. Sci.,* No. 1647, 1982.
107. **Andersen, O. G. N.,** Primary production associated with sea ice at Godhavn, Disko, West Greenland, *Ophelia,* 16, 205, 1977.
108. **Apollonio, S.,** Chlorophyll in Arctic sea ice, *Arctic,* 18, 118, 1965.
109. **Hoshiai, T. and Fukuchi, M.,** Sea ice colored by ice algae in a lagoon, Lake Saroma, Hokkaido, Japan, *Nankyoku Shiryo,* 71, 113, 1981.
110. **Palmisano, A. C. and Sullivan, C. W.,** Sea ice microbial communities (SIMCO). I. Distribution, abundance, and primary production of ice microalgae in McMurdo Sound, Antarctica in 1980, *Polar Biol.,* 2, 171, 1983.
111. **Hoshiai, T. and Kato, M.** Ecological notes on the diatom community of the sea ice in Antarctica, *Bull. Mar. Biol. St. Asamushi, Tohoku Univ.,* 10, 221, 1961.
112. **Hoshiai, T.,** Proliferation of ice algae in the Syowa Station area, Antarctica, *Mem. Nat. Inst. Polar Res. Ser. E,* 34, 1, 1981.
113. **Buinitsky, V. Kh.,** Organic life in sea ice, in *Polar Oceans,* Dunbar, M. J., Ed., The Arctic Institute of North America, Calgary, 1977, 301.
114. **Ackley, S. F., Taguchi, S., and Buck, K. R.,** Primary productivity in sea ice of the Weddell Region, Report 78-19, Cold Regions Research and Engineering Laboratory, Hanover, 1978.

Chapter 4

CHEMICAL COMPOSITION AND BIOCHEMISTRY OF SEA ICE MICROALGAE

M. J. McConville

TABLE OF CONTENTS

I. INTRODUCTION

Annual sea ice in the Arctic and Antarctic contains favorable microhabitats within the ice (brine channels, liquid pockets) or on the bottom surface of the ice which are colonized by microalgae. These ice communities constitute an important source of primary production in high latitude marine ecosystems, particularly during spring when phytoplankton growth is severely dampened. Information on the chemical composition and biochemistry of ice algae is useful for field ecology investigations and for providing an insight into the adaptive strategies of species living in this habitat.

Estimation of the physiological state of microalgae by measurement of the cellular ratios of several defined chemical components may provide an indirect measure of growth rate. Changes in the ratios between cellular carbon, nitrogen (carbohydrate, protein), adenylates, and pigments appear to reflect changes in growth rates and are commonly used to determine the physiological state of microalgae.[1] This approach generally gives qualitative estimates of growth, although quantitative estimates are possible given a detailed knowledge of the relationship between the chemical composition of the algae and various environmental factors.[2] While such an understanding has not yet been achieved with ice algae, measurement of the physiological state of ice communities is of interest for determining the viability of overwintering populations and in defining which environmental parameters are limiting for growth during the ice algal blooms. This approach is also promising in view of the difficulties encountered in directly measuring primary production from ^{14}C-assimilation rates. For example, perturbation of algal samples during the melting of ice cores or photoinhibition of shade-adapted cells may introduce serious errors into the results of laboratory-based experiments.[3,4] Alternatively, incubations *in situ* are only feasible with epontic or bottom ice communities and are still underestimates due to incomplete mixing and loss of the tracer through the ice.[5,6]

Certain chemical markers in microalgae may also be useful for demonstrating the relationship between different trophic levels.[7-9] The most commonly used chemical markers in microalgae are fatty acids which are highly conserved (low catabolism) in the predator. Many polyunsaturated fatty acids are only biosynthesized *de novo* by photosynthetic organisms, making them essential dietary factors in animals. This approach has been used to determine the dietary preferences of two euphausiids in the Southern Ocean.[9]

Adaptations of microalgae to the relatively extreme ice habitat are likely to be manifested by changes at the molecular level. In view of the long evolutionary time periods over which the ice biota has evolved,[10] these changes are of intrinsic interest with regard to the general adaptability of eukaryotes. While there are few studies on the chemical composition of ice algae, the data suggest that their gross chemical composition is similar to closely related species from other habitats (planktonic, benthic). This is in agreement with strategies of adaptation found in other microorganisms living in extreme environments, in which the biological activity and stability of various macromolecules are retained by relatively minor changes in their primary and secondary structure.

In this chapter, the sparse literature on the chemical composition of ice algae is reviewed and discussed in relation to the physiological ecology and adaptive strategies of these organisms. In some cases, where applicable, data on the gross chemical composition, (e.g., pigment composition) of particular algal groups found in sea ice have been derived from studies on related taxa from planktonic communities.

II. FACTORS INFLUENCING CHEMICAL COMPOSITION

A. Physical Parameters

Several authors have suggested that the relative growth rate ($\mu:\mu_{max}$) rather than the absolute growth rate is the physiologically relevant parameter influencing cellular chemical composition.[1,2,11] In the sea ice habitat, temperature appears to be the main factor controlling the maximum growth rate (μ_{max}) of microalgae. This is demonstrated by the consistent observation that the optimum temperature for growth of ice algae (approximately 4 to 10°C),[4] is appreciably higher than the *in situ* temperatures (approximately −2 to 0°C). Seasonal variation of temperature in the sea ice is small, although in this suboptimal temperature range, a small change in temperature may significantly affect μ_{max}.[4,12] In the ice diatom *Nitzschia sublineata* Hasle (*Fragilaria sublinearis* Van Heurck) the temperature optima for photosynthesis and respiration are not coincident and at low temperatures respiration, as well as growth rate decreases relative to the photosynthetic rate.[13] This appears to be an important physiological requirement for ice algae, as the optimum irradiance for growth decreases with temperature enabling net production at very low irradiances.[4] The effect of other physical parameters such as salinity may be superimposed in influencing μ_{max}.

Relative growth rates will vary with changes in light intensity and nutrient availability. As extreme seasonal variation occurs in the length of the photoperiod and intra ice irradiances, variation in the relative growth rate (and hence cellular chemical composition) of ice species would be expected to occur. Higher photosynthetic rates may not be reflected in higher growth rates due to limitations imposed on μ_{max} by temperature or salinity. Thus, an increase in photosynthetic capacity relative to growth rate may involve an increased flow of carbon into low molecular weight metabolites and storage products.[14] At light intensities approaching the compensation point (I_c; irradiance at which production and respiration are equal and opposite) photoassimilated carbon would be used solely to maintain basic metabolic requirements. This is manifested as a higher proportion of cellular protein relative to polysaccharide and lipid.[14-16] Decreases in the relative growth rate due to nutrient limitation are probably not as important as in phytoplankton communities due to the initially high levels of macronutrients in the ice.[17]

B. Species Composition

An understanding of the species composition of ice communities is essential in determining their energy content or nutritive value. Similarly, relating the chemical composition of communities to their physiological state requires a knowledge of the constituent species and how environmental factors regulate their cellular composition and growth rates. During the periods of maximum algal proliferation in the ice, pennate diatoms are characteristically dominant. However different algal groups may become abundant in different ice communities or temporally with a particular ice habitat. Changes in community structure are frequently observed during the development of bottom ice communities.[6,18] This often involves an increased species diversity, particularly flagellated species that may become dominant.[6,19] A further consideration in assessing the chemical composition and physiological state of natural communities is the relative cell size of different species.[20] Large cells will tend to contribute disproportionately to the overall chemical composition, while high concentrations of small cells that may occasionally dominate ice communities,[6] will be underrepresented.

III. PROTEIN

A. Cellular Composition

Cellular protein levels in microalgae appear to be conserved over a wide range of relative growth rates and usually constitute a major fraction of their organic dry weight. In nine planktonic diatoms, protein represented between 30 and 52% of the organic dry weight during exponential growth.[21] When the growth rate is reduced under suboptimal environmental conditions (such as low light intensities) enhanced relative incorporation of ^{14}C-label into the protein fraction is frequently observed.[14] In addition, protein synthesis may continue at appreciable rates during the dark period, usually at the expense of ^{14}C-incorporation into polysaccharide.[15,16,22] There appears to be a negative correlation between percent ^{14}C-incorporation into protein and irradiance.[22,23]

In contrast to patterns of ^{14}C-incorporation, the cellular protein content of diatoms is relatively invariant under different conditions of temperature,[24] light intensity,[22] and nutrient limitation.[21] This reflects the functional roles of proteins (structural, enzymic) in the cell which need to be maintained at certain levels for cellular metabolism. Variation of cellular protein was found to occur over a diel cycle in *Skeletonema costatum* (Greville) Cleve, although to a much lesser extent than cellular carbohydrate.[26] Diel variation was temperature dependent and at low temperatures changes in cellular protein were not significant. The tendency to conserve cellular protein levels has led several authors to suggest that measurement of protein-^{14}C (rate of protein synthesis) may provide a better indication of microalgal growth than total ^{14}C uptake.[14,16,27] If the cellular protein content is known, absolute growth rate can be estimated.

B. Strategies of Low Temperature Adaptation

Microalgae, like all other microorganisms and plants are ectotherms and are unable to maintain a cellular temperature that is different from that of the environment. Thus, microalgae living at temperatures between 0°C and −2°C in the ice might be expected to display lower metabolic rates simply as a consequence of the physical effects of low temperature on enzymes and metabolic reactions.

Microalgae in polar regions exhibit higher metabolic rates than would be expected based on the levels of thermal energy in the system, although complete compensation does not seem to have been achieved.[11,29] Microorganisms which grow at low temperatures have traditionally been termed psychrophiles, although further terminology is now being used depending on their temperature optima. Psychrophilic organisms may be defined as those having an optimum growth temperature <15°C, a maximum growth temperature of approximately 20°C, and a minimum growth temperature of <1°C. Microorganisms which grow at around 0°C, but which have maximum growth temperatures above 20°C, are considered to be psychrotrophic (facultative psychrophiles).[30] It follows that psychrophiles are specifically adapted to growth at low temperatures while psychrotrophes are merely cold tolerant. This difference may be reflected in different biochemical strategies to low temperature adaptation. While the growth characteristics of only a few sea ice diatoms are known, most species appear to be psychrophilic,[4] although some may be psychrotrophic.[31] Psychrophilic bacteria have also been found to dominate polar waters[32] and at low temperatures are able to outgrow psychrotrophic bacteria.[33] Thus, it would appear that low unvarying temperatures of polar waters apply a selective pressure on microorganisms which have developed growth characteristics closely fitting the environmental conditions.

Plants and microorganisms exposed to low temperatures appear to utilize a number of adaptive strategies at the molecular level. The biochemical basis of low temperature adaptation in ectotherms has been previously reviewed.[34-36] It is reasonable to suppose that the principles that have emerged from these studies will also apply to microalgae at low temperatures. Several strategies involving compensatory changes in enzyme activity have been found in ectotherms facing low ambient temperatures. These may involve changes in the amounts of existing enzyme forms, utilization of variant forms of a given enzyme, or modification of the intracellular environment affecting enzyme activities.[34] The strategies that have evolved appear to depend on the thermal stability of the environment as well as the evolutionary time periods involved.[34,35]

There are many examples of quantitative increases in the amount of cellular protein in response to growth at low temperatures and it has been suggested that levels of limiting enzymes are increased at low temperatures to compensate for their lower activity.[34] Increases in the enzymes involved in carbon dioxide fixation have been hypothesized to occur in several microalgae in response to lower growth temperatures.[37,38] This hypothesis was based primarily on observed increases in cellular protein levels at low temperatures.[37-39] However, concomitant changes in the activities of specific enzymes with changes in cellular protein are not always observed.[24,38] While some species may indeed stabilize metabolic reaction rates by regulating the concentration of rate limiting enzymes, it is probably not a general phenomenon, especially in organisms that have evolved over long periods at low temperatures. Such an approach has certain limitations, as protein synthesis is an energetically costly process and increased enzyme production at low temperatures would be highly demanding.[34] This would be an important consideration for microalgae growing at very low photon flux densities in the sea ice; Richardson et al.[40] have proposed that shade-adapted cells may reduce their basic metabolic requirements by reducing the level of photosynthetic enzymes maintained. Metabolic compensation through increased enzyme concentrations seems to be common in ectotherms that must adapt to different temperatures on a short term as well as seasonal time scale.[34,35]

It is possible that homologous enzymes from organisms living in different thermal environments, (e.g., equatorial vs. polar latitudes) may have different thermostabilities and activities. For species that have had a long evolutionary time span to adapt to low temperatures, catalytic efficiencies, but not ligand-binding affinities (K_m), are found to change in a temperature compensatory manner.[34-36] Thus, cold-adapted ectotherms appear to have enzymes that are more efficient (lower free energy of activation, $\Delta G\#$), less heat stable, and less temperature sensitive than homologous enzymes from warm-adapted ectotherms.[35] These characteristics are evident in the physiological responses of psychrophilic algae and may partially account for their ability to grow only over a relatively narrow range of temperatures. The presence of temperature-labile components in the ice diatom *Nitzschia sublineata* is demonstrated by the rapid destruction of protein-pigment complexes at temperatures that are normal for species of temperate origin.[13]

Enzyme activity may also be maintained by modification of the intracellular environment at low temperatures rather than by direct modification of the enzyme itself. Modification of the intracellular environment may involve changes in the intracellular pH, in levels of positive and negative allosteric modulators, or in the lipid environment adjacent to membrane bound proteins.[35] Burch and Marchant[41] suggest that modification of the intracellular environment may be required for the polymerization of tubulin and microtubule function at low temperatures. In many plant cells depolymerization of microtubules occurs just above 0°C.[42] However two ice flagellates, *Chlamydomonas* sp. and *Pyramimonas gelidicola* McFadden, Moestrup and Wetherbee, retained their peripheral microtubule cytoskeleton and flagellar function down to −14°C and −8°C respectively.[41] A similar low temperature resistance occurs in a species of *Dunaliella* found in Antarctic hypersaline lakes.[41] Under hyperosmotic conditions, glycerol is accumulated by *Dunaliella* and *Chlamydomonas* species, while mannitol appears to be accumulated in *Pyramimonas* sp. (Table 4). Polyols, such as glycerol and sucrose, stabilize microtubules and enhance tubulin polymerization in vivo,[43] and a similar function has been suggested to occur in these algae at subzero temperatures.[41] Whether polyols, such as mannitol, or other cellular compounds stabilize cellular microtubules as well has not been determined. The nucleotide guanosine triphosphate (GTP) also stabilizes tubulin protein from low temperature denaturation in vitro.[44]

IV. PIGMENTS

The pigment composition of microalgae is influenced by both the intensity and spectral composition of incident light. Within the sea ice habitat there may be considerable spatial heterogeneity with regard to light, although in general, the regions of highest algal prolif-

Table 1
MAJOR MICROALGAL GROUPS FOUND IN ICE COMMUNITIES AND THEIR DOMINANT LIGHT HARVESTING PIGMENTS[40]

	Chlorophylls						
Taxon	a	b	c_1	c_2	β carotene	Major xanthophyll	Biliproteins
Bacillariophyceae	+		+	+	+	Fucoxanthin	
Chrysophyceae	+		+	+	+	Fucoxanthin	
Prymnesiophyceae	+		+	+	+	Fucoxanthin	
Dinophyceae	+			+	+	Peridinin	
Cryptophyceae	+			+		Alloxanthin	+
Chlorophyceae	+	+			+	Lutein	
Prasinophyceae	+	+				(Siphonoxanthin)[a]	

[a] Found in some members of the Prasinophyceae.

eration are characterized by very low photon flux densities. The degree to which light is attenuated through ice depends on its physical properties (brine volume, bubble density, and crystal structure) as well as the thickness of the overlying snow cover.[45] The spectral extinction coefficient (degree of attenuation of photons by absorption and scattering) for sea ice shows that for wavelengths greater than 600 nm, attenuation is increased considerably compared to shortwave radiation (blue green region).[46] In general, the extinction coefficients for all wavelengths decrease as the ice begins to melt.[46] Ice melt often occurs concomitantly with the removal of the overlying snow layer resulting in much higher bottom ice irradiances during this period. In the following section, the distribution and properties of the major pigment complexes of ice microalgae will be discussed as well as environmental variables causing phenotypic variation.

A. Distribution and Properties

The dominant light harvesting pigments of the various microalgal groups found in the ice communities are presented in Table 1. During the periods of microalgal proliferation in the ice, species containing chlorophyll c_1 and c_2 in addition to chlorophyll *a* are characteristically dominant. In these algae, carotenoids, such as fucoxanthin, constitute a major proportion of the pigment composition. Towards the end of the spring bloom, as light intensities and photoperiod increase, species belonging to the Chlorophyceae and Prasinophyceae (green algal line) become more significant.[18,19] These species contain chlorophyll *a* and *b*, while carotenoids play a minor role in their light harvesting pigment composition.

The in vivo absorption spectra of the dominant ice microalgal group (Bacillariophyceae) is well suited to the available light conditions in the ice. Relatively high absorption occurs in the blue to green region (450 to 550 nm) due to chlorophyll *c* and fucoxanthin pigments.[47] Chlorophyll *c* contains two spectrally distinct components, c_1 and c_2, which occur in the Bacillariophyceae, Chrysophyceae, and Prymnesiophyceae (Haptophyceae).[48,49] Chlorophyll *c* absorbs predominantly in the blue region of the spectrum with very little absorption in the red,[48] in contrast to chlorophyll *a* and *b* which have major absorption peaks in the red. This is demonstrated by comparing the ratio of absorbance in the soret band (~ 450 nm) to that in the red (628 nm) band. For chlorophyll *c* this ratio is around 10, compared to about 1 and 2.8 for chlorophyll *a* and *b* respectively.[50] Chlorophyll *c* is thus, the most blue light adapted chlorophyll pigment.[40] In addition, the fluorescence spectra for c_1 and c_2 extend into the region of chlorophyll *a* absorption making it suitable for transferring trapped light energy for photosynthesis.[48] In the Dinophyceae and Cryptophyceae only c_2 is present.[49] The ratio of chlorophylls *a* to *c* was found to vary between 0.35 and 0.57 in 13 ice diatom species (2°C, 24 μE m^{-2} s^{-1}*).[4] This range is similar to that observed in other diatoms.[49]

The chlorophyll c_1-c_2 containing species also possess β carotene and fucoxanthin as their major xanthophylls. These light harvesting pigments have similar absorption spectra and are efficient in transferring absorbed light energy to chlorophyll *a*.[51] Their in vivo absorption maximum is in the blue region (450 to 550 nm),[47] making them important in maximizing the absorption of under-ice irradiance. A similar function would be performed by peridinin in dinoflagellates, while in some members of the Prasinophyceae, absorption in the blue green region may be facilitated by the presence of siphonoxanthin.[47] This xanthophyll is absent in the only member of the Prasinophyceae (*Pyramimonas gelidicola*) so far examined from the ice.[52] In addition to these pigments which have shown a light harvesting function, most of the microalgae contain a number of xanthophyll pigments. The distribution of xanthophylls in ice diatoms is variable, with diatoxanthin and diadinoxanthin being the main nonlight harvesting carotenoids in some, but not all, species.[52,53] Ratios of total carotenoids to chlorophyll *a* ranged from 1.12 to 1.4 in bottom ice diatoms from McMurdo sound (2°C and 24 μE m^{-2} sec^{-1}).[4] Higher values (1.6 to 2.2) were found in a natural population of *Navicula glaciei* Van Heurck sampled during early spring when *in situ* irradiances were very low.[54] The structure and distribution of algal carotenoids has been discussed by Goodwin.[55,56]

Cryptophytes are frequently found in the ice or in populations living beneath the ice. Cryptophytes, unlike other phycobiliprotein-containing algal groups, only contain phycoerythrin or phycocyanin, but not both.[57] These biliproteins absorb in the 540 to 565 nm and 610 to 640 nm range respectively and are effective in utilizing shortwave radiation.[57] Transfer of energy is thought to occur either directly to chlorophyll *a* or indirectly via chlorophyll c_1.[57]

B. Regulation of Pigment Composition

Microalgae living in or under ice are generally considered to be shade adapted and characteristically have low compensation light intensities (I_c).[4,58-60] This is more pronounced when measured at low temperatures approaching those found *in situ*.[4,59] Shade adaptation is thought to involve alteration of the photosynthetic apparatus at several levels and may include quantitative and qualitative changes in pigment composition, changes in the activities of enzymes involved in carbon fixation, and changes in the photosynthetic-irradiance profiles.[61]

Increases in the level of photosynthetic pigments are consistently observed in microalgae exposed to low photon flux densities, although the degree of variation is species specific.[40,62,63] These changes have been interpreted as signifying increases in either the size or the number of photosynthetic units in the cell.[40,61] Bunt found appreciable increases in the cellular chlorophyll *a* of several ice diatoms grown at low light intensities.[4] In *Nitzschia sublineata*, this increase was linearly related to the log of irradiance. After adaptation to the new irradiance, the chlorophyll *a* content increased by a factor of 7.5 between 2.4 and 72 μE m^{-2} sec^{-1} (1°C).[4] This response is more pronounced than in planktonic diatom species,[64] suggesting that ice species display considerable phenotypic adaptation to different irradiances.

Increased pigment levels are frequently enhanced under low intensity blue-green light relative to when white light is used[47] and may be accompanied by changes in the relative composition of light harvesting pigments. However, Jeffrey[47] has suggested that in algae containing carotenoids and chlorophyll *c* as accessory light harvesting pigments, complementary chromatic adaptation (selective increase in blue-green light absorbing pigments in cells grown under blue-green light) probably does not occur.

Temperature had a marked influence on the time course over which cells altered their chlorophyll *a* content at different light intensities. In cultures of *Nitzschia sublineata* grown at 2°C then transferred to 5°C, there was an immediate and rapid increase in cellular chlo-

* All values for light measurements were converted to μE m^{-2} sec^{-1} with the conversion factors used by Richardson et al.[40]

rophyll *a* at low light intensities.[4] In contrast, cells transferred to -1.6°C showed much slower accumulation of chlorophyll *a* levels. A similar temperature-dependent rate effect has been observed in other diatoms and at low temperatures, cells are unable to respond to short term (<24 hr) variations in irradiance.[26] Hitchcock[26] suggested that this was due to the depressant effects of temperature on the enzyme-mediated reactions of synthesis and degradation of cellular chlorophyll pools.

In addition to influencing the time scale over which pigment changes occur, temperature also appears to influence the cellular chlorophyll *a* levels directly. Increases in temperature from -1.6 to 3°C were accompanied by a twofold increase in chlorophyll *a* in *Nitzschia sublineata* grown at 72 μE m^{-2} sec^{-1}.[4] However, there was no change over this temperature range when cells were grown at irradiances near their I_c. Temperature-dependent changes in the ratios of other photosynthetic pigments also occurred in several ice diatoms at different irradiances.[4] It is difficult to conclude whether these changes signify adaptive adjustment of cellular pigments to optimize harvesting of available light or whether the enzymes involved in synthesis and turnover of different pigments merely differ in their temperature optima.

Many studies on ice communities have attempted to estimate microalgal biomass by measurement of chlorophyll *a* levels in the ice. In some cases these values are converted to cellular carbon on the basis of ratios observed in cultured species. Bunt[4] found that in the bottom ice microalgal bloom in McMurdo Sound, ratios of carbon to chlorophyll *a* ranged from 24 to 59. He concluded that estimation of standing stocks of plant cell carbon based on pigment data in ice covered waters should be accepted with caution.

Variations in the attenuation of light through the ice (due mainly to differences in overlying snow cover and the degree of selfshading in the community) would largely account for the spatial heterogeneity of irradiance in the ice and the associated variability in chlorophyll levels. During periods of algal proliferation, selfshading is undoubtedly important with the commensurate increase in photosynthetic pigments of shade-adapted cells. In Antarctic bottom ice communities, Sullivan et al.[65] observed average standing stocks of 129 ± 90 mg chlorophyll *a* m^{-2}. In some areas, chlorophyll *a* densities approached the theoretical maximum of 400 mg m^{-2}.[65] In the Arctic ice communities, where lower values of around 30 mg chlorophyll *a* m^{-2} are found during the period of maximum algal proliferation,[6] selfshading would still be very important. Community structure is relatively stable during the ice blooms and it is likely that cells would be able to achieve maximum pigment adaptation to the particular light regime they experience.

Overestimation of chlorophyll *a* standing crops in the ice may occur due to the presence of pigment degradation products. The main degradation products of chlorophyll *a* are chlorophyllide *a* (minus a phytol chain) and the pheopigments, pheophytin *a* and pheophorbide *a* (minus magnesium), although several other derivatives have also been resolved.[66,67] Pigment degradation products are frequently associated with senescent populations and have been used as indicators of the physiological state of natural phytoplankton communities.[66-68] Hoshiai[69] found that while pheopigments accounted for less than 2% and 20% of the measured total chlorophyll *a* in the spring and autumn ice blooms respectively, they represented more than 80% of the chlorophyll *a* pigment in a remnant overwintering population. He suggested that the presence of chlorophyll *a* (after correcting for pheopigments) indicated that these cells survived overwintering in the ice and constituted a seeding population for the spring ice bloom.[69] The fluorescence technique used to measure chlorophyll *a* in this study gives the sum of chlorophyll *a* and chlorophyllide *a* and may grossly overestimate the level of photosynthetically active pigment.[68] Recent techniques based on chromatographic separation by thin layer and high performance liquid chromatography (TLC and HPLC),[67] of pigment degradation products may be used to provide more information on the physiological state of overwintering populations in addition to improving the estimation of standing crops in the ice.

Table 2
FATTY ACID DISTRIBUTION (AS PERCENT TOTAL LIPID FATTY ACIDS) OF 4 SEA ICE MICROALGAE

Fatty acid	*Navicula glaciei*[54]	*Fragilaria* sp.[52] (−1°C)	*Stauroneis amphioxys*[92] (3°C)	*Stauroneis amphioxys*[92] (20°C)	*Pyramimonas gelidicola*[52] (−1°C)
14:0[a]	2.1	11.5	4.9	5.6	1.1
16:0	20.5	14.0	11.0	10.8	27.0
16:1	32.9	29.2	21.3	21.7	7.3
16:2	3.5	3.2	6.6	5.5	0.8
16:3	2.8	1.9	3.5	11.1	—
16:4	6.5	6.5	6.6	1.3	17.1
18:0	1.0	0.5	2.4	1.4	3.5
18:1	2.0	2.1	2.7	2.2	5.7
18:2	—	1.1	—	—	2.1
18:3	2.9	3.6	1.0	2.7	13.4
18:4	2.9	11.4	5.0	2.8	5.2
20:3	0.2	—	—	—	7.6
20:4	19.3	14.5	24.3	26.1	—
22:5	—	—	—	0.2	2.8
22:6	2.1	1.1	3.2	1.9	—
Minor components[c]	1.3	0.2	7.3	3.8	2.4
db/mol[d]	1.87	1.92	2.29	2.24	1.89

Note: *Navicula glaciei* was sampled from a natural population in the coastal tidecrack region during winter. The other ice algae were cultured at −1°C, 3°C, or 20°C under 16.5 $\mu E\ m^{-2}\ sec^{-1}$ irradiance.

[a] C:n, C is the number of carbon atoms in the chain, n is the number of double bonds.
[b] Not detected.
[c] Mainly saturated C_{12} and C_{15} acids and $>C_{20}$ unsaturated acids.
[d] Average number of double bonds per molecule of fatty acid.

V. LIPIDS

A. Cellular Composition

The major lipid components of microalgae are triglycerides, sulfonoquinovosyl diglyceride, monogalactosyl diglyceride, digalactosyl diglyceride, and the phosphoglycerides; phosphatidylcholine, phosphatidylethanolamine, phosphatidyglycerol, and phosphatidylinositol.[70] All of these lipid classes contain esterified fatty acids which account for approximately half of the total lipid.[71] Most studies on microalgae tend to give the total fatty acid composition after solvent extraction[72] and there is comparatively little information on the fatty acid distribution among different lipid classes.[71,73]

The total fatty acid analysis of four sea ice algae is given in Table 2. The predominance of 16:0, 16:1, and 20:5 in the three diatoms is similar to that typically found in other diatom species.[70-72] Significant differences were evident between the ice diatoms and other species in the production of high levels of C_{18} polyunsaturates, particularly 18:4 Δ6,9,12,15. This was most evident in the *Fragilaria* sp. which also had higher levels of 14:0 distinguishing it from the other ice species. The presence of decosahexaenoic acid (22:6) in all the ice diatoms is also unusual and has not been previously reported in other microalgae. The fatty acid analysis of *Pyramimonas gelidicola* is quite different from the diatoms and shows a distribution characteristic of some other green algal chemotypes in having high 16:0, 16:4, and 18:3 acids.[52]

The fatty acid distribution of exponentially growing cultures of *Navicula glaciei* (not given) differed from that found in the natural population by having a higher proportion of unsaturated fatty acids[52] (2.71 vs. 1.87 double bonds per fatty acid molecule). Volkman et

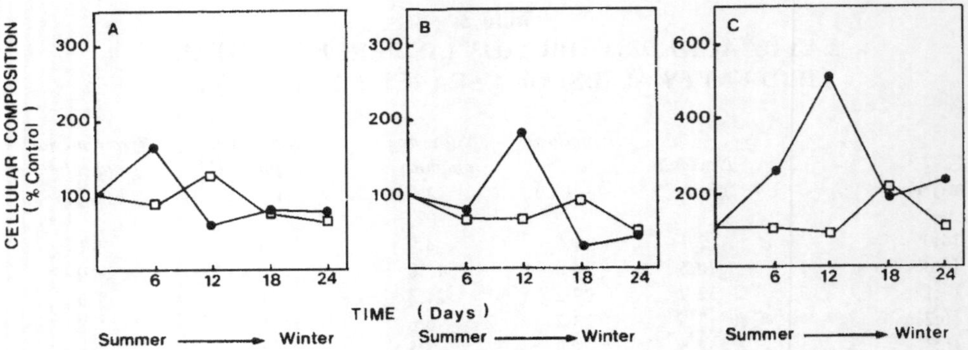

FIGURE 1. Changes in the cellular composition of carbohydrates (●) and lipid (□) in three sea ice diatom clones during a simulated summer-winter transition (as percent of cellular levels in control cultures maintained under summer conditions). During the transition, batch cultures of one Arctic sea ice diatom (A) and two clones of an Antarctic sea ice diatom (*Nitzschia cylindrus* (Grunow) Hasle) (B, C) were subjected to changes in irradiance (24 $\mu E\ m^{-2}\ sec^{-1}$ to darkness), temperature (0 to $-2°C$), and salinity (20 to 39‰) over a 30-day period that mimicked the onset of polar winter. (From Palmisano, A. C. and Sullivan, C. *J. Phycol.*, 18, 489, 1982. With permission.)

al.[74] found reduced 20:5 acids in old diatom cultures and a similar marked reduction in this fatty acid in the natural population may reflect the presence of senescent or resting cells.[52]

B. Regulation of Lipid Storage

The total lipid content of rapidly growing microalgae varies from 1 to 45% of the cell dry weight.[75] Average values for the percent of dry weight in diatom lipids range from 16 to 24%.[75,76] Both the quantitative and qualitative lipid composition of algae are influenced by environmental factors such as irradiance,[71,77,78] temperature,[79,80] and nutrients.[76,81] However, changes are not always consistent and may be species specific.[76]

The main storage lipids of diatoms and other microalgae are triglycerides (esters of glycerol in which each of the three hydroxyl groups are esterified with a fatty acid). In diatoms, triglycerides contain a high content of 16:1 and C_{20} polyenoic acids.[73] A shift to triglyceride synthesis has been found when cells are grown at high irradiances.[73] The low triglyceride content of a natural population of *Navicula glaciei* during winter is probably the consequence of very low irradiances in the snow/ice habitat at the time of sampling.[54] Under these conditions, the synthesis of polar lipids, which are major components of cellular membranes, is maintained.[54,73] While lipids may be utilized as short term reserves in some microalgae, this does not appear to occur in diatoms.[81] No significant diel periodicity has been observed in diatom lipids for example and appreciable accumulation of lipid occurs only under conditions in which growth and division have been limited for several generations.[22,76,81]

Pronounced lipid storage is generally observed in microalgae during nutrient limitation.[76] Werner has suggested that the appearance of oil droplets in natural populations of diatoms is probably attributable to $Si(OH)_4$ deficiency,[81] although similar accumulations may occur during prolonged nitrate deficiency.[76] The accumulation of oil droplets in many of the bottom ice diatoms towards the end of the spring ice bloom,[18,82] may indicate that these cells are nutrient limited. While nutrient levels within the sea ice are generally nonlimiting, there is some evidence to suggest that as extensive melting occurs throughout the ice, nutrients such as nitrate may become limiting.[83,84]

Accumulation of cellular lipid also may occur at low temperatures, although generalizations are difficult. In three sea ice diatoms, enhanced lipid levels were found toward the end of a simulated summer to winter transition in which temperature and irradiance were decreased (Figure 1). Similarly, high ^{14}C incorporation into lipids occurs in some,[80] but not all polar phytoplankton when temperatures are lowered (Section VI).[12,25,80]

C. Effects of Temperature on Membrane Lipid Composition

The lipid bilayers of membranes are largely fluid at physiological temperatures. However, at lower temperatures, membrane lipids undergo a reversible change of state from a fluid, liquid-like state, to an ordered state of fatty acyl chains.[86] The temperature at the midpoint of this transition is called the transition temperature and is a function of membrane lipid composition.[86] The state of cellular membranes is critical for the proper functioning of a large number of membrane attached enzymes, as well as for the regulation of fluxes of metabolites and ions through the cell.[86] It is frequently assumed that in ectothermic organisms such as bacteria and plants, some sort of regulation of the lipid phase transition occurs in response to lower temperatures. This is thought to involve changes in fatty acid composition (degree of unsaturation, chain length, etc.), as well as changes in the lipid composition, (i.e., sterol, phospholipid composition) of cellular membranes.[87] In a strain of *Chlamydomonas nivalis* (Bauer) Wille, for example, which actively photosynthesizes below 0°C in the snow habitat, the degree of unsaturation of the bulk lipid extract was appreciably higher than in a mesophilic species of *Chlamydomonas*.[88] Unequivocal evidence that these changes in the bulk lipid lead to adaptive changes in membrane fluidity is limited and there are numerous examples in microorganisms where the degree of unsaturation of fatty acids does not increase at lower growth temperatures.[89-92] While changes in the fatty acid composition of the ice diatom *Stauroneis amphioxys* Gregory occur at different culture temperatures, the overall degree of unsaturation remained the same (Table 2). The main changes were found between the polyenoic acids 16:3, 18:3, and 20:4 and their biosynthetically related fatty acids in the 16:4, 18:4, and 20:5 series. At 20°C there was a marked decrease in the ratio of 16:3 Δ6,9,12 to 16:4 Δ6,9,12,15 suggesting that the desaturase involved in the last desaturation step is temperature sensitive, although less pronounced changes occurred with the other biosynthetically related pairs (Table 2). The absence of an overall increase in the number of double bonds per fatty acid or in the chain length suggests that thermoregulation of cellular membranes in this species does not involve alterations in the synthesis of constituent fatty acids. This is not unexpected in view of the relatively high degree of unsaturation of microalgal fatty acids. Orcutt and Patterson,[71] found an average of 1.69 ± 0.79 double bonds per fatty acid in 11 diatom species. As most of the polar membrane lipids (phosphoglactolipids) have a higher proportion of polyunsaturated fatty acids,[73,93] the degree of unsaturation of membrane fatty acids would be appreciably higher. In the four sea ice algae studied, the average number of double bonds per fatty acid ranged from 1.9 to 2.3 and it is likely that their membrane lipids are sufficiently unsaturated to maintain membrane fluidity over a wide range of temperatures. This does not preclude the possibility that other changes in lipid composition or structure, not involving alteration of fatty acids composition, may be occurring at lower temperatures.

The sterol composition of *Stauroneis amphioxys* was similar to that found in other diatom species. The major sterol was 24-methylcholesta-5,22E-dien-3β-ol (79%), while cholesterol was also present.[92] As previously mentioned, the presence of sterols in the membrane may result in increased lipid fluidity. This effect is most pronounced below the transition temperature.[86] At 3°C there was a 50% increase in the sterol content of *S. amphioxys* relative to fatty acids. Increased sterol content at lower temperatures has also been found in *Ochromonas danica* Pringsheim. El-Betouhim et al.[94] suggested that the altered sterol content at different temperatures may explain the ability of this species to grow over a wide temperature range. Whether the changes in sterol level are involved in the adaptive thermoregulation of cellular membranes in ice algae requires further study.

The stresses accompanying extracellular ice formation (osmotic dehydration, intracellular ice formation) differ from those found at low temperatures[28] and there is evidence to suggest that adaptive strategies producing freezing resistance are different from those found in response to low temperatures.[95] In higher plant cells, accumulations of phospholipids rather

Table 3
MAJOR STORAGE CARBOHYDRATES OF SOME MICROALGAL CLASSES FOUND IN SEA ICE COMMUNITIES (SEE TEXT FOR REFERENCES)

Taxon	Storage carbohydrate
Bacillariophyceae	Chrysolaminarin
Chrysophyceae	Chrysolaminarin
Prymnesiophyceae	Chrysolaminarin
Chlorophyceae	Starch
Prasinophyceae	Starch
Dinophyceae	Starch
Cryptophyceae	Starch

than specific alterations in the composition of fatty acids have been correlated with increased resistance to freezing.[95] A similar augmentation of polar lipids in some *Chlorella* spp. was thought to increase freezing resistance by reducing the chance of intracellular ice formation.[91]

VI. CARBOHYDRATES

Microalgal carbohydrates form important structural components of the cell, accumulate as storage reserves, and also constitute an important fraction of the low molecular weight metabolite pool.

A. Reserve Carbohydrates

The major reserve carbohydrates characteristically synthesized by different ice taxa are shown in Table 3. The principal reserve polysaccharide in diatoms is chrysolaminarin, which has been extracted and characterized from a number of diatom species. It is also an important polysaccharide in three sea ice diatoms that have been examined, *Navicula glaciei, Stauroneis amphioxys,* and *Nitzschia* sp.[31] Structural investigations show that chrysolaminarin contains only glucose which is (1→3)-β-D-linked, with some branching through C(0)6.[96,98] Chrysolaminarin extracted from *Skeletonema costatum* contained, in addition, branching at C(0)2 which has not been previously observed in algal reserve polysaccharides.[98] Chrysolaminarin differs from laminarin produced by brown algae (Phaeophyceae) in lacking terminal mannitol groups and having a shorter average chain length.[96] It is readily extracted from cultured diatoms or natural populations with dilute acid and constitutes from 4 to 80% of the organic dry matter.[21] Chrysolaminarin is also the main storage reserve carbohydrate in the Chrysophyceae[99] and has been tentatively identified in the Prymnesiophyceae on the basis of cytochemical reactions.[100]

Starch is the main reserve carbohydrate produced by green algae (Chlorophyceae, Prasinophyceae), as well as the Dinophyceae and Cryptophyceae.[101] These carbohydrates contain predominantly (1 → 4)-∝ and (1 → 6)-∝-linked D-glucosyl residues and resemble the starches or amylopectins of higher plants. Starch is normally a mixture of two polymers, amylose and amylopectin, which differ mainly in their degree of branching and polymerization. The chemistry of algal starches has been reviewed elsewhere.[101]

Reserve polysaccharides are, by definition, accumulated when photoassimilated carbon is in excess of cellular metabolic needs. Thus, there is variation in the cellular carbohydrate content depending on the photosynthetic rate and physiological state of the cells. Low cellular levels of carbohydrate are generally observed when the photosynthetic rate is low relative to growth rate. This is evident at low irradiances and in rapidly growing nutrient-sufficient populations. Under these conditions higher relative levels of cellular protein are commonly found. Several authors have suggested that changes in the ratio of protein to carbohydrate, which is largely dependent on changes in reserve glucan in diatoms, may be a sensitive and convenient parameter for characterizing the physiological state of natural populations.[20,102] Whitaker and Richardson[54] estimated the total carbohydrate of a natural ice population of *Navicula glaciei* as 18% of the dry weight and calculated a protein-carbohydrate ratio of between 1.22 and 3.3. This ratio is high and is comparable to the ratios found in other diatoms grown at low irradiances.[22] The *N. glaciei* population was sampled during the Antarctic winter when irradiances are very low and the high protein-carbohydrate ratios suggest that accumulation of reserve carbohydrate is minimal. Levels of storage lipid (triacylglycerides) were also low in these cells, indicating that they were light limited and that all the photoassimilated carbon was being used to maintain basal metabolic rates.

Accumulation of reserve polysaccharide is frequently observed during the day followed by a decrease at night.[26,103-105] The magnitude of this diel variation is reduced at low light intensities and low temperatures.[26] The declining levels of cellular glucan in the dark are usually associated with continued protein synthesis and thus, chrysolaminarin appears to be rapidly mobilized as a short term storage product.[22,104] Recently, enzymes capable of catabolizing (1 → 3)-β-D-glucans have been found in several marine planktonic diatoms.[106] These enzymes hydrolyze monomer glucose from the nonreducing end of (1 → 3)-β-D-glucans, which can then be directly utilized during respiration. Maximum glucanase activity coincided with periods of chrysolaminarin accumulation.[106]

Accumulation of reserve polysaccharides is also observed when the growth rate is reduced relative to the photosynthetic capacity of the cells. This may occur at suboptimal temperatures,[26,58] high salinities,[107] or during nutrient limited growth,[108] and is necessarily dependent on light intensity. Marked accumulation of cellular carbohydrate occurred in three sea ice diatoms during a simulated summer to winter transition (Section V, Figure 1).[85] Whether the change in cellular carbohydrate was due to altered chrysolaminarin levels was not determined. Both temperature and salinity appear to act synergistically with light in influencing photosynthetic rate and carbohydrate metabolism in these diatoms. During the summer to winter transition, the accumulation of carbohydrate was associated with a decline in growth rate relative to photosynthesis. The higher relative levels of photosynthesis appear to have been facilitated by depressed respiration rates as the temperature was lowered. This is indicated by the finding that the decline in ^{14}C assimilation (measuring between net and gross photosynthesis) was faster when only light intensity was decreased without concomitant decrease in temperature. The subsequent decline in cellular carbohydrate during the summer to winter transition was associated with a sharp fall in photosynthetic rate and was suggested by the authors as being evidence of utilization of reserve carbohydrate as a carbon source for metabolism.[85] The increase in cellular lipid accompanying the decline in carbohydrate (Figure 1) suggests that the available carbon was merely being transferred from one storage pool to another. It is difficult to conclude whether the changes in chemical composition are solely attributable to the changes in temperature, light, and salinity and not to those commonly observed in batch cultures.[108]

B. Cell Wall Carbohydrates

In contrast to reserve polysaccharides, the composition of cell wall carbohydrates is relatively independent of the physiological state of the cell. The chemistry of microalgal cell wall carbohydrates has been reviewed by Painter.[109] The cell walls of most of the ice algae contain a siliceous component in addition to organic material. The siliceous frustule of diatoms is surrounded by an organic casing that appears to be derived from the silicalemma membrane.[110] The organic casing contains mainly polysaccharide and protein, as well as a minor component of lipid, that are all closely intermeshed. This material may account for between 16 to 90% of the cell wall.[110] The monosaccharide composition of diatom cell wall polysaccharides differs between species and also during growth.[111,112] Glucosamine and glucuronic acid, as well as six monosaccharides (fucose, xylose, mannose, galactose, glucose, and rhamnose), have been found in appreciable levels.[102,112,113] Of the several distinct polysaccharide fractions that have been extracted, two homogeneous sulfated polysaccharides have been characterized: one, a glucuronomannan (glucuronic acid and mannose only) sulfate, has been found in several different diatom species;[97,110,113] while a fucosan sulfate (containing fucose, mannose, galactose, and an unidentified deoxy sugar) has been found in *Navicula pelliculosa* Hilse.[110] The level of these polysaccharides in the cell wall generally increases after silica deposition ceases.[93]

Silicon is present as polymerized silicic acid in diatom frustules and in most planktonic diatoms constitutes from 10 to 30% of the total dry weight.[114] The silicon content within a species is dependent on environmental factors such as light, temperature, population density, and orthosilicic acid concentrations.[114] While there are no studies on the silica content of sea ice diatoms at present, the dominant species in the bottom ice communities characteristically appear to be heavily silicified.[3] The chemical analysis of a natural population of *Navicula glaciei* revealed that 33% was ash weight, which was assumed to be predominantly frustular material.[54]

The cell wall of members of the Volvocales, typified by *Chlamydomonas* spp., are noncellulosic and consist of a number of crystalline glycoproteins.[115] These glycoproteins contain high levels of hydroxyproline which are linked to arabinose and small amounts of galactose.[115] Comparatively few detailed studies have been made on the cell wall material of other unialgal groups. Many of these groups contain surface scales that may be silicified (Chrysophyceae) or unmineralized (Prasinophyceae), while the matrix material is frequently charged heteropolysaccharide (pectic material).[109] Fibrillar cellulosic material may be present in the Chrysophyceae and Dinophyceae.[109]

C. Nature and Role of Low Molecular Weight Carbohydrates

This group consists of a variety of monosaccharides and sugar polyols that are synthesized as early storage products of photosynthesis.[116] They may function as storage carbohydrates, as well as being intermediates in macromolecular biosynthesis and respiration.[117] Consequently, cellular levels vary with changes in demand and supply. In some algal groups the intracellular levels of particular polyols or sugars are regulated by the external osmotic pressure (Table 4) and there is good evidence that they are important in the osmoregulation of these algae.[118]

Microalgae living in the sea ice may be subjected to osmotic stress during freezing and melting. Fluctuations in salinity are most pronounced in the bottom layers of the ice where oscillatory changes in salinity occur in brine channels over relatively short time periods.[119] These species are generally euryhaline, sustaining appreciable growth rates over a wide range of salinities.[120] Very high tolerances to osmotic stress are evident in several ice flagellates (*Chlamydomonas* sp., *Pyramimonas gelidicola*) which are able to grow at salinities ranging from 5 to 140‰.[121] Closely related euryhaline species from other environments accumulate organic solutes in response to high osmotic stress and decrease their concentration when the stress is relieved.[118] Small increases in salinity frequently involve nonspecific changes in various metabolites and salts, while larger stresses always select a dominant solute.[122] The solutes are usually polyols, although the imino acid proline is characteristically accumulated by diatoms.[107,123,124] Schobert[124] has suggested that proline accumulation is a general response to osmotic stress in all diatoms, although the details of proline metabolism appear to vary between different species. These solutes may all be rapidly synthesized and equally rapidly metabolized or leaked from the cell in order to raise or lower intracellular concentrations. The extent to which they act in a purely colligative fashion to restore osmotic

Table 4
LOW MOLECULAR WEIGHT METABOLITES ACCUMULATED DURING HYPEROSMOTIC STRESS IN SEVERAL MICROALGAL GROUPS

Taxon	Metabolite	Ref.
Bacillariophyceae		
Cylindrotheca fusiformis Reimann and Lewin	Mannose	126
Cyclotella meneghiniana Kützing	Proline	107
Cyclotella cryptica Reimann, Lewin and Guillard	Proline	123
Phaeodactylum tricornutum Bohlin	Proline	124
Chlorophyceae		
Dunaliella spp.	Glycerol	125
Chlamydomonas spp.	Glycerol	118
Chlorella emersonii Shihira and Krauss	Proline, sucrose	127
Prasinophyceae		
Platymonas suecica Kylin	Mannitol	128
Pyramimonas sp.	Mannitol	118
Asteromonas gracilis Artari	Glycerol	129
Chrysophyceae		
Ochromonas malhamensis Pringsheim	Isofloridoside[a]	130

[a] O-α-D-galactopyranosyl-(1 – 1)-glycerol.

balance is uncertain. In many organisms the change in solute (polyol/proline) only represents a fraction of the external osmolar change.[107] This suggests that either other solutes are being increased or that the solute primarily accumulated has some additional protective function. Glycerol, for example, protects enzymes from inhibition and denaturation during prolonged water stress, as well as acting as an osmoticant.[122] Because glycerol maintains enzyme activity under conditions of high extracellular salt concentrations, it is called a compatible solute.[125] Information about the enzyme protective capacity of proline and other polyols is limited. Evidence so far suggests that they do have such a role and under relatively mild water stress conditions function as low grade compatible solutes.[122,124]

D. Pathways of Carbon Assimilation in Polar Microalgae

The distribution of ^{14}C among photosynthetic end products has been used as an extension of the ^{14}C productivity technique to give an indication of the physiological state of natural phytoplankton communities.[15] The results described below for the bottom ice community were obtained near the Australian Antarctic Station of Davis (68°35′S, 78°05′E) during mid-December 1982.[104] The percent ^{14}C-label in the protein, lipid, polysaccharide, and metabolite fractions is given in Table 5. These results show that most of the photosynthetically fixed ^{14}C was distributed in the polysaccharide and metabolite pools. High relative ^{14}C incorporation into the metabolite pool was also observed by Barlow and Henry[12] in Southern Ocean phytoplankton growing at low temperatures (<0°C). They concluded that at these temperatures, synthesis of the major macromolecular classes was reduced and consequently low molecular weight metabolites accumulated. However, most of the label in the ice algae was direct into chrysolaminarin (extracted in both the polysaccharide and metabolite pool), while protein synthesis was consistently low.[104] This pattern of carbon assimilation is characteristic of stationary phase populations which are nutrient limited,[20,108] and may indicate that the decline in the spring ice bloom during December is due to nutrient deficiency in the ice habitat.

The low relative incorporation of ^{14}C into the lipid fraction of ice algae is consistent with that found in other phytoplankton from polar waters.[12,25] This result differs from the study of Smith and Morris[80] who found that certain phytoplankton populations in the Southern

Table 5
DISTRUBUTION OF ^{14}C AMONG PHOTOSYNTHETIC END-PRODUCTS (AS PERCENT TOTAL ^{14}C FIXED) IN A BOTTOM ICE COMMUNITY, EAST ANTARCTICA

Date	Temp	I	t	P	Protein	Lipid	Polysaccharide	Metabolites
Dec. 11, 82	−1.0	12	3.0	0.66	14	18	26	42
Dec. 12, 82	−1.2	—	3.5	0.84	12	20	36	32
Dec. 14, 82	−1.0	26	3.0	0.59	10	17	46	27
Dec. 15, 82	−0.9	18	4.0	0.98	11	16	45	28

Note: ^{14}C incubations were performed *in situ* using a perspex chamber (500 mℓ capacity) similar to that used by Schrader et al.[5] After incubations were terminated, subsamples (100 mℓ) were filtered onto glass fiber filters and fractionated using the procedure of Li et al.[23] to give a lipid, protein, polysaccharide, and metabolite pool. Units: Temp, temperature (°C); I, irradiance ($\mu E\ m^{-2}\ sec^{-1}$); t, length of incubation (h); P, photosynthetic rate (mg C mg Chl $\underline{a}^{-1}\ h^{-1}$).

Ocean incorporated a high proportion of their photosynthetically fixed carbon into lipid (up to 80%). Enhanced lipid synthesis was most marked in areas experiencing lowest ambient temperatures (− 1°C) and light intensities.[80] It was suggested that prominent lipid synthesis was induced by prolonged exposure to these conditions during the Antarctic winter. However, the percent incorporation of ^{14}C into lipid was not appreciably influenced by either light intensity or temperature in other polar species. Li and Platt[25] have suggested that these differences in the labeling patterns of high latitude phytoplankton may reflect variation in species composition rather than a response to the extreme environmental conditions.

VII. EXTRACELLULAR PRODUCTS

The nature of extracellular material produced by microalgae is diverse and includes actively secreted high molecular weight mucilages, as well as a low number of molecular weight metabolites. This latter group are characteristically early products of photosynthesis and may be passively released.

A. Polysaccharide Mucilages

Many of the dominant ice algae produce extensive extracellular polysaccharide mucilages which form a type of matrix material between cells in the bottom ice communities and are responsible for the tendency of these cells to aggregate into clumps.[3,18,84] The rheological properties of these polysaccharides (solubility and gel forming characteristics) are dependent on their chemical composition and may vary markedly between different species and also under different environmental conditions in the same clone.[131,132] The extracellular insoluble mucilages can be visualized by using stains, (e.g., toluidine blue, alcian blue, or ruthenium red) that bind charged groups such as carboxylic acid or sulfate.

The extent of mucilage secretion and its degree of morphological organization varies among different ice microalgae, particularly among different pennate diatom species. Several species secrete mucilage pads that are restricted to either one or both ends of the frustule and which are involved in adhering cells to each other, (e.g., *Nitzschia frigida* Grunow) or to other cells (e.g., *Synedra* sp.). Other diatoms, when isolated into unialgal culture, produce extensive mucilaginous sheets to which cells are firmly attached, (e.g., *Stauroneis amphioxys, Navicula glaciei).* Finally, some diatom genera found in the ice, (e.g., *Berkeleya, Gomphonema*)[18,84] produce morphologically complex mucilages, forming stalks or encapsulating tubes that hold the cells together in true colonies. In addition to diatom species, members of the Chlorophyceae and Prymnesiophyceae frequently spend at least part of their life cycles in the ice as palmelloid colonies. Palmelloid colonies of *Phaeocystis pouchetii*

Table 6
CHEMICAL COMPOSITION OF EXTRACELLULAR MUCILAGES PRODUCED BY ICE MICROALGAE (SEE TEXT FOR REFERENCES

	Monosaccharide composition[a]			
Monosaccharide	***Stauroneis amphioxys***	***Navicula glaciei***	***Berkeleya rutilans***	***Phaeocystis pouchetii***
Arabinose				5
Fucose	7	12		
Galactose	25	25		5
Glucose	7	2		32
Mannose	9	3	57	27
Rhamnose	9	11		21
Ribose				3
Xylose	11	33	4	3
2-*O*-Methyl-xylose	5			
3,6-Anhydrohexose			19	
Uronic acid	17	11	19	5
Sulfate[b]	16	4	22	
Protein	+[c]	+	+	

[a] % Weight of total carbohydrate.
[b] % Weight of total mucilage.
[c] Protein found associated with mucilage polysaccharide.

(Hariot) Lagerheim, for example, are frequently dominant in many areas of the Antarctic pack ice.[19] The colonies are generally 1 to 2 mm in diameter, of irregular shape, with cells being arranged as small clusters of subcolonies within the mucilage. *Chlamydomonas* sp. may also form colonies in the ice and proliferate in a gelatinous matrix.

The chemical composition of mucilages from four ice algae is presented in Table 6. The diatom mucilages are charged heteropolysaccharides containing uronic acid residues and half ester sulfate groups. The composition of neutral and acidic residues in these polysaccharides varies markedly among species. In the *Stauroneis amphioxys* mucilage polysaccharide, the uronic acid is glucuronic acid,[133] while in the mucilage tubes of *Berkeleya rutilans* (Trentepohl) Grunow, mannuronic acid is the only uronic acid present.[134] The *Stauroneis amphioxys* mucilage also contains the methylated sugar 2-*O*-methyl-xylose that has not been previously reported in algal polysaccharides.[133] Studies on the structure of these and other diatom extracellular mucilages by methylation analysis suggests that they are highly branched polymers.[133-135]

The extracellular polysaccharides produced by *Phaeocystis pouchetii* (Prymnesiophyceae) in natural communities and in culture have been examined.[109,136] One component extracted from the palmelloid colonies in a natural community was a proteoglycan containing ester sulfate and residues of galacturonic acid, glucosamine, galactose, arabinose, xylose, and rhamnose.[109] In cultures of *P. pouchetii,* a range of polysaccharides and oligosaccharides was also released which have been partially characterized (Table 6).[136] Only about 3% of the total extracellular material had a molecular weight in the order of 100,000 or more, while most of the material (approximately 70%) fell in the 700 to 5000 molecular weight range as determined by Sephadex chromatography.[136] Some of this material may have been derived from intracellular reserve material, which in the Prymnesiophyceae is thought to be a (1→3)-β-glucan. This would explain the high glucose levels of the extracellular material of this clone compared to the warm water clone. Similarly, in axenic cultures of *Stauroneis amphioxys* and *Navicula glaciei*, a significant proportion of the soluble extracellular labeled material (approximately 95%) has a molecular weight of less than 6000. A variable fraction (up to 40%) of this material was identified as chrysolaminarin.[133] Greatest release of glucan occurred when intracellular levels were highest during stationary phase growth.[133] The presence of chrysolaminarin in the extracellular material has also been found in cultures of *Coscinodiscus nobilis* Grunow.[113]

The insolubility of algal extracellular mucilages in their native state frequently depends on the proportion of charged groups which can cross-link with divalent cations (Mg^{2+}, Ca^{2+}) in sea water.[132,134] Hydrophobic residues such as 3,6-anhydrohexose may also contribute to the overall insolubility.[134] The branched polysaccharide chains are cross-linked to form a network that can immobilize a large volume of water. Consequently the dry weight of a mucilage gel may be less than 2% of the fresh weight.[132] Increased sulfation of diatom mucilage polysaccharides has been associated with firmer gel properties and in habitats where shear forces are likely to be increased, there is a tendency to higher sulfation.[137] Correspondingly, the mucilages from *Navicula glaciei*, which characteristically colonize levels in the ice subjected to low shear forces, have a low sulfate content, while the mucilage tubes of *Berkeleya rutilans*, which are frequently attached to the under ice surface (higher shear forces), are highly substituted with sulfate.

The synthetic rate and rheological properties of extracellular polysaccharides are frequently dependent on culture age or stage in the life cycle of the microalgae.[131,136] In cultures of the ice diatoms *Stauroneis amphioxys* and *Navicula glaciei*, both the solubility and adhesive properties of the mucilage sheets vary with the physiological state of the cells, becoming successively more soluble and less "sticky" as cells enter stationary phase.[133] Similar changes in the mucilage polysaccharide of *Nitzschia frustulum* Kützing have also been found.[131] In the cold water clone of *Phaeocystis pouchetii* secretion of extracellular polysaccharide was dependent on whether the cells were in colonial or single cell stages. In the colonial, palmelloid state, extracellular polysaccharide represented between 16 to 63% of the total photoassimilated carbon, while only 3% of the photosynthate was released by single cells.[136]

The ubiquitous occurrence of extracellular mucilages associated with ice microalgae suggests that they are functionally important. As in benthic species, epiphytic and colonial species in the ice depend upon mucilage secretions for attachment to either the substratum (ice) or to each other. Attachment of diatoms to ice crystals has been observed in both spring and autumn bottom ice communities.[69] Similarly, these cells readily adhere to glass slides that are submerged in the layer of microalgal proliferation.[31,69] Attachment of diatoms to a substratum is usually a biphasic sequence,[137] similar to that characteristically observed in bacteria.[138] The adhesive polysaccharide involved in the initial adhesion may differ from the polysaccharides subsequently synthesized. Several stalk forming genera, (e.g., *Gomphonema, Licmophora*), are initially anchored to the substratum by a pad of polysaccharide in which the only charged groups are carboxyl groups. Subsequently produced stalk material is comprised of polysaccharides containing both carboxyl groups and half ester sulfate.[137] In some cases photosynthetic production of oxygen causes formation of gas bubbles within the mucilage matrix.[18] These bubbles, by making the algal aggregates positively buoyant, may be important in maintaining the cells in the ice habitat. This process of oxygen-rich bubble accumulation and retention is enhanced in cold waters which are generally supersaturated with oxygen.[139]

In addition to attachment, these mucilages may possibly function in ameliorating the immediate environment of the cell. Through the cation exchange properties of the charged polysaccharide, certain ions beneficial for growth may be selectively bound.[140] Protection from adverse fluctuations in salinity,[141] as well as freezing injury, may also be facilitated by extracellular mucilages. Some higher plants produce extracellular mucilages that are effective inhibitors of ice crystallization in plant tissues.[142] The xylan mucilages extracted from hardy species of rye had little effect on the freezing point of the solution, but retarded freezing by disrupting ice crystal structure and interfering with liquid diffusion.[143] By preventing adhesion of growing ice crystals to the cell wall, these extracellular mucilages minimize the chance of intracellular ice formation and cell death.[144]

Production of extracellular mucilages is also found in Antarctic limpets *(Patinigera polaris*, Hombron and Jacquinot, 1841) that become frozen into the inshore sea ice.[145] When trapped in the ice the limpets secrete an encapsulating mucilage up to 11 mm thick. In addition to lowering the interstitial fluid content between cells and thereby avoiding extracellular ice formation, this mucilage may play a more active role in depressing the freezing point.[145] This is analagous to the activity of soluble antifreeze glycoproteins found in the blood serum of several polar fish.[146] These glycoproteins have been proposed to interact at the ice-liquid interface, through formation of hydrogen bonds between the carbohydrate moiety and water molecules in the ice crystal lattice.[147] This interaction inhibits further ice growth until the temperature is lowered by approximately 0.8°C below the normal freezing point.[147] Whether the extracellular mucilages of ice algae function either nonspecifically, to interfere with ice crystallization or in a more specific manner (freezing point depression), is presently being investigated.

B. Low Molecular Weight Metabolites

In addition to the synthesis and secretion of high molecular weight polysaccharide, a variable amount of the rapidly labeled photosynthate is released to the medium. In the bottom ice community this release may account for 7 to 17% of the total labeled photosynthate.[18] This is similar to the release observed in phytoplankton communities in polar,[148] temperate,[149,150] and tropical waters.[151]

In the few studies that have attempted to characterize this soluble extracellular pool, it has been found to contain amino acids,[149] monosaccharides,[152] and frequently glycollate.[153] However, no evidence has been found for release of glycollate in cultures of ice diatoms grown under conditions equivalent to those found *in situ*.[31] An interesting metabolite secreted/released by *Phaeocystis pouchetii* is acrylic acid (CH_2·CH·COOH).[136,154] In cultures of the cold water clone approximately 2% of the extracellular material was acrylic acid.[136] This compound has been found to have inhibitory effects on bacterial growth and in rapidly growing cultures, the encapsulating mucilage surfaces are free of epibacteria.[154] This has also been observed in healthy colonies in the ice, although colonies that appear senescent are heavily covered with epibacteria.[31]

The rate of extracellular release has been found to be closely related to photosynthetic rate.[149,155] Mague et al.[149] have suggested that passive loss occurs continuously from the intracellular pool of soluble low molecular weight metabolites (amino acids, monosaccharides, etc.). This pool is in dynamic equilibrium with the biosynthesis and catabolism of cellular macromolecules. As photosynthesis increases above a certain rate, this intracellular pool becomes saturated and an increasing proportion of soluble metabolites are released. When cell division is reduced relative to photosynthesis (due to nutrient limitation or low temperature), the surplus photosynthate may initially be accumulated as reserve material followed by release into the media. When cultures of a subarctic cryptomonad, grown at low temperatures, were exposed to saturating irradiance, excess photosynthate was both stored and released.[58] Under these conditions the release of excess photoassimilated carbon constitutes an overflow mechanism which may be energetically cheaper than the synthesis of membrane bound storage organelles.[156]

Within the brine channel environment these released metabolites may be retained at higher levels than would normally occur in open waters. They may constitute an important source of carbon for heterotrophic organisms in the ice communities.

VIII. CONCLUSIONS

1. The gross chemical composition and pathways of carbon assimilation of ice algae are similar to related species from other marine ecosystems.
2. Morphologically and chemically, ice algae most closely resemble benthic microalgal species. The colonial or epiphytic habit of many of the dominant diatom species is facilitated by the secretion of sulfated heteropolysaccharide mucilages.
3. The fatty acids of ice algae are highly unsaturated and alterations in the degree of unsaturation of fatty acids do not appear to be involved in the thermoregulation of cellular membranes. The presence of decosahexaenoic acid (22:6) in three sea ice diatoms, has not been previously reported in other diatoms.
4. The cellular levels of reserve material (chrysolaminarin, triglycerides) are markedly influenced by environmental factors and may reflect the physiological state of diatom dominated ice communities. Accumulations of reserve material are not observed under winter conditions.
5. Considerable variability in cellular pigment composition may result from the spatial heterogeneity of under ice irradiances. The ratios of various pigments and their degradation products provide a measure of cell viability in the communities.
6. A similar approach to that suggested for phytoplankton communities, using the cellular ratios of different chemical constituents to measure the physiological state of microalgae may be useful for improving estimates of primary production in the ice community.

ACKNOWLEDGMENTS

It is a pleasure to thank Dr. T. Bacic and Mr. C. Mitchell for their valuable suggestions during the preparation of the manuscript and Drs. A. E. Clarke and R. Wetherbee for critical review of the manuscript. Logistic support for work at Davis Station was provided by the Australian Antarctic Division.

REFERENCES

1. **Sakshaug, E.,** Problems in the methodology of studying phytoplankton, in *The Physiological Ecology of Phytoplankton,* Morris, I., Ed., Blackwell Scientific, Oxford, 1980, 57.
2. **Goldman, J. C.,** Physiological processes, nutrient availability, and the concept of relative growth rate in marine phytoplankton ecology, in *Primary Productivity in the Sea,* Falkowski, P. G., Ed., Plenum Press, New York, 1980, 179.
3. **Bunt, J. S. and Lee, C. C.,** Observations within and beneath Antarctic sea ice in McMurdo Sound and the Weddell Sea, 1967—1968, Methods and Data, Tech. Rep. 69-1, Institute of Marine Sciences, University of Miami, Miami, 1969.
4. **Bunt, J. S.,** Some characteristics of microalgae isolated from Antarctic sea ice, *Antarct. Res. Ser.,* 11, 1, 1968.
5. **Schrader, G. C., Horner, R., and Smith, G. F.,** An improved chamber for in situ measurement of primary productivity of Arctic sea ice algae, *Can. J. Fish. Aquat. Sci.,* 39, 522, 1982.
6. **Horner, R. and Schrader, G. C.,** Relative contributions of ice algae, phytoplankton, and benthic microalgae to primary production in nearshore regions of the Beaufort Sea, *Arctic,* 35, 485, 1982.
7. **Sargent, J. R. and Whittle, K. J.,** Lipids and hydrocarbons in the marine food web, in *Analysis of Marine Ecosystems,* Longhurst, A. R., Ed., Academic Press, New York, 1981, 491.
8. **Lee, R. F., Nevenzel, J. C., and Lewis, A. G.,** Lipid changes during life cycle of marine copepod, *Euchaeta japonica* Murukawa, *Lipids,* 9, 891, 1974.
9. **Bottino, N. R.,** The fatty acids of Antarctic phytoplankton and euphausiids. Fatty acid exchange among trophic levels of the Ross Sea, *Mar. Biol.,* 27, 197, 1974.

10. **Dunbar, M. J.,** The evolution of polar ecosystems, in *Adaptations within Antarctic Ecosystems*, Llano, G. A., Ed., Gulf Publ., Houston, 1977, 1063.
11. **Li, W. K. W.,** Temperature adaptation in phytoplankton: cellular and photosynthetic characteristics, in *Primary Productivity in the Sea*, Falkowski, P. G., Ed., Plenum Press, New York, 1980, 259.
12. **Barlow, R. G. and Henry, H. L.,** Patterns of carbon assimilation in phytoplankton from the Southern Ocean, *Fish. Bull. S. Afr.*, 16, 25, 1982.
13. **Bunt, J. S., van Owens, O. H., and Hoch, G.,** Exploratory studies on the physiology and ecology of a psychrophilic marine diatom, *J. Phycol.*, 2, 96, 1966.
14. **Morris, I.,** Paths of carbon assimilation in marine phytoplankton, in *Primary Productivity in the Sea*, Falkowski, P. G., Ed., Plenum Press, New York, 1980, 139.
15. **Morris, I., Glover, H. E., and Yentsch, C. S.,** Products of photosynthesis by marine phytoplankton: the effect of environmental factors on the relative rates of protein synthesis, *Mar. Biol.*, 27, 1, 1974.
16. **Morris, I. and Skea, W.,** Products of photosynthesis in natural populations of marine phytoplankton from the Gulf of Maine, *Mar. Biol.*, 47, 303, 1978.
17. **Meguro, H., Ito, K., and Fukushima, H.,** Ice flora (bottom type): a mechanism of primary production in polar seas and the growth of diatoms in sea ice, *Arctic*, 20, 114, 1967.
18. **McConville, M. J. and Wetherbee, R.,** The bottom-ice microalgal community from annual ice in the inshore waters of East Antarctica, *J. Phycol.*, 19, 431, 1983.
19. **Bunt, J. S.,** Microalgae of the Antarctic pack ice zone, in *Symposium on Antarctic Oceanography*, Currie, R. I., Ed., W. Heffer & Sons Ltd., Cambridge, 1968, 198.
20. **Haug, A., Myklestad, S., and Sakshaug, E.,** Studies on the phytoplankton ecology of the Trondheimsfjord. I. The chemical composition of phytoplankton populations, *J. Exp. Mar. Biol. Ecol.*, 11, 15, 1973.
21. **Myklestad, S.,** Production of carbohydrates by marine planktonic diatoms. I. Composition of nine different species in culture, *J. Exp. Mar. Biol. Ecol.*, 15, 261, 1974.
22. **Terry, K. L., Hirata, J., and Laws, E. A.,** Light-limited growth of two strains of the marine diatom *Phaeodactylum tricornutum* Bohlin: chemical composition, carbon partitioning and the diel periodicity of physiological processes, *J. Exp. Mar. Biol. Ecol.*, 68, 209, 1983.
23. **Li, W. K. W., Glover, H. E., and Morris, I.,** Physiology of carbon assimilation by *Oscillatoria thiebautii* in the Caribbean Sea, *Limnol. Oceanogr.*, 25, 447, 1980.
24. **Li, W. K. W. and Morris, I.,** Temperature adaptation in *Phaeodactylum tricornutum* Bohlin: photosynthetic rate compensation and capacity, *J. Exp. Mar. Biol. Ecol.*, 58, 135, 1982.
25. **Li, W. K. W. and Platt, T.,** Distribution of carbon among photosynthetic end-products in phytoplankton of the Eastern Canadian Arctic, *J. Phycol.*, 18, 466, 1982.
26. **Hitchcock, G. L.,** Diel variation in chlorophyll *a*, carbohydrate and protein content of the marine diatom *Skeletonema costatum*, *Mar. Biol.*, 57, 271, 1980.
27. **Li, W. K. W. and Harrison, W. G.,** Carbon flow into the end-products of photosynthesis in short and long incubations of a natural phytoplankton population, *Mar. Biol.*, 72, 175, 1982.
28. **Franks, F.,** Biophysics and biochemistry of low temperatures and freezing, in *Effects of Low Temperature on Biological Membranes*, Morris, G. J. and Clarke, A., Eds., Academic Press, New York, 1981, 3.
29. **Yentsch, C. S.,** Some aspects of the environmental physiology of marine phytoplankton: a second look, *Oceanogr. Mar. Biol. Ann. Rev.*, 12, 41, 1974.
30. **Morita, R. Y.,** Psychrophilic bacteria, *Bacteriol. Rev.*, 39, 144, 1975.
31. **McConville, M. J.,** unpublished data, 1985.
32. **Herbert, R. A.,** Low temperature adaptation in bacteria, in *Effects of Low Temperature on Biological Membranes*, Morris, G. J. and Clarke, A., Eds., Academic Press, New York, 1981, 41.
33. **Harder, W. and Veldkamp, H.,** Competition of marine psychrophilic bacteria at low temperatures, *Antonie van Leeuwenhoek; J. Microbiol. Serol.*, 37, 51, 1971.
34. **Hochachka, P. W. and Somero, G. N.,** *Strategies of Biochemical Adaptation*, W. B. Saunders, Philadelphia, 1973, 179.
35. **Horwitz, B. A. and Hettinger, D. R.,** Cold induced enzymatic adjustments in ectotherms and homeotherms, in *Comparative Mechanisms of Cold Adaptation*, Underwood, L. S., Tieszen, L. L., Callahan, A. B., and Folk, G. E., Eds., Academic Press, New York, 1979, 91.
36. **Somero, G. N. and Low, P. S.,** Temperature: a shaping force in protein evolution, *Biochem. Soc. Symp.*, 41, 33, 1976.
37. **Jorgensen, E. G.,** The adaptation of plankton algae II. Aspects of the temperature adaptation of *Skeletonema costatum*, *Physiol. Plant.*, 21, 423, 1968.
38. **Morris, I. and Farrell, K.,** Photosynthetic rates, gross patterns of carbon dioxide assimilation and activities of ribulose diphosphate carboxylase in marine algae grown at different temperatures, *Physiol. Plant.*, 25, 372, 1971.
39. **Rhee, G.-Y. and Gotham, I. J.,** The effects of environmental factors on phytoplankton growth: temperature and the interactions of temperature with nutrient limitation, *Limnol. Oceanogr.*, 26, 635, 1981.

40. **Richardson, K., Beardall, J., and Raven, J. A.,** Adaptation of unicellular algae to irradiance: an analysis of strategies, *New Phytol.*, 93, 157, 1983.
41. **Burch, M. D. and Marchant, H. J.,** Motility and microtubule stability of Antarctic algae at sub-zero temperatures, *Protoplasma*, 115, 240, 1983.
42. **Melkonian, M., Kroger, K. H., and Marquardt, K. G.,** Cell shape and microtubules in zoospores of the green alga *Chlorosarcinopsis gelatinosa* (Chlorosarcinales): effects of low temperature, *Protoplasma*, 104, 283, 1980.
43. **Shelanski, M. L., Gaskin, F., and Cantor, C. R.,** Microtubule assembly in the absence of added nucleotides, *Proc. Natl. Acad. Sci. U.S.A.*, 70, 765, 1973.
44. **Weisenburg, R. C., Borisy, G. C., and Taylor, E. W.,** The colchicine binding protein of mammalian brain and its relation to microtubules, *Biochemistry*, 7, 4466, 1968.
45. **Thomas, C. W.,** On the transfer of visible radiation through sea ice and snow, *J. Glaciol.*, 4, 481, 1963.
46. **Perovich, D. K. and Grenfell, T. C.,** Laboratory studies of the optical properties of young sea ice, *J. Glaciol.*, 27, 331, 1981.
47. **Jeffrey, S. W.,** Algal pigment systems, in *Primary Productivity in the Sea*, Falkowski, P. G., Ed., Plenum Press, New York, 1980, 33.
48. **Jeffrey, S. W.,** Properties of two spectrally different components in chlorophyll *c* preparations, *Biochim. Biophys. Acta*, 177, 456, 1969.
49. **Jeffrey, S. W.,** Preparation and some properties of crystalline chlorophyll c_1 and c_2 from marine algae, *Biochim. Biophys. Acta*, 279, 15, 1972.
50. **Meeks, J. C.,** Chlorophylls, in *Algal Physiology and Biochemistry*, Stewart, W. D. P., Ed., Blackwell Scientific, Oxford, 1974, 161.
51. **Govindjee, and Braun, B. Z.,** Light absorption, emission and photosynthesis, in *Algal Physiology and Biochemistry*, Stewart, W. D. P., Ed., Blackwell Scientific, Oxford, 1974, 346.
52. **Gillan, F. T.,** Lipids of aquatic ecosystems, Ph.D. thesis, University of Melbourne, Parkville, 1981.
53. **Bunt, J. S.,** Analysis of algal pigments by thin layer chromatography, *Nature (London)*, 203, 1261, 1964.
54. **Whitaker, T. M. and Richardson, M. G.,** Morphology and chemical composition of a natural population of an ice-associated Antarctic diatom *Navicula glaciei*, *J. Phycol.*, 16, 250, 1980.
55. **Goodwin, T. W.,** Carotenoids and biliproteins, in *Algal Physiology and Biochemistry*, Stewart, W. D. P., Ed., Blackwell Scientific, Oxford, 1974, 176.
56. **Goodwin, T. W.,** Isoprenoid distribution and biosynthesis in flagellates, in *Biochemistry and Physiology of Protozoa*, Vol. 1, Levandowsky, M. and Hutner, S. H., Eds., Academic Press, New York, 1979, 91.
57. **Gantt, E.,** Phycobiliproteins of Cryptophyceae, in *Biochemistry and Physiology of Protozoa*, Vol. 1, Levandowsky, M. and Hutner, S. H., Eds., Academic Press, New York, 1979, 121.
58. **Morgan, K. C. and Kalff, J.,** Effect of light and temperature interactions on growth of *Cryptomonas erosa* (Cryptophyceae), *J. Phycol.*, 15, 127, 1979.
59. **Priddle, J.,** The production ecology of benthic plants in some Antarctic lakes. II. Laboratory physiology studies, *J. Ecol.*, 68, 155, 1980.
60. **Platt, T., Harrison, W. G., Irwin, B., Horne, E. P., and Gallegos, C. L.,** Photosynthesis and photoadaptation of marine phytoplankton in the Arctic, *Deep-Sea Res.*, 29, 1159, 1982.
61. **Falkowski, P. G.,** Light-shade adaptation in marine phytoplankton, in *Primary Productivity in the Sea*, Falkowski, P. G., Ed., Plenum Press, New York, 1980, 99.
62. **Yoder, J. A.,** Effect of temperature on light limited growth and chemical composition of *Skeletonema costatum* (Bacillariophyceae), *J. Phycol.*, 15, 362, 1979.
63. **Verity, P. G.,** Effects of temperature, irradiance, and daylength on the marine diatom *Leptocylindrus danicus* Cleve. I. Photosynthesis and cellular respiration, *J. Exp. Mar. Biol. Ecol.*, 55, 79, 1981.
64. **Falkowski, P. G. and Owens, T. G.,** Light-shade adaptation: two strategies in marine phytoplankton, *Plant Physiol.*, 66, 592, 1980.
65. **Sullivan, C. W., Palmisano, A. C., Kottmeier, S., and Moe, R.,** Development of the sea ice microalgal community in McMurdo Sound, *Antarct. J. U.S.*, 17, 155, 1982.
66. **Jensen, A. and Sakshaug, E.,** Studies on the phytoplankton ecology of the Trondheimsfjord. II. Chloroplast pigments in relation to abundance and physiological state of the phytoplankton, *J. Exp. Mar. Biol. Ecol.*, 11, 137, 1973.
67. **Jeffrey, S. W.,** An improved thin-layer chromatographic technique for marine phytoplankton pigments, *Limnol. Oceanogr.*, 26, 191, 1981.
68. **Jeffrey, S. W.,** Profiles of photosynthetic pigments in the ocean using thin-layer chromatography, *Mar. Biol.*, 26, 101, 1974.
69. **Hoshiai, T.,** Seasonal change of ice communities in the sea ice near Syowa Station Antarctica, in *Polar Oceans*, Dunbar, M. J., Ed., The Arctic Institute of North America, Calgary, 1977, 307.
70. **Darley, W. M.,** Biochemical composition, in *The Biology of Diatoms*, Werner, D., Ed., Blackwell Scientific, Oxford, 1977, 198.

71. **Orcutt, D. M. and Patterson, G. W.,** Sterol, fatty acid and elemental composition of diatoms grown in chemically defined media, *Comp. Biochem. Physiol.*, 15B, 579, 1975.
72. **Chuecas, L. and Riley, J. P.,** Component fatty acids of the total lipids of some marine phytoplankton, *J. Mar. Biol. Assoc. U.K.*, 49, 97, 1969.
73. **Opute, F. I.,** Lipid and fatty acid composition of diatoms, *J. Exp. Bot.*, 25, 823, 1974.
74. **Volkman, J. K., Eglinton, G., and Corner, E. D. S.,** Sterols and fatty acids of the marine diatom *Biddulphia sinensis, Phytochemistry*, 19, 1809, 1980.
75. **Aaronson, S., Berner, T., and Dubinsky, Z.,** Microalgae as a source of chemicals and natural products, in *Algal Biomass Production and Use*, Shelef, G. and Soeder, C. J., Eds., Elsevier/North-Holland, Amsterdam, 1980, 575.
76. **Shifrin, N. S. and Chisholm, S. W.,** Phytoplankton lipids: interspecific differences and effects of nitrate, silicate and light-dark cycles, *J. Phycol.*, 17, 374, 1981.
77. **Iwamoto, H., Yonekawa, G., and Asai, T.,** Fat synthesis in unicellular algae. I. Culture conditions for fat accumulation in *Chlorella* cells, *Bull. Agric. Chem. Soc. Jpn.*, 19, 240, 1955.
78. **Fisher, N. S. and Schwarzenbach, R. P.,** Fatty acid dynamics in *Thalassiosira pseudonana* (Bacillariophyceae): implications for physiological ecology, *J. Phycol.*, 14, 143, 1978.
79. **Aaronson, S.,** Effect of incubation temperature on the macro-molecular and lipid content of the phytoflagellate *Ochromonas danica, J. Phycol.*, 9, 111, 1973.
80. **Smith, A. E. and Morris, I.,** Pathways of carbon assimilation in phytoplankton from the Antarctic Ocean, *Limnol. Oceanogr.*, 25, 865, 1980.
81. **Werner, D.,** Silicate metabolism, in *The Biology of Diatoms*, Werner, D., Ed., Blackwell Scientific, Oxford, 1977, 110.
82. **Bunt, J. S.,** personal communication, 1982.
83. **Grainger, E. H.,** The annual nutrient cycle in sea ice, in *Polar Oceans*, Dunbar, M. J., Ed., The Arctic Institute of North America, Calgary, 1977, 285.
84. **Cross, W. E.,** Under-ice biota at the Pond Inlet ice edge and in adjacent fast ice areas during spring, *Arctic*, 35, 13, 1982.
85. **Palmisano, A. C. and Sullivan, C. W.,** Physiology of sea ice diatoms. I. Response of three polar diatoms to a simulated summer-winter transition, *J. Phycol.*, 18, 489, 1982.
86. **Pringle, M. and Chapman, D.,** Biomembrane structure and effects of temperature, in *Effects of Low Temperature on Biological Membranes*, Morris, G. J. and Clarke, A., Eds., Academic Press, New York, 1981, 21.
87. **Alexandrov, V. Ya.,** *Cells, Molecules and Temperature: Conformational Flexibility of Macromolecules and Ecological Adaptation*, Springer-Verlag, Berlin, 1977, chap. 8. (Translated from Russian by V. N. Bernstan.)
88. **Morris, G. J., Coulson, G. E., Clarke, K. J., Grout, B. W. W., and Clarke, A.,** Freezing injury in Chlamydomonas: a synoptic approach, in *Effects of Low Temperatures on Biological Membranes*, Morris, G. J. and Clarke, A., Eds., Academic Press, New York, 1981, 285.
89. **Patterson, G. W.,** Effect of culture temperature on fatty acid composition of *Chlorella sorokiniana, Lipids*, 5, 597, 1970.
90. **Brown, C. M. and Minnikin, D. E.,** The effect of growth temperature on the fatty acid composition of some psychrophilic marine pseudomonads, *J. Gen. Microbiol.*, 75, IX, 1973.
91. **Morris, G. J. and Clarke, A.,** The cryo-preservation of *Chlorella*. IV. Accumulation of lipid as a protective factor, *Arch. Microbiol.*, 119, 153, 1978.
92. **Gillan, F. T., McFadden, G. I., Wetherbee, R., and Johns, R. B.,** Sterols and fatty acids of an Antarctic sea ice diatom *Stauroneis amphioxys, Phytochemistry*, 20, 1935, 1981.
93. **Coombs, J. and Volcani, B. E.,** Studies on the biochemistry and fine structure of silica shell formation in diatoms. Chemical changes in the wall of *Navicula pelliculosa* during its formation, *Planta*, 82, 280, 1968.
94. **El-Betouhim, T., Kahan, D., and Eckstein, B.,** Influence of temperature on the sterols of *Ochromonas danica, Comp. Biochem. Physiol.*, 58B, 243, 1977.
95. **Singh, J., De La Roche, I. A., and Siminovitch, D.,** Membrane augmentation in freezing tolerance of plant cells, *Nature (London)*, 257, 669, 1975.
96. **Beattie, A., Hirst, E. L., and Percival, E.,** Studies on the metabolism of the Chrysophyceae: comparative structural investigations on leucosin (chrysolaminarin) separated from diatoms and laminarins from the brown algae, *Biochem. J.*, 79, 531, 1961.
97. **Ford, C. W. and Percival, E.,** The carbohydrates of *Phaeodactylum tricornutum*. II. A sulphated glucuronomannan, *J. Chem. Soc.*, 1299, 7042, 1965.
98. **Paulsen, B. S. and Myklestad, S.,** Structural studies of the reserve glucan produced by the marine diatom *Skeletonema costatum* (Grev.). Cleve, *Carbohydr. Res.*, 62, 386, 1978.

99. **Archibald, A. R., Cunningham, W. L., Manners, D. J., and Stark, J. R.,** Studies on the metabolism of the Protozoa. X. The molecular structure of the reserve polysaccharides from *Ochromonas malhamensis* and *Peranema trichophorum*, *Biochem. J.*, 88, 444, 1963.
100. **Hibberd, D. J.,** Prymnesiophytes (Haptophyceae), in *Phytoflagellates*, Cox, E. R., Ed., Elsevier/North-Holland, Amsterdam, 1980, 273.
101. **Manners, D. J. and Sturgeon, R. J.,** Reserve carbohydrates of algae, fungi and lichens, in *Plant Carbohydrates, Vol. I*, Loewus, F. A. and Tanner, W., Eds., Springer-Verlag, Basel, 1982, chap. 12.
102. **Haug, A. and Myklestad, S.,** Polysaccharides of marine diatoms with special reference to *Chaetoceros* species, *Mar. Biol.*, 34, 217, 1976.
103. **Handa, N.,** Carbohydrate metabolism in the marine diatom *Skeletonema costatum*, *Mar. Biol.*, 4, 208, 1969.
104. **McConville, M. J., Mitchell, C., and Wetherbee, R.,** Patterns of carbon assimilation in a microalgal community from annual sea ice, East Antarctica, *Polar Biol.*, in press, 1985.
105. **Cohen, D. and Parnas, H.,** An optimal policy for the metabolism of storage materials in unicellular algae, *J. Theor. Biol.*, 56, 1, 1976.
106. **Myklestad, S., Djurhuus, R., and Mohus, A.,** Demonstration of *exo*-(β 1,3)-D-Glucanase activity in some planktonic diatoms, *J. Exp. Mar. Biol. Ecol.*, 56, 205, 1982.
107. **Schobert, B.,** The influence of water stress on the metabolism of diatoms. I. Osmotic resistance and proline accumulation in *Cyclotella meneghiniana*, *Z. Pflanzenphysiol.*, 74, 106, 1974.
108. **Myklestad. S. and Haug, A.,** Production of carbohydrates by the marine diatom *Chaetoceros affinis* var. *willei* (Gran) Hustedt. I. Effects of the concentration of nutrients in the culture medium, *J. Exp. Mar. Biol. Ecol.*, 9, 125, 1972.
109. **Painter, T. J.,** Algal polysaccharides, in *The Polysaccharides*, Vol. 2, Aspinall, G. O., Ed., Academic Press, New York, 1983, 195.
110. **Volcani, B. E.,** Cell wall formation in diatoms: morphogenesis and biochemistry, in *Silicon and Siliceous Structures in Biological Systems*, Simpson, T. L. and Volcani, B. E., Eds., Springer-Verlag, Basel, 1981, 157.
111. **Kates, M. and Volcani, B. E.,** Studies on the biochemistry and fine structure of silica shell formation in diatoms. Lipid components of the cell walls, *Z. Pflanzenphysiol.*, 60, 19, 1968.
112. **Hecky, R. E., Mopper, K., Kilham, P., and Degens, E. T.,** The amino acid and sugar composition of diatom cell walls, *Mar. Biol.*, 19, 323, 1973.
113. **Percival, E., Rahman, M. A., and Weigel, H.,** Chemistry of the polysaccharide of the diatom *Coscinodiscus nobilis*, *Phytochemistry*, 19, 809, 1980.
114. **Paasche, E.,** Silicon, in *The Physiological Ecology of Phytoplankton*, Morris, I., Ed., Blackwell Scientific, Oxford, 1980, 59.
115. **Roberts, K.,** Crystalline glycoprotein cell walls of algae: their structure, composition and assembly, *Philos. Trans. R. Soc. London, Ser. B*, 268, 129, 1974.
116. **Craigie, J. S., McLachlan, J., Ackman, R. G., and Tocher, C. S.,** Photosynthesis in algae. III. Distribution of soluble carbohydrates and dimethyl-β-propiothetin in marine unicellular Chlorophyceae and Prasinophyceae, *Can. J. Bot.*, 45, 1327, 1967.
117. **Bieleski, R. L.,** Sugar alcohols, in *Plant Carbohydrates I. Intracellular Carbohydrates*, Vol. 13A, Loewus, F. A. and Tanner, W., Eds., Springer-Verlag, Basel, 1982, chap. 12.
118. **Hellebust, J. A.,** Osmoregulation, *Annu. Rev. Plant Physiol.*, 27, 485, 1976.
119. **Eide, L. E. and Martin, S.,** The formation of brine drainage features in young sea ice, *J. Glaciol.*, 14, 137, 1975.
120. **Grant, W. S. and Horner, R.,** Growth responses to salinity variation in four Arctic ice diatoms, *J. Phycol.*, 12, 180, 1976.
121. **McFadden, G.,** personal communication, 1982.
122. **Brown, A. D.,** Microbial water stress, *Bacteriol. Rev.*, 40, 803, 1976.
123. **Liu, M. S. and Hellebust, J. A.,** Regulation of proline metabolism in the marine centric diatom, *Cyclotella cryptica*, *Can. J. Bot.*, 54, 949, 1976.
124. **Schobert, B.,** The influence of water stress on the metabolism of diatoms. III. The effect of different nitrogen sources on proline accumulation, *Z. Pflanzenphysiol.*, 85, 463, 1977.
125. **Borowitzka, L. J. and Brown, A. D.,** The salt relations of marine and halophilic species of the unicellular green alga, *Dunaliella*. The role of glycerol as a compatible solute, *Arch. Microbiol.*, 96, 37, 1974.
126. **Paul, J. S.,** Osmoregulation in the marine diatom *Cylindrotheca fusiformis*, *J. Phycol.*, 15, 280, 1979.
127. **Greenway, H. and Setter, T. L.,** Accumulation of proline and sucrose during the first hours after transfer of *Chlorella emersonii* to high NaCl, *Aust. J. Plant Physiol.*, 6, 69, 1979.
128. **Hellebust, J. A.,** Effects of salinity on photosynthesis and mannitol synthesis in the green flagellate *Platymonas suecica*, *Can. J. Bot.*, 54, 1735, 1976.
129. **Ben-Amotz, A. and Grunwald, T.,** Osmoregulation in the halotolerant alga *Asteromonas gracilis*, *Plant Physiol.*, 67, 613, 1981.

130. **Kauss, H.,** Turnover of galactosylglycerol and osmotic balance in *Ochromonas*, *Plant Physiol.*, 52, 613, 1973.
131. **Allan, G. C., Lewin, J., and Johnson, P. G.,** Marine polymers. IV. Diatom polysaccharides, *Bot. Mar.*, 15, 102, 1972.
132. **Franks, F.,** Solvent interactions and the solution behaviour of carbohydrates, in *Polysaccharides in Food*, Blanshard, J. M. V. and Mitchell, J. R., Eds., Butterworths, London, 1979, 33.
133. **McConville, M. J.,** in preparation, 1985.
134. **Paulsen, B. S., Haug, A., and Larsen, B.,** Structural studies of a carbohydrate-containing polymer present in the mucilage tubes of the diatom *Berkeleya rutilans* (Trent.) Grun., *Carbohydr. Res.*, 66, 103, 1978.
135. **Smestad, B., Haug, A., and Myklestad, S.,** Production of carbohydrate by the marine diatom *Chaetoceros affinis* var. *willei* (Gran) Hustedt. III. Structural studies of the extracellular polysaccharide, *Acta Chem. Scand. Ser. B*, 28, 662, 1974.
136. **Guillard, R. R. L. and Hellebust, J. A.,** Growth and the production of extracellular substances by two strains of *Phaeocystis pouchetii*, *J. Phycol.*, 7, 330, 1971.
137. **Daniel, G. F., White, R., and Chamberlain, A. H. L.,** Attachment of freshwater and marine diatoms, *Br. Phycol. J.*, 17, 231, 1982.
138. **Fletcher, M.,** The attachment of bacteria to surfaces in aquatic environments, in *Adhesion of Microorganisms to Surfaces*, Ellwood, D. C., Melling, J., and Rutter, P. R., Eds., Academic Press, New York, 1979, 87.
139. **Parker, B. C., Simmons, G. M., Jr., Wharton, R. A., Jr., Seaburg, K. G., and Love, F. G.,** Removal of organic and inorganic matter from Antarctic lakes by aerial escape of blue green algal mats, *J. Phycol.*, 18, 72, 1982.
140. **Huntsman, S. and Sloneker, J. H.,** An exocellular polysaccharide from the diatom *Gomphonema olivaceum*, *J. Phycol.*, 7, 261, 1971.
141. **Drum, R. W.,** Light and electron microscope observations on the tube dwelling diatom *Amphipleura rutilans* (Trentepohl) Cleve, *J. Phycol.*, 5, 21, 1969.
142. **Olien, C. R. and Smith, M. N.,** Protective systems that have evolved in plants, in *Analysis and Improvement of Plant Cold Hardiness*, Olien, C. R. and Smith, M. N., Eds., CRC Press Inc., Boca Raton, Fla., 1981, 61.
143. **Olien, C. R.,** Interference of cereal polymers and related compounds with freezing, *Cryobiology*, 2, 47, 1965.
144. **Olien, C. R. and Smith, M. N.,** Ice adhesions in relation to freeze stress, *Plant Physiol.*, 60, 499, 1977.
145. **Hargens, A. R. and Shabica, S. V.,** Protection against lethal temperatures by mucus in an Antarctic limpet, *Cryobiology*, 10, 331, 1973.
146. **DeVries, A. L.,** Glycoproteins as biological antifreeze agents in Antarctic fishes, *Sciences*, 172, 1152, 1971.
147. **Feeney, R. E. and Yeh, Y.,** Antifreeze proteins from fish bloods, *Adv. Protein Chem.*, 32, 191, 1978.
148. **Horne, A. J., Fogg, G. E., and Eagle, D. J.,** Studies in-situ of the primary production of an area of inshore Antarctic Sea, *J. Mar. Biol. Assoc. U. K.*, 49, 393, 1969.
149. **Mague, T. H., Friberg, E., Hughes, D. J., and Morris, I.,** Extracellular release of carbon by marine phytoplankton, a physiological approach, *Limnol. Oceanogr.*, 25, 262, 1980.
150. **Ignatiades, L.,** Studies on the factors affecting the release of organic matter by *Skeletonema costatum* (Grev.) Cleve in field conditions, *J. Mar. Biol. Assoc. U. K.*, 53, 923, 1973.
151. **Samuel, S., Shah, N. M., and Fogg, G. E.,** Liberation of extracellular products of photosynthesis by tropical phytoplankton, *J. Mar. Biol. Assoc. U. K.*, 51, 793, 1971.
152. **Ittekkotl, V.,** Variations of dissolved organic matter during a plankton bloom: qualitative aspects based on sugar and amino acid analysis, *Mar. Chem.*, 11, 143, 1982.
153. **Fogg, G. E.,** The ecological significance of extracellular products of phytoplankton photosynthesis, *Bot. Mar.*, 26, 3, 1983.
154. **Sieburth, J. McN.,** Acrylic acid, an "antibiotic" principle in *Phaeocystis* blooms in Antarctic waters, *Science*, 132, 676, 1960.
155. **Verity, P. G.,** Effects of temperature, irradiance and daylength on the marine diatom *Leptocylindrus danicus* Cleve. II. Excretion, *J. Exp. Mar. Biol. Ecol.*, 55, 159, 1981.
156. **Wangersky, P. J.,** Production of dissolved organic matter, in *Marine Ecology*, Vol. 4, Kinne, O., Ed., John Wiley & Sons, New York, 1978, 115.

[illegible]

Chapter 5

GROWTH, METABOLISM, AND DARK SURVIVAL IN SEA ICE MICROALGAE

Anna C. Palmisano and Cornelius W. Sullivan

TABLE OF CONTENTS

I. INTRODUCTION

Sea ice microbial communities (SIMCO), three types, have been distinguished based on their location within the ice column.[1,2] Surface ice communities have been found in melt pools, infiltration ice, and at the ice edge in the Antarctic,[3-5] and in the Arctic.[6,7] Interior ice communities at intermediate depths were reported within both annual sea ice[1,8] and in multi-year[9] ice in the Antarctic, although it is unclear whether these cells are actively metabolizing while in the ice. Bottom ice communities are located both in the diffuse lower layers of hard congelation ice in the Arctic and Antarctic,[6,10-13] as well as in the unconsolidated ice platelet layer in the Antarctic.[14-16] It may be more appropriate to consider the congelation and platelet communities as separate because the physicochemical nature of these habitats and the species composition[17] may be quite different. As a habitat for marine microorganisms, sea ice provides a unique set of physical and chemical conditions to which physiological processes of associated biota have adapted. Physical parameters that may affect growth, metabolism, and survival of microalgae include light (intensity, spectral composition, photoperiod, and seasonality), temperature, nutrients, and salinity.

Despite low temperatures and low, episodic irradiance in the ice habitat, microalgae attain relatively high annual standing crops ranging from 4.6 to 30 mg chlorophyll *a* m^{-2} in the Arctic,[6,11,18-23] and 1.37 to 309 mg chlorophyll *a* m^{-2} in the Antarctic.[1,4,5,8,15,24] High standing crops may partly be a function of reduced grazing due to the inaccessibility of ice microalgae to pelagic herbivores. The success of microalgae in colonizing and growing in sea ice also reflects metabolic adaptations to this environment. What kinds of physiological adaptations allow the proliferation of microalgae in sea ice? Although this question remains largely unanswered, we review here much of the work published to date on growth, metabolism, and dark survival in sea ice microalgae.

II. GROWTH

Relatively few estimates have been made of growth rates in natural populations or laboratory cultures of sea ice microalgae. *In situ* estimates of ice algal growth rates have been made by determining algal cell numbers in ice samples at two different time points during a spring bloom. Increases in algal numbers with time reported in sea ice microbial communities[6,15] may actually be the result of algal accumulation rather than cell division. Growth of the sea ice itself can lead to entrapment of algae from the water column[25,26] resulting in an overestimate of growth rate. Conversely, loss of algae due to ice ablation might lead to an underestimate of growth rate. This problem can be avoided either by sampling during a period of constant ice thickness or by sampling repeatedly from a given depth in the ice that is unaffected by addition and loss of algae at the ice-seawater interface. Rates of ice growth are difficult to predict and may vary with season, year, or location. Losses of algae due to grazing at the ice-seawater interface by ice-associated herbivores or cryopelagic grazers[27] may also lead to an underestimate of *in situ* growth rate.

With these limitations in mind, we consider the *in situ* growth rates estimated by Bunt et al.[15,28] Algal cells associated with platelet ice and interstitial water in McMurdo Sound, Antarctica, were sampled with scuba techniques. Bunt[28] found that the algal populations during the 1962-63 summer season increased from 2.3×10^6 to 40.6×10^6 cells ℓ^{-1} over an 18-day period. Assuming the increase was solely the result of cell division, this represents a specific growth rate (μ) of 0.12 day^{-1} or a doubling time of 6 days:

$$\mu = \frac{\ln N - \ln N_o}{t - t_o} \qquad (1)$$

where N and N_o are the numbers of cells at time t and t_o, respectively. Using the data of

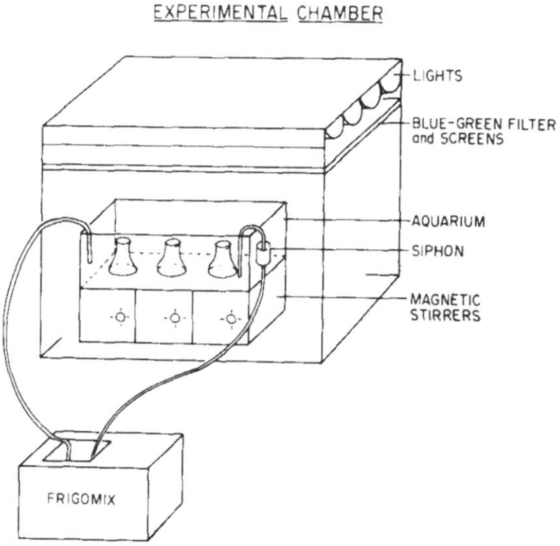

FIGURE 1. Incubator for the growth of sea ice diatoms. See text for explanation. (From Palmisano, A. C. and Sullivan, C. W., *J. Phycol.*, 18, 489, 1982. with permission.)

Bunt and Lee,[15] a range of specific growth rates of 0.08 to 0.21 day^{-1} (doubling times of 3 to 9 days) can be calculated for the spring bloom in McMurdo Sound in 1967. The maximum growth rate found by Holm-Hansen et al.[29] for Antarctic phytoplankton was 0.33 day^{-1}; in most samples, maximum rates were closer to 0.1 day^{-1}. Using data from algae grown over a wide range of temperatures, Eppley[30] described maximum growth rate as a function of growth temperature by the empirically derived equation $\log \mu = 0.0275T - 0.070$, where T is temperature in degrees Celsius. The lowest temperature for which growth data were included was +2°C. An extrapolation of the curve to −1.8°C predicts a maximum growth rate of 0.76 day^{-1},[29] a value four to ten times higher than most *in situ* estimates of ice algal growth rates.[15]

Sea ice diatoms may be cultured in seawater-based media contained in a low temperature ethylene-glycol water bath fitted with fluorescent lights (Figure 1). Palmisano and Sullivan[2] mimicked the spectral quality of under-ice light using blue-green Plexiglas filters. Plastic screening was used to adjust light intensity available to cultures, which were gently stirred. They found an average doubling time of 2.5 days ($\mu = 0.28\ day^{-1}$) for three sea ice diatoms including two clones of *Nitzschia cylindrus* (Grunow) Hasle. Cells were grown at 46 $\mu E\ m^{-2}\ sec^{-1}$, 0°C and 20‰ in modified f/2 medium to mimic summer conditions under annual sea ice. This estimate of growth rate should not be considered maximal, because growth was light-limited at this irradiance. By growing a culture of *Fragilaria sublinearis* Van Heurck at 7°C after adaptation to higher light intensities (about 220 $\mu E\ m^{-1}\ sec^{-1}$), Bunt[31] obtained growth rates equivalent to 1.16 day^{-1}.

Morita[32] defined psychrophiles as organisms having an optimal temperature for growth at about 15°C or lower. Bunt[31] demonstrated that *F. sublinearis* was a truly psychrophilic diatom; it had a temperature optimum for growth of about 7°C. Although optimum temperature for growth of *F. sublinearis* does not coincide with the ambient sea ice temperature of −1.8°C or less, low temperature adaptation is evident. Adaptations to overcome the depression of metabolic rate by low temperatures may involve structural modifications of enzymes and biological membranes.[33,34]

Water temperatures under annual sea ice in McMurdo Sound are relatively constant with a yearly mean and range of −1.81 ± 0.09°C.[35] Although low temperatures in polar waters limit algal growth, temperature variation cannot explain the seasonal changes in ice algal population densities nor the patchiness in the spatial distribution and abundance of the standing crop of ice algae in McMurdo Sound.[24]

Steep temperature gradients exist across sea ice: bottom ice temperatures resemble underlying seawater, while surface ice approaches air temperatures which may be as low as −50°C. Ice temperatures may also indirectly affect algal growth by determining the volume and concentrations of brine inclusions.[36] The brine bathes the ice microalgal cells and provides an aqueous environment necessary for growth. Little is known of the freezing tolerance of sea ice microalgae. Bunt[37] found that some species were viable after storage at −10°C for 1 week, but storage at −20°C was uniformly fatal. The effect of temperature on the growth and mortality of microalgae in sea ice will determine, in part, the vertical distribution of viable algae in the ice column.

Light has been proposed as the primary factor influencing growth of the bottom ice algal community.[2,23,37,38] Under-ice irradiance, attenuated in its passage through snow cover, sea ice, and the ice algal community, is typically < 1% of the surface down-welling irradiance.[39,40] Sullivan and Palmisano[13] reported under-ice irradiance between 0.3 to 13.0 μE m^{-2} sec^{-1} during the 1980 ice algal bloom. Similarly, Horner and Schrader[23] found that irradiance of 5 to 18 μE $m^{-2} sec^{-1}$ reached the top of the algal layer in a nearshore region of the Beaufort Sea. Light appears to be a critical environmental variable affecting bottom ice microalgal growth and metabolism not only during the spring bloom, but throughout the year. Bunt and Lee[15] found the standing crop size of ice algae was a function of snow cover. A site with surface snow yielded 1076 mg C m^{-2}, while a snow-free site yielded 520 mg C m^{-2} in McMurdo Sound in 1967, but differences in ice thickness at these two sites makes interpretation difficult. Palmisano et al.[41] and Sullivan et al.[40] conducted a light perturbation experiment in which snow cover over 1.5 m of annual sea ice was manipulated to reduce under-ice irradiance. At the end of the growth season in McMurdo Sound in 1981, a standing crop of 100 mg chlorophyll *a* m^{-2} was found under 1.5 cm of natural snow cover in the control quadrat, while only 5.5 mg chlorophyll *a* m^{-2} was found under 100 cm of snow in the experimental quadrat. Underwater observations by scuba divers[42] suggest a negative correlation between algal standing crop and surface snow cover.

Nutrients in sea ice are generally considered to be high.[12,15,21] Inorganic and organic nutrients may be supplied to ice microalgae by bacterial nutrient regeneration,[43] penetration of nutrients from underlying sea water, desaltation during ice formation, or their combined effects.[11] Antarctic waters are comparatively rich in nutrients, presumably due to upwelling, with 40 to 90 μM silicon, 10 to 20 μM nitrate, and 1 to 2 μM phosphate.[29] In sea ice, nutrients may become concentrated in brine pockets as solutes are excluded from the ice matrix.[44] It is possible that dense growth of microalgae during bloom conditions may create localized nutrient depletion in brine pockets resulting in nutrient limitation of growth. Moreover, if ice algae have a low affinity for essential nutrients, they may be nutrient-limited despite high nutrient concentrations. In cultures of Antarctic phytoplankton, Jacques[45] reported a low affinity for silicon with transport constants (Ks) between 12 and 22 μg-at ℓ^{-1}, so Antarctic phytoplankton may be silicon-limited. Some of the phytoplankton he tested, such as *Nitzschia kerguelensis* (O'Meara) Hasle, are also known to occur in the ice. To test for nutrient limitation in sea ice algae, the transport characteristics (Ks and V_{max}) of the algae for major nutrients (silicate, nitrate, ammonia, and phosphate) and the availability of these nutrients in the liquid brine surrounding the algae need to be determined.

Sea ice microalgae may be exposed to large changes in salinity. As sea ice is formed, brine pockets become trapped between the interstices of ice crystals.[36] Sullivan[9] found the freezing point depression of interstitial brine from 1-year ice collected in October at a depth of 1 meter to be −15.2°C, which corresponds to a salinity approximately five times more concentrated than seawater. Trapped brine migrates through interconnecting drainage channels in sea ice, leaving the upper layer with a reduced salinity. As snow and the upper layer of ice melt in early summer, low salinity meltwater percolates through the sea ice and may form a freshwater lens at the ice-seawater interface. Although Grant and Horner[46] found that growth rates of four sea ice diatoms were relatively constant over salinities ranging from 10 to 50‰, local brine pockets may reach salinities low enough to inhibit algal growth in the summer.

The environment of the surface ice algal communities in melt pools and at the ice edge is quite different from the bottom ice community. While irradiance is high, temperature and nutrient availability may fluctuate. Algae growing in surface melt pools may be subjected to repeated freezing and thawing of their growth milieu. At an ice edge zone in the eastern Bering Sea, a bloom was initiated as the water column stabilized and light penetration increased.[47] Nutrient distributions suggested that depletion of nitrate was the primary factor limiting the spring algal bloom at the ice edge.

One environmental factor influencing algal growth in bottom, interior, or surface ice communities that is rarely considered is the ice substrate itself. Some ice algae are chain-forming pennate diatoms that are found attached to ice crystals. This attachment confers the simple advantage that ice algae remain where more light is available. The nature of attachment of algae to ice crystals is not known, but a special structural adaptation or exopolymeric substance[48] may be involved.

III. PHOTOSYNTHESIS

Microalgae growing in the bottom of annual sea ice have a remarkable ability to photosynthesize at light levels commonly below 1% of the surface down-welling irradiance. Estimates of primary production in ice have been hindered by the difficulties inherent in quantitative sampling of bottom ice and *in situ* incubations under sea ice. In Antarctic sea ice studies by Bunt et al.[15,28] scuba divers collected samples by forcing sample bottles through the ice matrix trapping crystals and interstitial water. Ice crystals represented about 20% of the volume of the sample. Primary productivity was determined in the laboratory on interstitial water alone, a procedure that would grossly underestimate total productivity because most of the ice algae are attached to ice crystals. Bunt and Lee[15] estimated a peak fixation rate of 3.82 mg C m^{-3} hr^{-1} for interstitial water or a chlorophyll *a* specific photosynthesis of 0.4 mg C mg chl' a^{-1} hr^{-1} at 1800 μW cm^{-2} (about 100 μE m^{-2} sec^{-1}) and −1.8°C. Palmisano and Sullivan[24] also used scuba divers to collect the loose ice matrix at the ice-sea water interface. They incubated the entire sample in a chamber designed to mimic both the temperature and the quantity and quality of light available to bottom ice algae (Figure 1). Under-ice light was measured with a QSI underwater scalar irradiance meter (Biospherical Instruments, La Jolla, Calif.). The chamber light intensity was then adjusted to match the *in situ* irradiance. They estimated primary productivity of ice algae to be 0.31 mg C mg chl a^{-1} hr^{-1} at 13 μE m^{-2} sec^{-1}. Microautoradiographs showed that pennate diatoms forming stellate colonies (Figure 2A) and chains of *Amphiprora* sp. (Figure 2B), the dominant species in this community, were heavily labeled with $H^{14}CO_3$. Rates of primary production in this study may have been influenced by slight changes in salinity resulting from the melting of a small amount of ice crystals during the 24 hr incubation at −2°C. Schrader et al.[49] have designed a combined sampler-incubation chamber for *in situ* measurements of primary production in sea ice. Although successfully employed in their Arctic studies, such samplers may be limited by problems of nonuniform distribution of isotope in the sample and possible isotope loss. Their use is also dependent on the nature of the ice-seawater interface. We found that this particular sampler was not satisfactory in the platelet ice layer in Antarctica. Using *in situ* incubation chambers, Horner and Schrader[23] obtained primary productivity estimates up to 2.64 mg C m^{-3} hr^{-1} during peak bloom conditions in ice in the Beaufort Sea. Using their data from Stefansson Sound in May, the mean chlorophyll *a* specific photosynthesis was 0.50 mg C mg chl^{-1} *a* hr^{-1}.

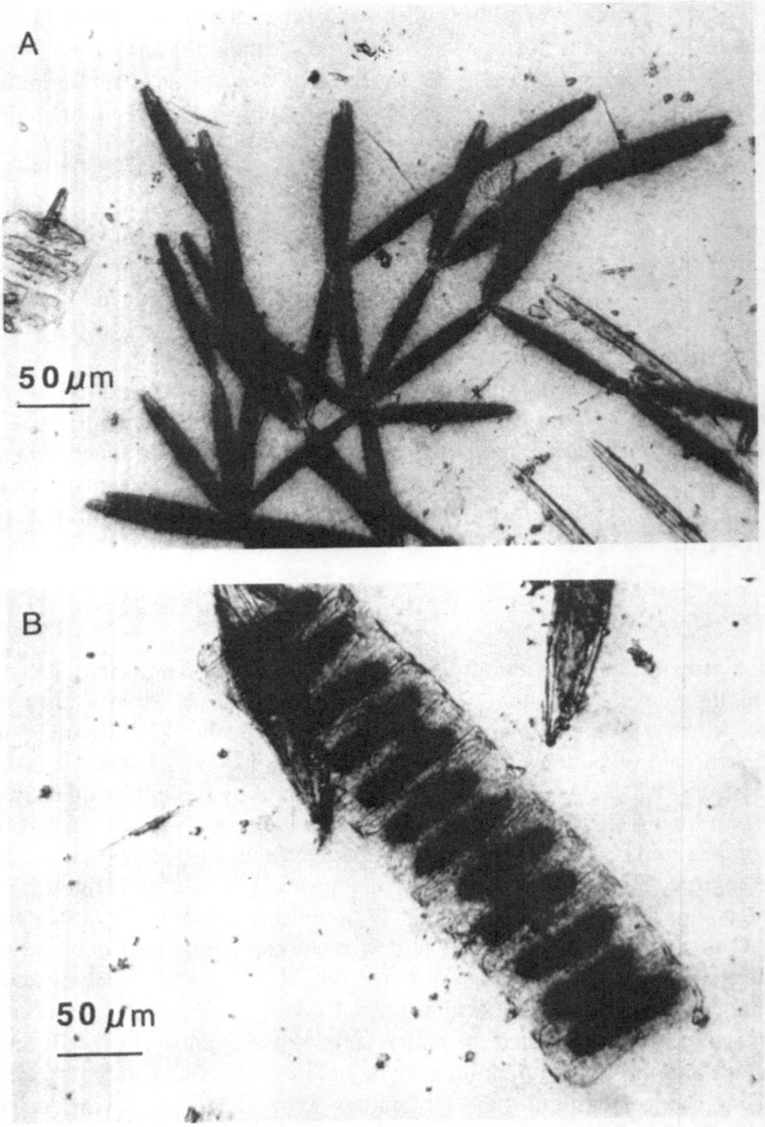

FIGURE 2. Microautoradiographs of sea ice diatoms labeled with $H^{14}CO_3$ under simulated *in situ* incubation conditions. (A) *Nitzschia stellata* Manguin; (B) *Amphiprora* sp.

Comparisons of primary productivity estimates in ice algae between studies pose some difficult problems. There is a lack of consistency both in collection methods and incubation procedures. This is partly the result of variation in the types of sea ice that may occur in the Arctic and Antarctic. A unit that can efficiently sample both congelation and platelet ice has not been designed. Problems also exist in the expression of primary production in different studies. Some investigators have chosen to normalize data to chlorophyll *a* because of the patchiness in biomass that occurs in the sea ice community. Some report productivity per unit volume meltwater in samples (often collected at an unspecified depth in the ice column). Others integrate productivity over the depth of the ice column and report production on an areal basis; this allows comparisons of ice standing crops to other ecosystems. A comparison of productivity in sea ice to productivity in other marine ecosystems is difficult to make because so few assimilation numbers (photosynthesis per unit chlorophyll *a* at light saturation) have been determined for natural communities or clonal cultures.

The ice algal community is generally considered to be "shade adapted." Bunt[50] found that photosynthesis by natural populations of ice microalgae in McMurdo Sound was light-saturated at 100 fc (about 22 μE m^{-2} sec^{-1}) and photoinhibited at 1100 fc (about 240 μE m^{-2} sec^{-1}). Assimilation numbers for the ice community were less than 0.15 mg C mg chl a^{-1} hr^{-1}. By comparison, the under-ice plankton dominated by *Phaeocystis* was light-saturated at about 300 fc. Bunt[31] later confirmed his observations of shade adaptation in laboratory studies with the ice diatom *Fragilaria sublinearis*. Cells grown at −2°C and 13 fc resulted in an assimilation number of only 0.37. Cultivating *F. sublinearis* at 5.0°C and 1000 fc increased the assimilation number to 1.2. Preliminary studies by Palmisano and Sullivan[51] indicated that the photosynthetic capacity (P_{max}) for natural populations of sea ice algae in McMurdo Sound was only 0.5 mg C mg chl a^{-1} hr^{-1}, a value less than one fourth that of temperate phytoplankton. Sea ice microalgae had an I_k (irradiance at which photosynthesis is light-saturated) of only 19 μE m^{-2} sec^{-1}, a value which compares with the lowest I_k's reported for microalgae.[52-54] The mean photosynthetic efficiency (the initial slope of the P(I) curve or α) was 0.04 mg C mg chl a^{-1} hr^{-1} (μE m^{-2} sec^{-1})$^{-1}$. These data confirm Bunt's[50] observations that ice algae are very shade adapted. Sea ice microalgae could serve as a model for the study of photoadaptation in nature.

Little attention has been given to two important aspects of light affecting primary production in bottom ice algae: photoperiod and spectral composition of under-ice light. Littlepage[35] concluded from insufficient data that under-ice irradiance failed to show a diel periodicity in the summer months. Sullivan et al.,[40] noted a pronounced diel change in under-ice irradiance during the spring bloom, ranging from 2.9 μE m^{-2} sec^{-1} at 1200 hr to 0.5 μE m^{-2} sec^{-1} at 2300 hr on November 14, 1981, despite continuous 24 hr surface downwelling irradiance of 1500 μE m^{-2} sec^{-1}. Diel rhythms in photosynthesis may exist in sea ice diatoms similar to those reported in temperate phytoplankters.[55] Unfortunately, most investigators have not taken this into account in their incubation procedures either in the laboratory or *in situ*. Short term *in situ* incubations during the morning or afternoon hours may greatly overestimate daily productivity if photosynthetic rate is assumed to be constant throughout 24 hr.

Light under sea ice, falls in the blue green region. Maykut and Grenfell[39] found the maximum transmission of light through Arctic sea ice was between 400 to 550 nm, regardless of surface conditions, ice thickness, and age. Spectral composition of light has been shown to affect photosynthesis in diatoms. Jeffrey and Vesk[56] demonstrated an increase in chlorophyll concentration and chloroplast number in a temperate marine diatom in blue-green light. Moreover, blue-green light appeared to enhance carbon fixation relative to cultures grown in white light. Because of the difficulty of performing *in situ* incubations under sea ice, estimates of primary production in ice algae are often done in the laboratory. However, laboratory incubations often fail to mimic the spectral quality of light under sea ice.

Microalgae undergoing shade adaptation typically increase their cellular pigment concentrations.[57-59] Cellular chlorophyll *a* increased logarithmically in the sea ice diatom *Fragilaria sublinearis* grown at −2°C when light was decreased from 300 to 3 fc with decreasing light intensity.[31] Small changes in temperature resulted in large changes in carotenoid: chlorophyll *a* ratios that varied among four sea ice diatom cultures, possibly due to separate temperature optima in the synthesis of xanthophyll precursors.[31]

Fogg[38] suggested that a low compensation intensity is the principal physiological attribute enabling algae to grow in and under ice. Bunt[31,50] estimated a compensation intensity of 2.5 fc (about 0.5 μE m^{-2} sec^{-1}) for sea ice microalgae in the platelet layer of McMurdo Sound sea ice and 3 to 6 fc for laboratory cultures. In an *in situ* experiment, Palmisano et al.[41] found that microalgae in bottom sea ice were clearly capable of net photosynthesis at irradiance as low as 2.5 μE m^{-2} sec^{-1}. Phytoplankton in the Southern Ocean were able to photosynthesize at depths corresponding to 0.1% surface downwelling irradiance.[29]

While bottom ice algae are considered "shade adapted," microalgae in surface melt pools and infiltration ice are considered "sun adapted". Burkholder and Mandelli[4] found that surface microalgal communities had a maximum photosynthetic rate at 18 klux (about 400 μE m^{-2} sec^{-1}). They determined an assimilation number of 2.6 mg C mg chl a^{-1} hr^{-1} for this community. Algae dominating the brash ice at the highly productive ice edge zones appear to be sun adapted. At the Bering Sea ice edge at the time of ice breakup, Alexander and Chapman[7] found pockets of dense algal communities associated with brash ice. Photosynthesis by this community was greatest at surface light intensity (about 1500 μE m^{-2} sec^{-1}) and showed no photoinhibition.

Factors other than light may affect primary production in sea ice including temperatures,[50,60-62] nutrient status,[63] cell size,[64,65] and salinity.[50] Preliminary experiments by Bunt[50] showed that net carbon fixation by natural populations of sea ice microalgae was markedly temperature dependent, particularly between −1.5 and +5°C. A Q_{10} of about 20 can be calculated for Bunt's data in this temperature range. Photosynthesis by ice microalgae was optimal between 4 and 7°C and inhibited by temperatures greater than 15°C. Neori and Holm-Hansen[62] found that temperatures above 7°C caused a decrease in photosynthetic rate in Antarctic phytoplankton collected from seawater with ambient temperatures of −0.8 to +1.0°C. Although net carbon fixation by polar microalgae appears to be somewhat adapted to low temperatures, photosynthesis was not optimal at temperatures as low as ambient (−1.8°C or lower). These observations conform to a generalization made by Li[61] that photosynthetic temperature optima are typically higher than the ambient growth temperature of microalgae.

A preliminary study by Bunt[50] on the effect of salinity on photosynthesis yielded some surprising results. Photosynthesis by ice microalgae was strongly affected by salinity with an optimum at only 7.5 to 10‰. Under-ice phytoplankton dominated by *Phaeocystis* reached a maximum photosynthetic activity at only 15‰. Bunt's samples were collected in late December to early January, a period when increasing air temperatures melt snow and surface ice.

Meltwater percolates through the ice forming a low salinity lens at the ice-seawater interface. Bottom ice diatoms sampled by Bunt may have been preconditioned by low salinity meltwater and may not reflect the physiology typical of the spring bloom of ice diatoms in November. Ice algae are exposed to both high salinities resulting from brine exclusion during ice formation and low salinities as snow and ice meltwater percolate through the algal layer in summer. Such changes in salinity could have a profound effect not only on ice algal photosynthetic rate, but on cell viability. Individual species may respond in different ways depending on whether they are euryhaline or stenohaline.

Estimates of dissolved organic carbon (DOC) released by phytoplankton range from less than 5%[66] to 70%[67] of carbon fixed. Sharp[68] concluded that many of the higher values were overestimates due to cell breakage during filtration. More recent estimates of phytoplankton exudation are primarily in the range of 8 to 20% of total carbon fixed.[69-71] Preliminary estimates of exudation in sea ice microalgae range from 7%[72] to 33%.[73] Algal exudation contributes to the DOC available to microheterotrophs in the ice including bacteria, microflagellates, and diatoms.

IV. HETEROTROPHY

Diatoms show a nutritional diversity. They may be obligate autotrophs, facultative heterotrophs (mixotrophs), or obligate heterotrophs. Facultative heterotrophy is not uncommon in pennate diatom species, particularly those isolated from low light, organic rich environments such as salt marshes, and benthic sediments.[74,75] Substrates taken up by diatoms include sugars, amino acids, and organic acids.[76] Light may influence both the uptake and utilization of organic substrates by diatoms. Uptake of glucose by *Cyclotella cryptica* Reimann, Lewin, and Guillard, for example, was greatly enhanced in the dark.[77] Cooksey[74] found that metabolism of organic acids by the benthic diatom *Cocconeis diminuta* Pantocsek was affected by light conditions. Lipid was the major end product in the light, while CO_2 was the major product in the dark.

Heterotrophic uptake of organic substrates may play one or more roles in the nutrition of ice diatoms. First, it may provide carbon and energy for maintenance metabolism or growth during the winter months. To grow heterotrophically, cells must be able to utilize organic substrates both in respiratory pathways and in biosynthetic pathways necessary for cell division. Second, heterotrophy may supplement carbon and energy provided by photosynthesis during the spring bloom when light is attenuated by passage through snow and ice and by self-shading in dense ice algal populations. The possibility of algae under ice employing heterotrophic metabolism for winter survival was suggested by Rodhe[78] for subarctic lake phytoplankton and by Wilce[79] for Arctic seaweeds. Allen[80] thought it probable that Arctic sea ice diatoms survived the dark polar winter by heterotrophic metabolism.

The first experimental test of heterotrophy in natural populations of ice microalgae was performed by Horner and Alexander[12] in the Arctic. Ice core samples from the Chukchi Sea near Point Barrow, Alaska, were slowly melted, maintaining an ice-seawater environment until incubation. Melted samples were incubated with ^{14}C-glycine, ^{14}C-glucose, and ^{3}H-acetate at 4°C for 6 hr in the dark. Assimilation of these substrates was slow. The turnover time of glycine was 200 hr uncorrected for respiration. They concluded that heterotrophic metabolism by ice algae was negligible and that assimilation of added substrate was almost exclusively by bacteria. Their study was conducted in March when increased light levels may have already triggered a switch in sea ice microalgae to photosynthetic nutrition from heterotrophy. Horner and Alexander's[12] study was also limited in that the kinds and concentrations of natural organic substrates in their samples were not determined.

Bunt and Lee[81] found variable growth responses in four sea ice microalgae incubated at 861 to 1076 lux (about 20 $\mu E\ m^{-2}\ sec^{-1}$) with a variety of organic substrates. While growth of a chrysomonad flagellate was generally stimulated by organic substrates, *Chaetoceros fragilis* Meunier was often inhibited. Although their data suggested that organic substrates may, in some cases, influence growth in the light, dark incubation failed to suggest that heterotrophy was important in dark survival.

In a laboratory study, Palmisano and Sullivan[2] found that uptake of ^{14}C-glucose by three sea ice diatoms was greatly enhanced by dark incubation. Uptake of 1 μM glucose increased 60-fold in an isolate of *Nitzschia cylindrus* that had been incubated in the dark at -2°C for 6 days. Dark uptake of glucose provided less than 1% of the carbon needed for cell division, although it may be sufficient for maintenance metabolism.

In general, transport affinities of diatoms for organic substrates are low with Km's ranging from 0.002 to 1.0 mM.[76] The uptake of organic substrates by sea ice diatoms will depend to a large extent on the concentrations available in the interstitial brine. Recent data indicate that free amino acid concentrations in under-ice seawater are much higher than previously considered. Using HPLC, Manahan et al.[82] determined dissolved free amino acid concentrations of 0.5 to 1.5 μM in seawater during the spring bloom in McMurdo Sound. Because solutes are excluded from the crystal matrix during ice formation, actual concentrations of organics in brine pockets are probably even higher. High amino acid concentrations may allow facultatively heterotrophic diatoms to compete successfully with bacteria trapped in brine pockets for organic substrates, supplementing carbon and energy derived from photosynthesis during the spring bloom and providing essential growth factors. If high amino acid concentrations are present throughout the year, heterotrophic uptake may increase survival potential for certain species during the winter months.

Using natural populations of sea ice diatoms from the spring bloom in McMurdo Sound in 1981, Palmisano et al.[41] demonstrated the uptake of ^{14}C-serine at substrate concentrations close to ambient relative to formalin-killed controls. Samples were collected by divers at the ice-seawater interface from quadrats with 4 and 70 cm of snow cover. Ice-seawater slurry was incubated with radiolabel for 24 hr in the dark at $-2°C$. Serine uptake was evident in samples from both quadrats with rates ranging from 0.31 to 12.4 *nM* serine mg chlorophyll a^{-1} hr^{-1}.

To study heterotrophy in a community of microorganisms, one must be able to distinguish algal from bacterial uptake. In this study, two approaches were taken. First, samples were size-fractionated using 1.0 μm and 0.2 μm porosity Nucleopore filters.[83] Virtually all of the ^{14}C-serine was collected on a 1.0 μm porosity filter, indicating that free-living bacteria made a relatively small contribution to total uptake. Second, microautoradiographs revealed that several species of diatoms were responsible for ^{14}C-serine uptake including *Nitzschia sublineata* Hasle, a common ice diatom. Unlike *Amphiprora* spp., *N. sublineata* was not extensively colonized by attached bacteria that have been observed to take up a variety of radiolabeled organic substrates.[41]

We believe that future research on heterotrophy in sea ice microalgae needs to address the following questions:

1. What are the rates of utilization of organic substrates by the ice algal community?
2. Is substrate utilization species specific?
3. Can ice algae compete effectively with ice bacteria for substrates?
4. What portion of algal nutrition is provided by heterotrophy vs. photosynthesis?
5. Can heterotrophy support maintenance metabolism or even growth during the winter months?

V. DARK SURVIVAL

With the onset of polar winter, sea ice microalgae are subject to a prolonged period of total darkness. A critical percentage of the ice diatom population must survive the polar winter to seed the spring bloom. Survival of an organism under adverse conditions is important to the perpetuation of a species; when a favorable environment is encountered, the genome may be expressed in terms of growth and metabolism.[84] It is not clear whether sea ice algae overwinter in the ice, the water column, the benthos, or all three environments. Matheke and Horner[85] suggested that arctic ice algae may arrive in the spring from the benthos, while Bunt and Lee[15] and later others,[25,26] proposed that ice crystals formed in the water column carry attached microalgae upward to the growing annual ice. In the Arctic, Allen[80] observed dividing pennate diatoms in ice cores collected in December near Point Barrow. Hsiao[6] found microalgae in seasonal shorefast ice in the Canadian Arctic from late October when sea ice formed through early June. Horner[20] reported cells scattered throughout the ice in the winter. In the Antarctic, Bunt and Lee[15] found midwinter population densities of about 100 cells ℓ^{-1} in the bottom of annual sea ice at McMurdo Sound. Algae were observed both in sea ice and in the underlying water column in the Mirny region of Antarctica from April 1968 to January 1969.[86]

All these reports suggest a population of microalgae surviving the winter in the ice. Most rely on microscopic observations to determine viability. Microscopic observations of cell morphology, however, may not be sufficient to determine cell viability, especially in those cells exposed to prolonged darkness.[87] Morphological alterations can occur in cells left in darkness for only 1 week.[88] Viability can be better demonstrated by the use of vital stains,[89] by microautoradiography after incubation with suitable radiolabeled substrates, or by the subsequent growth of overwintered cells which are inoculated into culture media. If these cells are involved in the seeding of the spring bloom, they must be capable of initiating the growth of a new population of cells.

In 1982, a pigmented band of microalgae was found within 2-year sea ice from the New Harbor region of McMurdo Sound.[9] The band was 80 cm from the ice-seawater interface and it appeared to be a portion of the 1981 bloom to which an extra season of ice growth had been added. Using epifluorescence microscopy, samples stained with acridine orange and DAPI (4′, 6-diamidino-2-phenylindole 2HCl) revealed diatoms with vesicular cytoplasmic contents that weakly fluoresced. From this band, a pennate diatom was cultured in seawater-based medium, confirming that some cells in ice can maintain their viability over winter.

Sea ice microalgae overwintering in the ice, water column, or the benthos must survive periods of total darkness at temperatures of −1.8°C or less. Dark survival by temperate planktonic and benthic microalgae has been studied extensively.[87,90,91] Antia[92] found three temperate benthic diatoms capable of autotrophic growth after 12 to 36 months in the dark at +2°C. Similarly, nonaxenic cultures of sea ice diatoms kept in the dark at −2°C remained viable after 12 months.[93] Little is known of the physiological mechanisms of dark survival in sea ice diatoms. Bunt and Lee[15] found that three of four sea ice microalgae remained viable after 3 months of dark incubation in the absence of organic supplements. Palmisano and Sullivan[93] found that 1 to 100% of the populations of three sea ice diatoms were able to grow autotrophically after dark incubation for 5 months at −2°C. Preconditioning the cells with a simulated summer to winter transition (gradually decreasing light, temperature, and increasing salinity) or with a simple light to dark transition slightly enhanced survival. Furthermore, eight out of nine experiments showed a decrease in survival after 3 months followed by an increase, suggesting a possible recycling of nutrients by associated bacteria within the closed system of the incubation flask. Palmisano and Sullivan[2] proposed and tested three survival strategies: facultative heterotrophy; storage and utilization of carbon reserves at a reduced metabolic rate; and resting spore formation. These strategies are not considered to be mutually exclusive and may vary in importance between species. As heterotrophic metabolism has been discussed in a previous section, we will consider here the latter two strategies.

Endogenous stores of carbohydrates and lipids may supply carbon and energy required for maintenance metabolism during the winter months if utilized at reduced rates. The basal metabolic rate of sea ice diatoms in the dark at −2°C is likely to be very low. Bunt et al.[94] found that dark respiration by *Fragilaria sublinearis* at 3°C was less than 50% of the value at 10°C and presumably would be even lower at ambient water temperatures of −1.8°C. Anderson[88] demonstrated that the temperate diatom *Amphora coffeaeformis* (Agardh) Kützing formed resting cells at +4°C in the dark. Resting cell formation was characterized by an accumulation and utilization of lipids in the cytoplasm and cells respired at a rate less than 20% of light-grown cells.[95] Anderson[88] hypothesized that resting cell formation allowed the prolonged survival of *A. coffeaeformis* below the aphotic zone, an environment somewhat analogous to the polar winter in its low temperatures and darkness. Morgan and Kalff[96] suggested that the prolonged survival of a subarctic cryptomonad for 2 months in the dark at +10°C was due to the slow respiration of carbohydrate reserves. Palmisano and Sullivan[2] found an accumulation and subsequent depletion of cellular lipids and carbohydrates in sea ice diatoms exposed to conditions mimicking a summer to winter transition. All these observations suggest that the slow metabolism of stored carbon plays a role in providing maintenance energy for microalgal cells under low light or dark conditions.

Some diatoms form resistant resting spores. These are heavily silicified cells with a morphology and metabolism different from actively growing vegetative cells.[90,97] French and Hargraves[98] found that resting spores had higher cellular carbon and chlorophyll *a* and a lower respiration rate than their corresponding vegetative cells. The formation of resting spores has been suggested as an overwintering strategy for sea ice diatoms because resting spores can survive long periods of darkness.[90] This survival strategy has not been indicated for sea ice diatoms in general for the following reasons. First, few spore-forming diatoms are endemic to polar waters[99] with notable exceptions, including species of *Odontella,*[100] *Eucampia,*[100] *Chaetoceros,*[101] and *Thalassiosira.*[102] Second, ice diatoms are primarily pennate species and the occurrence of resting spores in pennates is rare.[99] Von Stosch and Fecher[103] listed only two pennate marine diatoms known to form resting spores. Recently, Sullivan et al.[40] reported an ice algal community in McMurdo Sound dominated by the centric diatom *Porosira pseudodenticulata* (Hustedt) Jousé. It is possible that local populations of centric ice diatoms may develop resting spores. Finally, resting spore formation is commonly induced by nitrogen deficiency rather than the absence of light.[99] Antarctic waters are considered to be nutrient rich,[30] and it is unlikely that growth is nutrient limited, at least at the ice-seawater interface. Doucette[104] observed resting spore formation in *Thalassiosira antarctica* Comber in response to increased salinity. Such increases in salinity may occur in brine pockets during sea ice formation in the fall. Overall, present evidence fails to suggest the importance of resting spores in winter survival of sea ice diatoms. Fryxell et al.[97] pointed out that the apparent absence of resting spores may be due to seasonality, geographic differences, or their nonrecognition by investigators.

VI. CONCLUSION

Annual sea ice is extensive in the polar regions, reaching up to 40 $\times$ 10^6 km^2 at its maximum extent. The algae associated with annual sea ice reach standing crops up to 309 mg chl a m^{-2}.[24] They must certainly play an important role in the trophodynamics of polar marine ecosystems. In the Arctic, Alexander[19] estimated that ice algae contributed 25 to 50% of the total primary productivity in the Beaufort Sea. Horner and Schrader[23] found that ice algae provided two thirds of the total productivity in nearshore regions of the Beaufort Sea during their spring bloom, while phytoplankton productivity provided one third, and benthic productivity was negligible. In the Antarctic, Ackley et al.[1] estimated that up to 21% of the productivity in the Weddell Sea was derived from sea ice microalgae. The high protein content of sea ice microalgae (C/N = 4)[15] makes them a rich food source for cryopelagic grazers[105] such as amphipods,[106] krill,[107] and larval fish.[27]

Sea ice microalgae offer an important opportunity to study physiological adaptation. Their adaptations to low light have implications for algae living in the aphotic zone, the benthos, and other low light environments. Their adaptations to low temperatures are of interest to biologists studying extraterrestrial systems.[108] Although past studies of ice algae have provided a data base, the lack of consistency in sampling methods, incubation techniques, and data reporting have limited comparisons between studies. Future research on sea ice microalgal physiology and ecology should attempt to examine metabolic rates either *in situ* or under conditions that accurately mimic the sea ice environment.

ACKNOWLEDGMENT

The authors wish to thank the NSF Division of Polar Programs through grants DPP-81-17237 and DDP-82-18752 for support during the conduct of research and writing of this chapter.

REFERENCES

1. **Ackley, S. F., Buck, K. R., and Taguchi, S.,** Standing crop of algae in the sea ice of the Weddell Sea region, *Deep-Sea Res.*, 26A, 269, 1979.
2. **Palmisano, A. C. and Sullivan, C. W.,** Physiology of sea ice diatoms. I. Response of three polar diatoms to a simulated summer-winter transition, *J. Phycol.*, 18, 489, 1982.
3. **Meguro, H.,** Plankton ice in the Antarctic Ocean, *Nankyoku Shiryo*, 14, 1192, 1962.
4. **Burkholder, P. R. and Mandelli, E. F.,** Productivity of microalgae in Antarctic sea ice, *Science*, 148, 872, 1965.
5. **Whitaker, T. M.,** Sea ice habitats of Signy Island (South Orkneys) and their primary productivity, in *Adaptations within Antarctic Ecosystems*, Llano, G. A., Ed., Gulf Publ., Houston, 1977, 75.
6. **Hsiao, S. I. C.,** Quantitative composition, distribution, community structure and standing stock of sea ice microalgae, *Arctic*, 33, 786, 1980.
7. **Alexander, V. and Chapman, T.,** The role of epontic algal communities in Bering Sea ice, in *The Eastern Bering Sea Shelf: Oceanography and Resources*, Vol. 2, Hood, D. W. and Calder, J. A., Eds., University of Washington Press, Seattle, 1981, 773.
8. **Hoshiai, T.,** Seasonal change of ice communities in sea ice near Syowa Station, Antarctica, in *Polar Oceans*, Dunbar, M. J., Ed., Arctic Institute of North America, Calgary, 1977, 307.
9. **Sullivan, C. W.,** unpublished data, 1982.
10. **Apollonio, S.,** The chlorophyll content of Arctic sea ice, *Arctic*, 14, 197, 1961.
11. **Meguro, H., Ito, K., and Fukushima, H.,** Ice flora (bottom type): a mechanism of primary production in polar seas and growth of diatoms in sea ice, *Arctic*, 20, 114, 1967.
12. **Horner, R. and Alexander, V.,** Algal populations in Arctic sea ice: an investigation of heterotrophy, *Limnol. Oceanogr.*, 17, 454, 1972.
13. **Sullivan, C. W. and Palmisano, A. C.,** Sea ice microbial communities in McMurdo Sound, Antarctica, *Antarct. J. U.S.*, 16, 126, 1981.
14. **Bunt, J. S. and Wood, E. J. F.,** Microalgae and Antarctic sea ice, *Nature (London)*, 199, 1254, 1963.
15. **Bunt, J. S. and Lee, C. C.,** Seasonal primary production in Antarctic sea ice at McMurdo Sound in 1967, *J. Mar. Res.*, 28, 304, 1970.
16. **Sullivan, C. W., Palmisano, A. C., Kottmeier, S., McGrath Grossi, S., Moe, R., and Taylor, G. T.,** The influence of light on the development and growth of sea ice microbial communities (SIMCOs) in McMurdo Sound, Antarctica, *Antarct. J. U.S.*, 18(5), 177, 1983.
17. **Grossi, S. M.,** unpublished data, 1982.
18. **Apollonio, S.,** Chorophyll in Arctic sea ice, *Arctic*, 18, 118, 1965.
19. **Alexander, V.,** Primary productivity regimes of the nearshore Beaufort Sea with reference to the potential role of ice biota, in *The Coast and Shelf of the Beaufort Sea*, Reed, J. C. and Sater, J. E., Eds., Arctic Institute of North America, Calgary, 1974, 609.
20. **Horner, R.,** Sea ice organisms, *Oceanogr. Mar. Biol. Ann. Rev.*, 14, 167, 1976.
21. **Grainger, E. H.,** The annual nutrient cycle in sea ice, in *Polar Oceans*, Dunbar, M. J., Ed., Arctic Institute of North America, Calgary, 1977, 285.
22. **Dunbar, M. J. and Acreman, J.,** Standing crops and species composition of diatoms in sea ice from Robeson Channel to the Gulf of St. Lawrence, *Ophelia*, 19, 16, 1980.
23. **Horner, R. and Schrader, G. C.,** Relative contributions of ice algae, phytoplankton, and benthic microalgae to primary production in nearshore regions of the Beaufort Sea, *Arctic*, 35, 485, 1982.
24. **Palmisano, A. C. and Sullivan, C. W.,** Sea ice microbial communities (SIMCO). I. Distribution, abundance, and primary production of ice microalgae in McMurdo Sound, Antarctica in 1980, *Polar Biol.*, 2, 171, 1983.
25. **Ackley, S. F.,** Ice scavenging and nucleation: two mechanisms for incorporation of algae into newly-formed sea ice, *Eos*, 63, 54, 1982.
26. **Garrison, D. L., Ackley, S. F., and Buck, K. R.,** A physical mechanism for establishing algal populations in frazil ice, *Nature (London)*, 306, 363, 1983.
27. **Bradford, J. M.,** Sea ice organisms and their importance in the Antarctic ecosystem, *N.Z. Antarct. Res.*, 1, 43, 1978.
28. **Bunt, J. S.,** Diatoms of Antarctic sea ice as agents of primary production, *Nature (London)*, 199, 1254, 1963.
29. **Holm-Hansen, O., El-Sayed, S. A., Franceschini, B. A., and Cuhel, R.,** Primary production and the factors controlling phytoplankton growth in the Antarctic seas, in *Adaptations within Antarctic Ecosystems*, Llano, G. A., Ed., Gulf Publ., Houston, 1977, 11.
30. **Eppley, R. W.,** Temperature and phytoplankton growth in the sea, *NOAA (U.S.) Fish. Bull.*, 71, 1063, 1972.
31. **Bunt, J. S.,** Some characteristics of microalgae isolated from Antarctic sea ice, *Antarct. Res. Ser.*, 11, 1, 1968.

32. **Morita, R. Y.**, Psychrophilic bacteria, *Bacteriol. Rev.*, 39, 144, 1975.
33. **Hochachka, P. and Somero, G.**, *Strategies of Biochemical Adaptation*, W. B. Saunders, Philadelphia, 1973, chap. 7.
34. **McConville, M. J.**, Chemical composition and biochemistry of sea ice microalgae, this volume, chap. 4.
35. **Littlepage, J. L.**, Oceanographic investigations in McMurdo Sound, Antarctica, *Antarct. Res. Ser.*, 5, 1, 1965.
36. **Weeks, W. F. and Lee, O. S.**, The salinity distribution in young sea-ice, *Arctic*, 15, 93, 1962.
37. **Bunt, J. S.**, Microbial productivity in polar regions, *Microbes and Biological Productivity*, Twenty-first Symposium of the Society for General Microbiology, 1971, 333.
38. **Fogg, G. E.**, Aquatic primary production in the Antarctic, *Philos. Trans. R. Soc. London, Ser. B*, 279, 27, 1977.
39. **Maykut, G. A. and Grenfell, T. C.**, The spectral distribution of light beneath first-year sea ice in the Arctic Ocean, *Limnol. Oceanogr.*, 20, 554, 1975.
40. **Sullivan, C. W., Palmisano, A. C., Kottmeier, S., and Moe, R.**, Development of the sea ice microbial community in McMurdo Sound, *Antarct. J. U. S.*, 17, 155, 1982.
41. **Palmisano, A. C., Kottmeier, S. T., Moe, R. M., and Sullivan, C. W.**, Sea ice microbial communities (SIMCO). IV. The effect of light perturbation on microalgae at the ice-seawater interface, *Mar. Ecol. Prog. Ser.*, 21, 37, 1985.
42. **Moe, R. M.**, unpublished observations, 1981.
43. **Sullivan, C. W.**, this volume, chap. 7. Sea Ice Bacteria: Reciprocal Interactions of the Organisms and Their Environment.
44. **Lake, R. A., and Lewis, E. L.**, Salt rejection by sea ice during growth, *J. Geophys. Res.*, 75, 583, 1970.
45. **Jacques, G.**, Some ecophysiological aspects of the Antarctic phytoplankton, *Polar Biol.*, 2, 27, 1983.
46. **Grant, W. S. and Horner, R.**, Growth responses to salinity variation in four Arctic ice diatoms, *J. Phycol.*, 12, 180, 1976.
47. **Niebauer, H. J., Alexander, V., and Cooney, R. T.**, Primary production at the eastern Bering Sea ice edge: the physical and biological regimes, in *The Eastern Bering Sea Shelf: Oceanography and Resources*, Vol. 2, Hood, D. W. and Calder, J. A., Eds., University of Washington Press, Seattle, 1981, 763.
48. **Geesey, G. G.**, Microbial exopolymers: ecological and economic considerations, *Am. Soc. Microbiol. News*, 48, 9, 1982.
49. **Schrader, G. C., Horner, R., and Smith, G. F.**, An improved chamber for in situ measurement of primary productivity of Arctic sea ice algae, *Can. J. Fish. Aquat. Sci.*, 39, 522, 1982.
50. **Bunt, J. S.**, Primary productivity under sea ice in Antarctic waters 2. Influence of light and other factors on photosynthetic activities of Antarctic marine microalgae, *Antarct. Res. Ser.*, 1, 27, 1964.
51. **Palmisano, A. C. and Sullivan, C. W.**, Photosynthesis-irradiance relationships in natural populations of sea ice algae from McMurdo Sound, Antarctica, *Eos*, 64, 1090, 1983.
52. **Beardall, J. and Morris, I.**, The concept of light intensity adaptation in marine phytoplankton: some experiments with *Phaeodactylum tricornutum*, *Mar. Biol.*, 37, 377, 1976.
53. **Chan, A. T.**, Comparative physiological study of marine diatoms and dinoflagellates in relation to irradiance and cell size. I. Growth under continuous light, *J. Phycol.*, 14, 396, 1978.
54. **Morris, I. and Glover, H. E.**, Physiology of photosynthesis by marine coccoid cyanobacteria — Some ecological implications, *Limnol. Oceanogr.*, 26, 957, 1981.
55. **Prezelin, B. B. and Ley, A. C.**, Photosynthesis and chlorophyll *a* fluorescence rhythms of marine phytoplankton, *Mar. Biol.*, 55, 295, 1980.
56. **Jeffrey, S. W. and Vesk, M.**, Effects of blue-green light on photosynthetic pigments and chloroplast structure in the marine diatom *Stephanopyxis turris*, *J. Phycol.*, 13, 271, 1977.
57. **Jorgensen, E. G.**, The adaptation of plankton algae. V. Variation in the photosynthetic characteristics of *Skeletonema costatum* cells grown at low light intensity, *Physiol. Plant.*, 23, 11, 1970.
58. **Halldal, P.**, The photosynthetic apparatus of microalgae and its adaptation to environmental factors, in *Photobiology of Microorganisms*, Halldal, P., Ed., John Wiley & Sons, New York, 1970, 17.
59. **Prezelin, B. B.**, The role of peridinin-chlorophyll *a*-proteins in the photosynthetic light adaptation of the marine dinoflagellate, *Glenodinium* sp., *Planta*, 130, 225, 1976.
60. **Steemann-Nielsen, E. and Jorgensen, E. G.**, The adaptation of plankton algae. I. General part, *Physiol. Plant.*, 21, 401, 1968.
61. **Li, W.**, Temperature adaptation in phytoplankton: cellular and photosynthetic characteristics, in *Primary Production in the Sea*, Falkowski, P., Ed., Plenum Press, New York, 1980, 259.
62. **Neori, A. and Holm-Hansen, O.**, Effects of temperature on rate of photosynthesis in Antarctic phytoplankton, *Polar Biol.*, 1, 33, 1982.
63. **Preszelin, B. B.**, Effects of light intensity on aging of the dinoflagellate *Gonyaulax polyedra*, *Mar. Biol.*, 69, 129, 1982.

64. **Taguchi, S.,** Relationship between photosynthesis and cell size of marine diatoms, *J. Phycol.*, 12, 185, 1976.
65. **Malone, T. C. and Neale, P. J.,** Parameters of light-dependent photosynthesis for phytoplankton size fractions in temperate estuarine and coastal environments, *Mar. Biol.*, 61, 289, 1981.
66. **Ignatiades, L.,** Studies on the factors affecting release of organic matter by *Skeletonema costatum* (Greville) Cleve in field conditions, *J. Mar. Biol. Assoc. U.K.*, 53, 937, 1973.
67. **Fogg, G. E.,** Extracellular products of phytoplankton and the estimation of primary production, *Rapp. P. V. Reun. Cons. Int. Explor. Mer*, 144, 56, 1958.
68. **Sharp, J. H.,** Excretion of organic matter by marine phytoplankton: Do healthy cells do it?, *Limnol. Oceanogr.*, 22, 381, 1977.
69. **Smith, W. O., Barber, R. T., and Huntsman, S. A.,** Primary production off the coast of northwest Africa: excretion of dissolved organic matter and its heterotrophic uptake, *Deep-Sea Res.*, 24, 35, 1977.
70. **Iturriaga, R.,** Phytoplankton photoassimilated extracellular products: heterotrophic utilization in marine environment, *Kiel. Meeresforsch. Sonderh*, 5, 318, 1981.
71. **Larsson, U. and Hagstrom, A.,** Fractionated phytoplankton primary production, exudate release and bacterial production in a Baltic eutrophication gradient, *Mar. Biol.*, 67, 57, 1982.
72. **McConville, M. J. and Wetherbee, R.,** The bottom-ice microalgal community from annual ice in the inshore waters of East Antarctica, *J. Phycol.*, 19, 431, 1983.
73. **Palmisano, A. C. and Sullivan, C. W.,** Physiological response of microalgae in the ice platelet layer to ambient low light conditions, *Proc. SCAR Symposium on Antarctic Biology*, Wilderness, South Africa, in press.
74. **Cooksey, K. W.,** Metabolism of organic acids by a marine pennate diatom, *Plant Physiol.*, 50, 1, 1972.
75. **Jolley, E. T., Jones, A. K., and Hellebust, J. A.,** A description of glucose uptake in *Navicula pelliculosa* (Bréb). *Arch. Microbiol.*, 109, 127, 1976.
76. **Hellebust, J. A. and Lewin, J.,** Heterotrophic nutrition, in *The Biology of Diatoms*, Werner, D., Ed., University of California Press, Los Angeles, 1977, 169.
77. **Hellebust, J. A.,** The uptake and utilization of organic substances by marine phytoplankters, in *Symposium on Organic Matter in Natural Waters*, Hood, D. W., Ed., Institute of Marine Science, University of Alaska, Fairbanks, 1970, 225.
78. **Rodhe, W.,** Can plankton production proceed during winter darkness in subarctic lakes?, *Verh. Int. Verein. Limnol.*, 12, 117, 1955.
79. **Wilce, R. T.,** Heterotrophy in Arctic sublittoral seaweeds: an hypothesis, *Bot. Mar.*, 10, 185, 1967.
80. **Allen, M. B.,** High-latitude phytoplankton, *Ann. Rev. Ecol. Syst.*, 2, 261, 1971.
81. **Bunt, J. S. and Lee, C. C.,** Data on the composition and dark survival of four sea-ice microalgae, *Limnol. Oceanogr.*, 17, 458, 1972.
82. **Manahan, D., Sullivan, C. W., and Delaca, T. E.,** unpublished data, 1982.
83. **Azam, J. F. and Hodson, R. E.,** Size distribution and activity of marine microheterotrophs, *Limnol. Oceanogr.*, 22, 492, 1977.
84. **Morita, R. Y.,** Microbial life in the deep sea, *Can. J. Microbiol.*, 26, 1375, 1980.
85. **Matheke, G. E. M. and Horner, R.,** Primary production of the benthic microalgae in the Chukchi Sea near Barrow, Alaska, *J. Fish. Res. Board Can.*, 31, 1779, 1974.
86. **Buinitsky, V. Kh.,** Organic life in sea ice, in *Polar Oceans*, Dunbar, M. J., Ed., Arctic Institute of North America, Calgary, 1977, 301.
87. **Smayda, T. J. and Mitchell-Innes, B.,** Dark survival of autotrophic planktonic marine diatoms, *Mar. Biol.*, 25, 195, 1974.
88. **Anderson, O. R.,** The ultrastructure and cytochemistry of resting cell formation in *Amphora coffeaeformis*, *J. Phycol.*, 11, 272, 1975.
89. **Hobbie, J. E., Daley, R. J., and Jasper, S. I.,** Use of Nuclepore filters for counting bacteria by fluorescence microscopy, *Appl. Environ. Microbiol.*, 33, 1225, 1977.
90. **Hargraves, P. E. and French, F. W.,** Observations on the survival of diatom resting spores, *Nova Hedwigia Beih.*, 53, 229, 1975.
91. **Antia, N. J. and Cheng, J. Y.,** The survival of axenic cultures of marine planktonic algae from prolonged exposure to darkness at 2°C, *Phycologia*, 9, 179, 1970.
92. **Antia, N. J.,** Effects of temperature on the darkness survival of marine microplanktonic algae, *Microb. Ecol.*, 3, 41, 1976.
93. **Palmisano, A. C. and Sullivan, C. W.,** Physiology of sea ice diatoms. II. Dark survival of three polar diatoms, *Can. J. Microbiol.*, 29, 157, 1983.
94. **Bunt, J. S., Owens, O., van H., and Hoch, G.,** Exploratory studies on the physiology and ecology of a psychrophilic marine diatom, *J. Phycol.*, 2, 69, 1966.
95. **Anderson, O. R.,** Respiration and photosynthesis during resting cell formation in *Amphora coffeaeformis* (Ag.) Kütz., *Limnol. Oceanogr.*, 21, 452, 1976.

96. **Morgan, K. and Kalff, J.**, The winter survival of an algal flagellate *Cryptomonas erosa* (Skuja), *Verh. Internat. Verein. Limnol.*, 19, 2734, 1975.
97. **Fryxell, G. A., Villareal, T. A., and Doucette, G. J.**, Diatom resting spores and Agulhas collections, *Antarct. J. U.S.*, 16, 128, 1981.
98. **French, F. W. and Hargraves, P. E.**, Physiological characteristics of plankton resting spores, *Mar. Biol. Lett.*, 1, 185, 1980.
99. **Hargraves, P. E. and French, F. W.**, Diatom resting spores: significance and strategies, in *Survival Strategies of the Algae*, Fryxell, G. A., Ed., Cambridge University Press, New York, 1983, 49.
100. **Hoban, M. A., Fryxell, G. A., and Buck, K. R.**, Biddulphioid diatoms: resting spore formation in Antarctic *Eucampia* and *Odontella*, *J. Phycol.*, 16, 591, 1980.
101. **Hasle, G. R.**, An analysis of the phytoplankton of the Pacific Southern Ocean, *Hvalråd. Skr.*, 52, 1, 1969.
102. **Fryxell, G. A., Villareal, T. A., and Hoban, M. A.**, *Thalassiosira scotia* sp. nov.: observations on a phytoplankton increase in early austral spring north of the Scotia Ridge, *J. Plankton Res.*, 1, 335, 1979.
103. **Von Stosch, H.-A. and Fecher, K.**, "Internal thecae" of *Eunotia soleirolii* (Bacillariophyceae): development, structure and function as resting spores, *J. Phycol.*, 15, 233, 1979.
104. **Doucette, G. J.**, Variability in patterns of resting spore formation in the diatom *Thalassiosira antarctica*, *Eos*, 63, 47, 1982.
105. **Carey, A. G. and Montagna, P. A.**, Arctic sea ice faunal assemblage: first approach to description and source of the underice fauna, *Mar. Ecol. Prog. Ser.*, 8, 1, 1982.
106. **Sagar, P.**, Life cycle and growth of the Antarctic gammarid amphipod *Paramoera walkeri* (Stebbing 1960), *J. R. Soc. N.Z.*, 10, 259, 1980.
107. **Hamner, W. M., Hamner, P. P., Strand, S. W., and Gilmer, R. W.**, Behavior of Antarctic krill *Euphausia superba:* chemoreception, feeding, schooling, and molting, *Science*, 220, 433, 1983.
108. **Squyres, S. W., Reynolds, R. T., and Colburn, D. S.**, On the habitability of Europa, *EOS*, 63, 1022, 1982.

Chapter 6

TAXONOMY OF SEA ICE MICROALGAE

Rita Horner

TABLE OF CONTENTS

I. INTRODUCTION

Although organisms in sea ice have been known and studied for nearly 150 years, diatoms were the only organisms found, or at least were the only ones reported, in nearly all of the earliest studies on ice algae.[1-10] Only Sutherland[11] and Nansen[12,13] reported finding organisms such as flagellates, infusoria, and bacteria in the ice in the Arctic, while Hooker[14] reported silicoflagellates from the Antarctic ice. Later, Heiden and Kolbe,[15] and Hendey[16] in the Antarctic, and Meister[17] in the Arctic found only diatoms in the ice. Even some recent investigators have reported finding only diatoms,[18,19] while some who have found other organisms chose to study only the diatoms.[20] In his review of Soviet ice algal studies, Usachev[21] said other algae were rarely found.

In the Arctic, a number of recent investigators[19,22-32] have spent considerable effort determining the species present and more than 200 species have been identified. Few of these, however, have been isolated and studied in culture (but see Grant and Horner;[33] Palmisano and Sullivan[34]). In the Antarctic, less effort has been spent identifying species and more on experimental studies. Some species lists are available for the Antarctic,[20,35-38] but except for Buinitsky et al.,[37] they are not extensive. El-Sayed[39] pointed out that of the 20 most abundant species at McMurdo Sound[35] and at Mirny,[36] not one was found at both sites. In the Antarctic, more individual species have been isolated and used in autecological studies,[34,40-46] often with identification being of secondary interest. The elegant morphological study of *Pyramimonas gelidicola* by McFadden et al.[44] is an important exception.

II. ORGANISMS FROM ARCTIC SEA ICE

Green and bluegreen algae, including *Scenedesmus quadricauda* de Brébisson, *Pediastrum boryanum* (Turpin) Meneghini, and the remains of *Anabaena* and *Aphanizomenon* were found in freshwater ice that formed in the Yenisey River and remnants of *Peridinium* and *Ceratium* were found in sea ice.[48] Most of the cells, including the diatoms, were dead (empty frustules) or only cell remnants, making it difficult to determine whether the cells actually lived and grew in the ice or were incorporated into the ice as dead cells. "Typical cryophiles" from the genera *Navicula, Nitzschia,* and *Gomphonema* with well-preserved cell contents were found in ice that formed in the spring.[21] Usachev[21] found 64 diatoms, 3 dinoflagellates, 1 silicoflagellate, and 5 ciliates, plus several kinds of spores and cysts. Of the diatoms, 24 were considered to be "typical cryophiles" that form blooms on the underside of the ice and have been reported from a number of Arctic ice collections. About 50% of the diatoms reported here were benthic species; only three diatoms were from freshwater.

Melt ponds located on the surface of sea ice near Barrow, Alaska, contained bacteria, fungi, and pigmented and nonpigmented flagellates, while the freshwater green algae *Chlorella salina* Butcher, *Oocystis* sp., and *Ulothrix* sp. were found in ponds as the ice surface melted.[47] In mid-July, a system of shallow ponds contained flagellates, diatoms, and ciliates. In late August, the marine planktonic species *Achnanthes taeniata* Grunow, *Fragilaria oceanica* Cleve, *Navicula* sp., and *Chaetoceros wighami* Brightwell were found in pools on the ice, having entered them through deep perforations in the ice.[47] The bottom ice community was not studied.

Meguro et al.[18,19] found 24 pennate diatom species and no other algae in sea ice from offshore in the Chukchi Sea. A preliminary paper on this work contained photomicrographs of the species present.[49]

Diatom species (28) and numerous unidentified pennate diatoms were identified from the nearshore Chukchi Sea.[23] In addition to pennate diatoms, several centric diatom species, euglenoids, dinoflagellates, cryptomonads, and unidentified pigmented and nonpigmented flagellates were found. Flagellates were the most abundant organisms in the ice in winter. Only seven pennate diatom species were the same as those found offshore[18,19] and the two dominant species in the nearshore area, *Nitzschia frigida* Grunow and *Navicula marina* Ralfs, were not reported from offshore. It is not known if these differences are real because there may be considerable variability in species composition both between years and over relatively narrow geographic limits. Large numbers of meiofauna were also reported in the ice from the nearshore area.[22]

In the nearshore Beaufort Sea near Prudhoe Bay, Alaska, 55 species or groups of species were identified, including 39 pennate diatoms, 4 centric diatoms, 3 dinoflagellates, and 6 unidentified flagellates, plus groups of unidentified pennate diatoms, dinoflagellates, and flagellates.[31] Only six species ever accounted for more than 10% of the cells counted. This community changed over time from one dominated by *Nitzschia cylindrus* (Grunow) Hasle in April and May to one with more species of pennate diatoms and increased abundance of *Navicula directa* (W. Smith) Ralfs and *Nitzschia frigida* in early June.[31] Unidentified flagellates also increased in number later in the bloom and were dominant during the winter.

Of the 101 taxa identified from sea ice samples collected in the Beaufort Sea and in McKinley Bay near the mouth of the Mackenzie River, 78 taxa were pennate diatoms and comprised 33 to 99% of the total number of cells counted.[50] *Nitzschia frigida* and *N. grunowii* Hasle were the most abundant species in all samples. The centric diatom *Chaetoceros septentrionalis* Østrup and microflagellates were also numerically important in the ice during the course of this study.

In the Canadian Arctic, where more emphasis has been placed on species identification than anywhere else, 196 species of microalgae were identified from the Eskimo Lakes, Eclipse Sound, and Frobisher Bay, including 189 diatoms, 1 green flagellate, 2 euglenoids, 2 dinoflagellates, and 2 chrysophytes.[25] Of these, only 20 species were also found near Prudhoe Bay.[31] Later, 227 species were found in Frobisher Bay alone.[29] From the earlier study[25] 68 species were not found at Frobisher Bay,[29] while 96 species from Frobisher Bay were not previously reported.[25] These numbers suggest that some species are present over a wide area of the North American Arctic, but many species may not be so widespread. Additional floristics studies are obviously needed. These studies also suggest that more attention should be paid to the identification of the organisms present, something that has been difficult to do, partly because of the scattered literature.

Poulin and Cardinal[51-53] reevaluated taxonomic information for more than 110 diatom species found in Manitounuk Sound, Hudson Bay. They described new taxa, rejected others, and added characteristics to earlier descriptions. These papers have line drawings, scanning electron micrographs, and useful reference lists.

In the Gulf of St. Lawrence, centric diatoms were reported to be more numerous than pennate diatoms,[26] but another study[54] found benthic diatoms to be more numerically dominant 1 year, with pelagic species, especially microflagellates, more numerous the next year. These studies again show the wide variability in species present that exists over a relatively narrow geographic range and indicate annual variation as well.

A checklist of marine phytoplankton and sea ice microalgae from Arctic Canada has recently been compiled from published and unpublished reports.[55] Of the 685 taxa listed, 193 diatoms, 3 chrysophytes, 2 euglenoids, 2 dinoflagellates, and 1 green alga were recorded from sea ice. Only 27 of the ice diatoms were centric species, with 6 species each from the genera *Chaetoceros* and *Thalassiosira*.

Elsewhere in the Northern Hemisphere, Saito and Taniguchi,[56] working in the northern Bering and southern Chukchi seas, listed 13 pennate diatoms that they called "ice plankton" or species that probably grew in sea ice. *Fragilaria islandica* Grunow, *F. striatula* Lyngbye, and *Nitzschia grunowii* were the dominant components. This study was based on water samples collected sometime after the ice receded from the study area. Alexander and Chapman[57] and Schandelmeier and Alexander[27] have provided the only lists of organisms actually found in ice from the Bering Sea. From ice cores collected in late May, they found 14 diatom species, 6 of which were centric species, while 7 of 15 diatoms from slush ice collected in late March were centric species.

In sea ice from Lake Saroma, Hokkaido, Takahashi[28] found 44 taxa, including diatoms, green algae, cryptomonads, dinoflagellates, euglenoids, chrysophytes, prymnesiophytes, and prasinophytes, plus scales of 5 unknown taxa that were found with the electron microscope.

In the Baltic, the prymnesiophyte *Chrysochromulina birgeri* Hällfors and Niemi has been described from seawater under the ice.[58] This organism has also been found in holes and fissures in the ice, in cracks along the shore where it formed blooms in the water above the ice, and in slushy snow along the shore. This species may not be a true member of the ice community, but it is closely associated with ice and the slushy snow, and blooms in water above the ice may be similar to some of the surface ice communities in the Antarctic. Also in the Baltic, Grøntved[59] reported large quantities of *Nitzschia frigida* from ice in nearshore waters.

III. ORGANISMS FROM ANTARCTIC SEA ICE

Hooker's[14] samples came from seawater, the "scum of the ocean", salp stomachs, penguin guano, pancake ice, snow floating on the water surface, and mud with some species being found in more than one of these environments. Other early lists were primarily of species found in the water column near pack ice and the species were considered to be neritic or ice edge forms.[15,60,61] Bunt and Wood[35] listed diatom species they found attached to ice crystals, as well as nonattached forms that were living in close association with the ice. Of the 11 dominant species, 5 were centric diatoms, including *Biddulphia weissflogii* Grunow, *Coscinodiscus subtilis* Ehrenberg, *Eucampia balaustium* Castracane, *Rhizosolenia alata* Brightwell, and *R. rostrata* Heiden and Kolbe. The dominant pennate diatoms were *Amphiprora kjellmanii* Cleve, *A. oestrupii* Van Heurck, *Fragilaria linearis* Castracane, *Nitzschia martiana* Agardh, *N. seriata* Cleve, and *Pleurosigma antarcticum* Heiden and Kolbe.

Burkholder and Mandelli[62] listed diatom species from the ice and also planktonic species found floating near the ice. The ice contained "a very small diatom", *Fragilaria curta* Van Heurck, and *Nitzschia seriata,* plus small numbers of other pennate diatoms, centric diatoms, a few dinoflagellates, small green flagellates, and *Phaeocystis pouchetii* (Hariot) Lagerheim. Centric diatoms were most common in the water column with *Fragilaria antarctica* Castracane and *Thalassiothrix antarctica* Schimper also being present. The same diatom species were often found in both the ice and seawater, but the proportion of small cells was always higher in the ice.

Whitaker[63] described a number of ice habitats from the South Orkney Islands and identified the characteristic species from each one, but only listed the most common species. *Navicula glaciei* Van Heurck was found in infiltration ice and slush, on the vertical walls of tide cracks, on the ice foot, and in congelation ice. *Nitzschia curta* (Van Heurck) Hasle, *N. lineata* Hasle, and *Thalassiosira tumida* (Janisch) Hasle were found on the vertical walls of tide cracks and in dense snow slush on seawater; *N. curta* was also found on the ice foot, the lower surface of fast ice, and in congelation ice. *Chaetoceros schimperianum* Karsten was found in congelation ice, while *Nitzschia closterium* (Ehrenberg) W. Smith formed dense concentrations in infiltration ice. Other *Nitzschia* species were common in congelation ice, loosely adhering to the undersurface of fast ice, in underwater ice, and in dense snow slush on seawater. Diatoms were present on the ice foot, shelf area, and vertical walls of icebergs, with *Biddulphia punctata* Greville being extremely abundant.

Buinitsky[36] reported 50 species of microalgae from ice collected from about 11 to 93°E along the Antarctic coast; of these, 48 species were diatoms. *Fragilariopsis cylindrus* (Grunow) Helmcke and Krieger was the most numerous in all samples. Other abundant species were *Thalassiosira dubia* Koslova, *Navicula glaciei, Gomphonema* sp., *Fragilariopsis curta* (Van Heurck) Hustedt, and *Nitzschia angustissima* Van Heurck. The four freshwater diatoms,

Eunotia lunaris (Ehrenberg) Grunow, *Achnanthes charcoti* Peragallo, *Gomphonema clevei* Fricke, and *Nitzschia acicularis* W. Smith, were found in ice collected close to shore and it was assumed that they came from freshwater lakes near Molodezhnaya Station. Later, 78 diatom species were found in the fast ice of Alasheyev Bay, including 32 centric and 47 pennate species.[37] Of these, pennate diatoms were generally more numerous than centric diatoms, with *Pleurosigma elongatum* W. Smith, *Amphiprora kjellmanii Fragilariopsis ritscherii* Hustedt, and *Nitzschia stellata* Manguin being the most abundant.

Variations in species composition were compared for the ice and water column at some stations in the Weddell Sea.[64] Pennate and centric diatoms, chrysophytes, chrysophyte spores, and choanoflagellates were among the organisms found. *Nitzschia cylindrus* was the dominant species in the ice and in the water column at stations in the drifting pack ice and in the ice at a fast ice station, but *N. curta* was most abundant in the water column at the fast ice station. Other pennate diatoms that were abundant in the ice were *Nitzschia delicatissima* Cleve, *N. subcurvata* Hasle, and *Synedra* sp. Centric diatoms that were numerous in the ice included *Chaetoceros gracilis* Schütt, *C. debilis* Cleve, *Dactyliosolen antarcticus* Castracane, and *Porosira pseudodenticulata* (Hustedt) Jousé. *Phaeocystis* sp., chrysophyte spores, and choanoflagellates were also present in the ice at the northern stations. The choanoflagellates were later described in detail.[65,66] Choanoflagelletes were also studied at Lützow-Holm Bay[67] where 15 species were found.

Vertical distribution has been described from sea ice at Lützow-Holm Bay[20] and at McMurdo Sound.[68] At Lützow-Holm Bay, eight diatom communities were identified and the relative numerical composition of the species in each community was determined. *Fragilaria* spp. and *Nitzschia* spp. were present in all communities and dominant in five of them. *Chuniella* sp., *Chaetoceros* spp., and *Coscinodiscus* spp. were also abundant.[20] At McMurdo Sound, algal abundances in the ice decreased with increasing distance from the ice-seawater interface and with surface snow cover. *Nitzschia stellata* was dominant in the bottom 5 cm regardless of snow cover, while *Navicula glaciei* and *Fragilaria islandica* var. *adeliae* Manguin were dominant in the upper section of the ice when little or no snow cover was present, and *Amphiprora* sp. and three centric diatoms were present at intermediate depths.[68]

Recently, the prasinophyte *Pyramimonas gelidicola* was isolated and described from sea ice at Prydz Bay, east Antarctica.[44] This organism has now been used for a number of physiological and biochemical experiments.[45,46]

IV. UNUSUAL ORGANISMS

In addition to the diatoms, dinoflagellates, and other relatively common taxa, a number of unusual organisms, often with unknown taxonomic affinities, have been reported from sea ice. Meunier[69] named and figured a number of these from ice collected in the Kara and Barents seas. Most were small and in some cases were relatively well-described and figured, but generally their affinities were unknown. Both flagellated and nonflagellated cells were seen and some of the nonflagellated forms were thought to be cysts. Undoubtedly some of these organisms have been seen by more recent workers and perhaps placed in taxonomic categories, but others have probably not been found again.

In 1965, Bursa[70] described an organism from sea ice samples collected at Devon Island in the Canadian Arctic that produced star-shaped, calcareous scales. He named this organism *Discoasteromonas calciferus* Bursa and considered it to be a living member of a group of organisms previously known only from their scales and thought to have been extinct since the beginning of the Pleistocene. These organisms were thought to be related to the coccolithophorids because of their calcareous scales. Bursa's[70] descriptions and figures are not clear and this work has generally met with skepticism, based partly on the paleontologist's assumption that discoasters are extinct. However, I found a similar organism in sea ice collected near Barrow, Alaska. My observations,[71,72] based on living cells found in sea ice and on cells growing in mixed cultures started from pieces of sea ice containing cells, tend to support some of Bursa's findings, and it is now thought that discoasters are dinoflagellates, probably belonging to the class Desmomonadales.[72,73] These organisms appeared to be restricted to sea ice as they have not been found in either the water column or bottom sediments, although both of these habitats have been intensively studied at Barrow.[74,75] It is not known whether these cells are in fact living members of the discoaster group.

Members of another group of little known marine organisms, the archaeomonads, have been found in sea ice in the Antarctic, as well as in surface sediments from the western Bering, Weddell, and Ross seas, and in water samples collected off the California coast and in the Weddell Sea.[76] Archaeomonads are thought to be siliceous, fossil, marine chrysophyte cysts that cannot be matched to living species[77] and thus, they have been assigned to the artificial family Archaeomonadaceae.[78] Their occurrence in polar and near polar waters and their distribution in modern sediments suggests that they may be useful guide fossils for sea ice environments.[76]

V. COMPARISONS

A. Ice-Water Column

To some extent, this comparison is tied to the fate of the ice algae when they leave the ice (Chapter 4). Usachev[21] pointed out that the ice bloom is different from the water column bloom. He found benthic, littoral forms with heavy mucilage for attachment to prevail in the ice. A few centric diatoms were also present. *Chaetoceros septentrionalis* with short setae that does not form chains was common in his ice samples. In contrast, species found in the water column were primarily planktonic species, such as *Rhizosolenia* spp., *Thalassiosira nordenskioeldii* Cleve, and dinoflagellates. The *Chaetoceros* species in the water column, including *Chaetoceros furcellatus* Bailey, *C. socialis* Lauder, *C. decipiens* Cleve, and *C. borealis* Bailey, all have long setae and form long chains. *Melosira arctica* (Ehrenberg) Dickie was cited as an ice species with heavy mucilage production, but this species is also commonly found in large masses drifting in the water column among ice floes.

In a number of other Arctic localities, including Barrow,[23,75,79] and Narwhal Island,[31] Alaska, the Eskimo Lakes,[80] and Frobisher Bay,[81] N.W.T., species differences separate the two blooms, with pennate diatoms being dominant in the ice and centric diatoms in the water column. A time lag may also occur between the release of the ice algae and the beginning of the phytoplankton bloom.

In the Bering Sea in late March, 50 to 67% of the diatoms found in the ice and seawater were centric species, with about half of these being found in both habitats. The seawater also contained dinoflagellates and microflagellates. By late May, many additional centric diatom species were found in the water column.[57]

In the Antarctic, centric diatoms are more numerous in the ice and they are also commonly found in the water column near drifting ice.[15,61-63] The origin of the centric diatoms in the water column has not been determined, but has been considered to be the ice.[62,82] More centric diatoms were found in the water column than in the sea ice in the Weddell Sea,[64] but pennate diatoms were numerically dominant in both habitats.

Phaeocystis was reported to be the dominant genus in the spring phytoplankton at McMurdo Sound,[34] but *P. pouchetii* has also been reported from ice on the western side of the Antarctic Peninsula.[62] Burkholder and Mandelli[62] also reported more centric diatoms from the water column and more pennate diatoms from the ice.

B. Arctic — Antarctic

It is difficult to compare the Arctic and Antarctic with regard to species present in sea ice because adequate floristics studies are lacking. Pennate diatoms are the most abundant group of organisms in the ice in both polar regions where the ice acts as an inverted benthic environment and the species present are often benthic forms. Only a few centric diatoms are commonly found in the ice, with more centric species occurring in the Antarctic than in the Arctic, perhaps because of the presence of a thick layer of unconsolidated, frazil ice crystals that allows more water to be present between ice crystals and provides a more favorable habitat for centric species.

A number of centric diatoms have been reported from sea ice, but they are rarely dominant. Watanabe,[38] however, found *Porosira pseudodenticulata* to be dominant at Langhovde (69°12′S, 39°37′E) and *Coscinodiscus furcatus* Karsten to be dominant at Cape Bird (77°13′S, 166°28′E). Some other centric diatoms reported to be numerically important in Antarctic sea ice include *Biddulphia weissflogii, Coscinodiscus subtilis, Eucampia balaustium, Rhizosolenia alata,* and *R. rostrata*. In the Arctic, centric diatoms rarely occur in large numbers. Grainger and Hsiao[29] found 3 to 7% centric diatoms in their samples from Frobisher Bay, but farther south in the Gulf of St. Lawrence, Dunbar and Acreman[26] found 43% centric diatoms. But even in the Gulf of St. Lawrence, pennate diatoms may be numerically dominant.[54]

In both polar regions where marginal ice zones have been studied, centric diatoms are dominant in the ice and water column, but it is not known whether the diatoms from the ice in these areas actually grew in the ice or were washed onto it by wave action. There is no information on the growth or survival of diatoms that may be washed onto the ice or how long they remain on the ice before possibly being washed back into the water column. It is especially difficult to determine the source of the diatoms found in slush ice when both ice and water are collected at the same time.

Several diatom species from sea ice have been reported from both polar regions, although bipolarity in marine phytoplankton has not been satisfactorily explained.[83] *Thalassiosira antarctica* Comber and *Porosira glacialis* (Grunow) Jørgensen are two of these bipolar species, but there is recent evidence that there may be morphological differences,[84,85] and *T. antarctica* has been separated into two varieties based on resting spore morphology.[84] *Nitzschia cylindrus* and *N. frigida* are two important additional ice species that have been reported from both polar regions.

There is not enough taxonomic information available about organisms from other algal groups to make comparisons.

A partial list of algal species found in sea ice is given in the Appendix and the Arctic and Antarctic are compared with regard to species present. The list has been complied from a number of references, but should not be considered complete. The Arctic species list was compiled using Hsiao's[55] extensive checklist for Arctic Canada as a base and other recent taxonomic lists have been added. The early papers by Cleve, Gran, etc., have not been searched, but see Van Heurck[10] for an extensive list from the early literature. The Antarctic list was compiled from many sources because so few taxonomic lists are available.

VI. CONCLUSIONS

Organisms from many algal classes, including the Bacillariophyceae, Chrysophyceae, Chlorophyceae, Cryptophyceae, Dictyochophyceae, Dinophyceae, Euglenophyceae, Prymnesiophyceae, Prasinophyceae, Craspedophyceae, and Cyanophyceae are found in sea ice. Diatoms (Bacillariophyceae) are usually the most abundant organisms in the ice, with pennate species being more numerous than centric species, but there are exceptions, particularly in the Antarctic and in the marginal ice zones where centric diatoms are often dominant. Microflagellates from a number of algal classes may dominate in winter.

Some species are found over wide geographic areas, while others have been reported only in particular habitats and a few species have been reported from both polar regions. Vertical distribution of species in sea ice is variable.

Some species have been found in both the ice and the water column near the ice, but the relationship between these two habitats with regard to the origin and fate of the cells in the ice is not known (see Chapter 4).

A number of unusual organisms have been found in the ice and in seawater near drifting ice, but their biological and taxonomic affinities are often unknown.

In some areas, species succession occurs during the course of the ice bloom, perhaps in response to changes in light conditions or nutrient concentrations.

As is evident from this brief discussion of the organisms found in sea ice, taxonomic studies are still necessary. In particular, additional floristics studies are needed; comparisons should be made between similar geographic areas. This is especially true in the Antarctic where species lists from McMurdo Sound and Mirny had no species in common. Organisms other than diatoms should be identified and studied. The identification and importance of small cells (<10 μm) should be determined using scanning electron microscope and epifluorescence techniques. Epifluorescence techniques should be used to determine the abundance and importance of nonphotosynthetic organisms. Culture studies should be combined with electron microscope studies to determine possible morphological differences in the bipolar species. The possibility of species succession should be determined in more areas, along with the environmental factors that cause it; this should include changes in vertical distribution patterns as well.

REFERENCES

1. **Dickie, G.,** On the algae found during the Arctic Expedition, *J. Linn. Soc. London, Bot.*, 17, 6, 1880.
2. **Cleve, P. T.,** Diatoms collected during the expedition of the *Vega, Vega-Exped. Vetenskap. Iakttag.*, 3, 455, 1883.
3. **Cleve, P. T.,** Microscopical examination of dust from drift-ice north of Jan Mayen, *Oefv. K. Vetenskapsakad. Foerh.*, 1900(4), 393, 1900.
4. **Cleve, P. T. and Grunow, A.,** Beiträge zur Kenntniss der arctischen Diatomeen, *K. Sven. Vetenskapsakad. Handl.*, 17(2), 1, 1880.
5. **Grunow, A.,** Die Diatomeen von Franz Josephs Land, *Akad. Wiss., Wien, Denkschr. Math.-Naturwiss. Kl. B*, 48, 53, 1884.
6. **Østrup, E.,** Marine Diatomeer fra østgrønland, *Medd. Groenl.*, 18, 395, 1895.
7. **Gran, H. H.,** Bacillariaceae vom kleinen Karajakfjord, *Bibl. Bot.*, 42, 13, 1897.
8. **Gran, H. H.,** Diatomaceae from the ice-floes and plankton of the Arctic Ocean, *Sci. Res. Norw. N. Polar Exped.*, 4(11), 3, 1904a.
9. **Gran, H. H.,** Die Diatomeen der arktischen Meere. I. Teil. Diatomeen des Planktons, *Fauna Arct.*, 3, 511, 1904b.
10. **Van Heurck, H.,** Diatomeés. Resultats du Voyage du S. Y. "Belgica" en 1897-9 sous le Commandement de A. de Gerlache de Gomery, Rapports Scientifiques: Botaniques, *Exped. Antarct. Belg.*, 5(2), 3, 1909.
11. **Sutherland, P. C.,** *Journal of a Voyage in Baffin's Bay and Barrow Straits, in the Years 1850-1851, Performed by H. M. Ships "Lady Franklin" and "Sophia," Under the Command of Mr. William Penny, in Searth of The Missing Crews of H. M. Ships "Erebus" and "Terror,"* 2 Vols., Longman, Brown, Green, and Longmans, London, 1852.
12. **Nansen, F.,** *Farthest North*, Vol. 1, Harper and Brothers, New York, 1897.
13. **Nansen, F.,** Protozoa on the ice-floes of the North Polar Sea, *Sci. Res. Norw. N. Polar Exped.*, 5(16), 1, 1906.
14. **Hooker, J. D.,** *The Botany of the Antarctic Voyage of H. M. Discovery Ships Erebus and Terror in the Years 1839-1843. I. Flora Antarctica*, J. Cramer, Weinheim, 1847 (reprint 1963), 503.
15. **Heiden, H. and Kolbe, R. W.,** Die marinen Diatomeen der deutschen Südpolar-Expedition 1901-1903, *Dtsch. Süd-Polar-Exped. 1901-1903*, 8(5), 450, 1928.
16. **Hendey, N. I.,** The plankton diatoms of the Southern Seas, *Discovery Rep.*, 16, 151, 1937.
17. **Meister, F.,** Mission du "Porquoi-Pas?" en 1929 sous le Commandement du Dr. J.-B. Charcot. Diatomes récolteés par R.-Ph. Dollfus un glace flottante, *Bull. Mus. Nat. Hist. Nat.*, (2nd series), 2, 329, 1930.
18. **Meguro, H., Ito, K., and Fukushima, H.,** Diatoms and the ecological conditions of their growth in sea ice in the Arctic Ocean, *Science*, 152, 1089, 1966.

19. **Meguro, H., Ito, K., and Fukushima, H.,** Ice flora (bottom type): a mechanism of primary production in polar seas and the growth of diatoms in sea ice, *Arctic*, 20, 114, 1967.
20. **Hoshiai, T. and Kato, M.,** Ecological notes on the diatom community of the sea ice in Antarctica, *Bull. Mar. Biol. St. Asamushi, Tohoku Univ.*, 10, 221, 1961.
21. **Usachev, P. I.,** The microflora of polar ice, *Tr. Inst. Okeanol., Akad. Nauk SSSR*, 3, 216, 1949 (in Russian).
22. **Horner, R. and Alexander, V.,** Algal populations in Arctic sea ice: an investigation of heterotrophy, *Limnol. Oceanogr.*, 17, 454, 1972.
23. **Alexander, V., Horner, R., and Clasby, R. C.,** Metabolism of Arctic sea ice organisms, Rep. R74-4, Institute of Marine Science, University of Alaska, Fairbanks.
24. **Foy, M. G. and Hsiao, S. I. O.,** Phytoplankton data from the Beaufort Sea, 1973-1975. Environment Canada, Fisheries and Marine Service Tech. Rep. No. 617, Ottawa, 1976.
25. **Hsiao, S. I. C.,** Quantitative composition, distribution, community structure and standing stock of sea ice microalgae in the Canadian Arctic, *Arctic*, 33, 768, 1980.
26. **Dunbar, M. J. and Acreman, J. C.,** Standing crops and species composition of diatoms in sea ice from Robeson Channel to the Gulf of St. Lawrence, *Ophelia*, 19, 61, 1980.
27. **Schandelmeier, L. and Alexander, V.,** An analysis of the influence of ice on spring phytoplankton population structure in the southeast Bering Sea, *Limnol. Oceanogr.*, 26, 935, 1981.
28. **Takahashi, E.,** Floristic study of ice algae in the sea ice of a lagoon, Lake Saroma, Hokkaido, *Mem. Natl. Inst. Polar Res., (Ser. E)*, 34, 49, 1981.
29. **Grainger, E. H. and Hsiao, S. I. C.,** A study of the ice biota of Frobisher Bay, Baffin Island 1979-1981, Canadian Manuscript Report of Fisheries and Aquatic Sciences No. 1647, Ottawa, 1982.
30. **Cross, W. E.,** Under-ice biota at the Pond Inlet ice edge and in adjacent fast ice areas during spring, *Arctic*, 35, 13, 1982.
31. **Horner, R. and Schrader, G. C.,** Relative contributions of ice algae, phytoplankton, and benthic microalgae to primary production in nearshore regions of the Beaufort Sea, *Arctic*, 35, 485, 1982.
32. **Booth, J. A.,** The epontic algal community of the ice edge zone and its significance to the Davis Strait ecosystem, *Arctic*, 37, 234, 1984.
33. **Grant, W. S. and Horner, R.,** Growth responses to salinity variation in four Arctic ice diatoms, *J. Phycol.*, 12, 180, 1976.
34. **Palmisano, A. C. and Sullivan, C. W.,** Physiology of sea ice diatoms. I. Response of three polar diatoms to a simulated summer-winter transition, *J. Phycol.*, 18, 489, 1982.
35. **Bunt, J. S. and Wood, E. J. F.,** Microalgae and Antarctic sea-ice, *Nature (London)*, 199, 1254, 1963.
36. **Buinitsky, V. Kh.,** The influence of microalgae on the structure and strength of Antarctic sea ice, *Oceanology (USSR)*, 8, 771, 1968.
37. **Buinitsky, V. Kh., Nikolayev, V. A., and Nikolayev, S. Ye.,** Microscopic algae in fast ice of Alasheyev Bay, *Probl. Arktiki Antark.*, 49, 87, 1977 (in Russian).
38. **Watanabe, K.,** Centric diatom communities found in the Antarctic sea ice, *Nankyoku Shiryo*, 74, 119, 1982.
39. **El-Sayed, S. Z.,** Biological aspects of the pack ice ecosystem, in *Proc. Symposium Antarctic Ice and Water Masses*, Deacon, G., Ed., Scientific Committee on Antarctic Research, Cambridge, 1971, 35.
40. **Bunt, J. S., Owens, O., van H., and Hoch, G.,** Exploratory studies on the physiology and ecology of a psychrophilic marine diatom, *J. Phycol.*, 2, 96, 1966.
41. **Bunt, J. S.,** Some characteristics of microalgae isolated from Antarctic sea ice, *Antarct. Res. Ser.*, 11, 1, 1968.
42. **Bunt, J. S. and Lee, C. C.,** Data on the composition and dark survival of four sea-ice microalgae, *Limnol. Oceanogr.*, 17, 458, 1972.
43. **Whitaker, T. M. and Richardson, M. G.,** Morphology and chemical composition of a natural population of an ice-associated Antarctic diatom *Navicula glaciei*, *J. Phycol.*, 16, 250, 1980.
44. **McFadden, G. I., Moestrup, Ø., and Wetherbee, R.,** *Pyramimonas gelidicola* sp. nov. (Prasinophyceae), a new species isolated from Antarctic sea ice, *Phycologia*, 21, 103, 1982.
45. **Gillen, F. T., McFadden, G. I., Wetherbee, R., and Johns, R. B.,** Sterols and fatty acids of an Antarctic sea ice diatom, *Stauroneis amphioxys*, *Phytochemistry*, 20, 1935, 1981.
46. **Burch, M. D. and Marchant, H. J.,** Motility and microtubule stability of Antarctic algae at sub-zero temperatures, *Protoplasma*, 115, 240, 1983.
47. **Bursa, A. S.,** Phytoplankton in coastal waters of the Arctic Ocean at Point Barrow, Alaska, *Arctic*, 16, 239, 1963.
48. **Usachev, P. T.,** Biological analysis of ice-floes, *C. R. Dokl. Acad. Sci. URSS*, 19, 645, 1938.
49. **Fukushima, H.,** Preliminary report on diatoms from colored sea ice in Point Barrow, Alaska, *Nankyoku Shiryo*, 24, 31, 1965. (In Japanese, English summary.)

50. **Pett, R. J., Vickers, G. P., Acreman, J. C., and Ethier, A. G.,** A seasonal study of the epontic and planktonic communities near the Issungnak artificial island and McKinley Bay in the Canadian Beaufort Sea. Contract ISZ81-00026 for Department of Supply and Services Canada by Arctic Laboratories Ltd., Syndey, 1983.
51. **Poulin, M. and Cardinal, A.,** Sea ice diatoms from Manitounuk Sound, southeastern Hudson Bay (Quebec, Canada). I. Family Naviculaceae, *Can. J. Bot.*, 60, 1263, 1982.
52. **Poulin, M. and Cardinal, A.,** Sea ice diatoms from Manitounuk Sound, southeastern Hudson Bay (Quebec, Canada). II. Naviculaceae, genus *Navicula*, *Can. J. Bot.*, 60, 2825, 1982.
53. **Poulin, M. and Cardinal, A.,** Sea ice diatoms from Manitounuk Sound, southeastern Hudson Bay, (Quebec, Canada). III. Cymbellaceae, Entomoneidaceae, Gomphonemataceae, and Nitzschiaceae, *Can. J. Bot.*, 61, 107, 1983.
54. **Demers, S., Therriault, J.-C., and Descolas-Gros, C.,** Biomasse et composition spécifique de la microflora des glaces saisonnières: influences de la lumière et de la vitesse de congélation, *Mar. Biol.*, 78, 185, 1984.
55. **Hsiao, S. I. C.,** A checklist of marine phytoplankton and sea ice microalgae recorded from Arctic Canada, *Nova Hedwigia*, 37, 225, 1983.
56. **Saito, K. and Taniguchi, A.,** Phytoplankton communities in the Bering Sea and adjacent seas. II. Spring and summer communities in seasonally ice-covered areas, *Astarte*, 11, 27, 1978.
57. **Alexander, V. and Chapman, T.,** The role of epontic algal communities in Bering Sea ice, in *The Eastern Bering Sea Shelf: Oceanography and Renewable Resources*, Vol. 2, Hood, D. W. and Calder, J. A., Eds., University of Washington Press, Seattle, 1981, 773.
58. **Hällfors, G. and Niemi, Å.,** A *Chrysochromulina* (Haptophyceae) bloom under the ice in the Tvärminne Archipelago, southern coast of Finland, *Memo. Soc. Fauna Flora Fenn.*, 50, 89, 1974.
59. **Grøntved, J.,** Phytoplankton studies. I. *Nitzschia frigida* Grun., an Arctic-Inner-Baltic diatom found in Danish waters, *K. Dan. Vidensk. Selsk. Biol. Medd.*, 18(12), 1, 1950.
60. **Hart, T. J.,** Phytoplankton periodicity in Antarctic surface waters, *Discovery Rep.*, 21, 261, 1942.
61. **Hasle, G. R.,** An analysis of the phytoplankton of the Pacific Southern Ocean, *Hvalråd. Skr.*, 52, 1, 1969.
62. **Burkholder, P. R. and Mandelli, E. F.,** Productivity of microalgae in Antarctic sea ice, *Science*, 149, 872, 1965.
63. **Whitaker, T. M.,** Sea ice habitats of Signy Island (South Orkneys) and their primary productivity, in *Adaptations within Antarctic Ecosystems*, Llano, G. A., Ed., Gulf Publ. Co., Houston, 1977, 75.
64. **Ackley, S. F., Taguchi, S., and Buck, K. R.,** Primary productivity in sea ice of the Weddell Region, Report 78-19, Cold Regions Research and Engineering Laboratory, Hanover, 1978.
65. **Buck, K. R.,** Morphology and distribution of the Acanthoecidae (Choanoflagellata) from the Weddell Sea during the austral summer, 1977, Report 80-16, Cold Regions Research and Engineering Laboratory, Hanover, 1980.
66. **Buck, K. R.,** A study of choanoflagellates (Acanthoecidae) from the Weddell Sea, including a description of *Diaphanoeca multiannulata* n. sp., *J. Protozool.*, 28, 47, 1981.
67. **Takahashi, E.,** Loricate and scale-bearing Protista from Lützow-Holm Bay, Antarctica. I. Species of the Acanthoecidae and the Centrohelida found at a site selected on the fast ice, *Nankyoku Shiryo*, 73, 1, 1981.
68. **Grossi, S. M. and Sullivan, C. W.,** The vertical distribution of algae in bottom sea ice, *EOS*, 64, 1090, 1983.
69. **Meunier, A.,** *Microplankton des Mers de Barents et de Kara, Duc d'Orleans Compagne Arctique de 1907*, Charles Bulens, Brussels, 1910.
70. **Bursa, A. S.,** *Discoasteromonas calciferus* n. sp., an Arctic relict secreting *Discoaster* Tak [sic] Sin Hok 1927, *Grana Palynol.*, 6, 147, 1965.
71. **Horner, R.,** *Discoasteromonas calciferus* Bursa, an unusual organism occurring in Arctic sea ice, *J. Phycol.*, 8, (Suppl.) 7, 1972.
72. **Horner, R.,** unpublished data, 1972.
73. **Bursa, A. S.,** Morphogenesis and taxonomy of fossil and contemporary Dinophyta secreting discoasters, in *Proc. II Planktonic Conference Roma, 1970*, Vol. 1, Farinacci, A., Ed., Edizioni Technoscienza, Rome, 1971, 129.
74. **Horner, R.,** Phytoplankton Studies in Coastal Waters near Barrow, Alaska, Ph.D. thesis, University of Washington, Seattle, 1969.
75. **Matheke, G. E. M. and Horner, R.,** Primary productivity of the benthic microalgae in the Chukchi Sea near Barrow, Alaska, *J. Fish. Res. Board Can.*, 31, 1779, 1974.
76. **Mitchell, J. G. and Silver, M. W.,** Modern archaeomonads indicate sea-ice environments, *Nature (London)*, 296, 437, 1982.
77. **Adam, D. P. and Mahood, A. D.,** Chrysophyte cysts as potential environmental indicators, *Bull. Geol. Soc. Am.*, 92, 839, 1981.
78. **Deflandre, G.,** Archaeomonadiceae, une famille nouvelle de Protistes fossiles a loge siliceux, *C. R. Acad. Sci.*, 194, 1859, 1932.

79. **Horner, R.,** Do ice algae produce the spring phytoplankton bloom in seasonally ice-covered waters? A review of recent literature, in *Proc. Seventh Symposium on Living and Fossil Diatoms*, Mann, D., Ed., J. Cramer, Braunschweig, 1984, 401.
80. **Grainger, E. H. and Evans, M. S.,** Seasonal variations in chlorophyll and nutrients in a Canadian Arctic estuary, *Estuaries*, 5, 294, 1982.
81. **Grainger, E. H.,** The annual nutrient cycle in sea-ice, in *Polar Oceans*, Dunbar, M. J., Ed., The Arctic Institute of North America, Calgary, 1977, 285.
82. **Walsh, J. J.,** Vertical distribution of Antarctic phytoplankton. II. A comparison of phytoplankton standing crops in the Southern Ocean with that of the Strait of Florida, *Limnol. Oceanogr.*, 14, 86, 1969.
83. **Hasle, G. R.,** The biogeography of some marine planktonic diatoms, *Deep-Sea Res.*, 23, 319, 1976.
84. **Fryxell, G. A., Doucette, G. J., and Hubbard, G. F.,** The genus *Thalassiosira:* the bipolar diatom *T. antarctica* Comber, *Bot. Mar.*, 24, 321, 1981.
85. **Villareal, T. A. and Fryxell, G. A.,** Bipolarity in the marine diatoms *Thalassiosira antarctica* Comber and *Porosira glacialis* (Grun.) Jorg., *EOS*, 63, 47, 1982.

Chapter 7

SEA ICE BACTERIA: RECIPROCAL INTERACTIONS OF THE ORGANISMS AND THEIR ENVIRONMENT

Cornelius W. Sullivan

TABLE OF CONTENTS

I. INTRODUCTION

The available literature on bacteria from polar oceans is small and information on sea ice associated bacteria is very scant. However, a renewed interest in studies of high latitude biological oceanography coupled with recent advances in microbial ecology call for an appraisal of bacteria associated with ice and their role in Arctic and Antarctic ecosystems.

Maykut[1] has thoroughly covered the physical characteristics of the ice microenvironment and Palmisano and Sullivan[2] have reiterated and extended the discussion of those physical and chemical features of the ice which probably have significant influence on the associated microbiota. It is reasonable to expect from thermal conditions alone that bacteria which are metabolically active within the ice microenvironment would have characteristics defined as psychrotrophic or psychrophilic by Morita.[3] One expects these organisms to be capable of growth and metabolism at ambient temperatures of 0 to −2°C. Yet even as recently as 1968, some microbiologists continued to describe microflora and their metabolic activities as extremely low or nonexistent in polar seas.[4,5]

Since that time, several investigators have demonstrated an abundance of bacterioplankton in polar oceans with growth rates and metabolic activities commensurate with those measured in temperate oceans.[6-8] There is no reason to believe that at least a portion of the seawater bacterial community also inhabits the ice, since sea ice presumably is inoculated or "seeded" primarily from the water column. One might also expect ice bacteria to be physiologically unique because of differences in the physical, chemical, and biological characteristics of the ice microenvironment compared with the water column. It is likely that a selection takes place for those bacterioplankton whose growth is permitted or even favored by high salinities, low temperatures (−2 to −50°C), surfaces (ice crystals, organic detritus, and algal cells), and high concentrations of dissolved and particulate organic matter.

From the current literature some information is available on distribution and abundance, morphological and taxonomic diversity, activity, growth, and associations of ice bacteria with ice algae. Areas in which ice bacteria are suspected to play an important role include:

1. Secondary microbial production mediated through the microbial loop
2. Remineralization and recycling of ice-associated organic matter
3. Maintenance of a balance in the ice microenvironment with regard to detoxification and oxygen consumption
4. Trace gas production
5. Ice nucleation and early stages of sea ice formation

II. TAXONOMY AND DIVERSITY

Bacterial isolates collected from the surface of Beaufort Sea ice were characterized as Gram-negative rods which occur as single cells. Approximately 20% of the bacteria were motile, 30% were pigmented, all were facultative anaerobes, and none could grow at 37°C. Thus, these bacteria were psychrotrophic[3] in their temperature/growth characteristics. Compared with water column and sediment bacteria they showed very limited abilities to utilize individual organic substrates[9] and all required vitamins and complex growth factors.[10] The complex growth factor requirements of these bacteria may result from the close association of bacteria with photosynthetically active microalgae in the sea ice environment.[11-13]

The diversity of bacterial populations from sea ice in the Beaufort Sea was determined by Kaneko et al.[14] They reasoned that methods used for determining community stress for higher organisms could be applied to bacterial populations. Biological communities of plants and animals characteristically contain few species with many individuals or many species with few individuals.[15] Species diversity is usually low in physically stressed environments such as polar regions. Atlas' group applied a taxonomic diversity indexe to bacterial populations to investigate their relationship to population size and population-independent factors. They used numerical taxonomic criteria to describe phenotypic clusters of bacteria which were considered equivalent to species. Numbers of taxonomic groups and individuals were used to calculate the Shannon diversity index.[16] Kaneko et al.[14] found that bacterial diversity increased with decreasing population size as in winter when water column bacteria were below 10^4 CFU ℓ^{-1} (CFU = colony-forming units). By contrast, summer showed lower diversity at similar population sizes. The 2°C temperature difference between winter and summer was considered insufficient to explain the differences, but the lack of light and phytoplankton photosynthesis during winter was postulated to be a severe stress on the bacteria since their absence might deprive the community of needed carbon and energy sources for metabolism and growth. Bacteria from sea ice samples were also less than 10^4 CFU ℓ^{-1} and no correlation between population size and diversity index was found. The authors[14] speculated that bacteria are frozen into the ice matrix and survive but do not grow, so that selection pressures are not operative. Since samples for bacteria were taken from surface ice, the no growth supposition is probably valid for winter-collected samples; however, bacteria at the ice surface were probably derived from populations frozen into the ice during autumn when late season plankton blooms may still occur.[17]

III. DISTRIBUTION AND ABUNDANCE

Ice bacteria have been reported from both Arctic[9,18-20] and Antarctic sea ice.[11,12,21-24] Some of these reports were based on samples taken within 20 km of shore, but Kaneko et al.,[9] Griffiths et al.,[19] and Dahlback et al.[20] sampled Arctic pack ice. Marra et al.[22] found bacteria in three samples from cores collected in pack ice and Miller et al.[24] reported bacterial concentration, biomass, and activity in 22 cores from western Weddell Sea pack ice in November. This suggests a wide distribution for ice bacteria since 12% of the world oceans are covered by sea ice.[25]

The vertical distribution of bacteria in sea ice has been described only for the McMurdo Sound and the Weddell Sea, Antarctica. Sullivan and Palmisano[12] found an abundant, diverse, and active bacterial community within annual sea ice and at the ice-seawater interface. The bacterial standing crop from 30 ice cores was 3.9×10^{11} bacteria m^{-2} and concentrations as high as 6×10^9 bacteria ℓ^{-1} were observed in ice meltwater. Vertical profiles of bacterial concentration revealed that the bottom 0.2 m, including the skeletal layer, of ice contained approximately 10 times higher biomass concentrations of bacteria than the upper portions of congelation ice. The observed difference in bacterial cell numbers and biomass in the lower and upper ice was postulated to result from the high to low gradients (going from bottom to top of annual ice) in temperature, organic nutrients, ice algae, and their metabolic products. By contrast, salinity increased from the ice-seawater to ice-air interface.

Bacterial concentrations in ice samples have been reported to vary from 10^4 to 10^6 bacteria $m\ell^{-1}$; however, since these crop sizes have not been reported on a m^{-2} basis, it is not possible to make valid biomass comparisons among various geographical areas. An exception to this is the report by Sullivan and Palmisano[12] who estimated an ice bacterial biomass of 9.8 mg C m^{-2} for McMurdo Sound. Marra et al.[22] reported that bacteria from three samples of Weddell Sea pack ice could represent about 700 mg C m^{-3}, while in the water column, bacteria contributed only 0.7 mg C m^{-3}. Such concentration estimates can be misleading when ice sample size or location in the ice are not reported. Recent investigations[24,26] in the Weddell Sea pack ice have shown a vertical profile for ice bacteria that differs significantly from those observed for fast ice and a bacterial biomass which ranged from 24.0 to 97.5 mg C m^{-2}. These studies, which included data from 22 ice cores, did not substantiate the high biomass values reported by Marra et al.[22] Yet, one must be cautious in making comparisons on such small data sets, since geographic, seasonal, and small scale distributions may be highly variable.

IV. *IN SITU* ACTIVITY

During investigations of microbial activity and bacterial concentration in the Beaufort Sea, Griffiths et al.[19] collected five samples of floating ice during the summer of 1975 and three samples of ice of unspecified origin during winter of 1976. In the latter three samples they found 1.3 to 2.2 $\times$ 10^8 cells ℓ^{-1} using the acridine orange direct count (AODC) technique. This bacterial concentration was not significantly different from that found in the adjacent seawater. The focus of the ice-related portion of this work was to determine the influence of freshwater input by melting pack ice on cell concentrations and microbial activity.[19] This approach was taken because it was believed that these processes might have a profound effect on nutrient recycling in surface water of the Beaufort Sea. The relative microbial activity (V_{max}) of the ice meltwater was measured using uniformly labeled ^{14}C-L-glutamic acid as substrate. Incubations were in the dark within 1°C of the *in situ* temperature. The V_{max} of summer ice samples ranged from 1 to 68 ng glutamic acid ℓ^{-1} hr^{-1}, while for winter ice, the values were 0.7 to 4 ng glutamic acid ℓ^{-1} hr^{-1}. A comparison of relative microbial activities in melted ice compared to seawater adjacent to the ice showed similar values. Mixtures of seawater and melted ice (1:1) showed no adverse effect on glutamic acid utilization. Thus, the input of freshwater from the melted pack ice reduced neither the numbers of bacteria nor their heterotrophic activity.[19]

Possible explanations were offered for the high relative bacterial activity in ice meltwater compared to seawater: (1) organisms that survive freezing are those that most actively utilize glutamic acid; (2) a significant number of marine bacteria survive freezing and are able to grow when the ice melts; and (3) many bacteria are not actually frozen, but are concentrated into saline pockets of water within the ice. The high salinity brine was hypothesized to offer some degree of freeze-injury protection. Griffiths et al.[19] suggested that the latter possibility seemed most likely. Unpublished data of Dr. R. M. Atlas, who found a patchy distribution of bacterial colony-forming units in ice core samples taken from the Beaufort Sea, were offered as supporting evidence.

V. GROWTH

A fundamental question with regard to the presence of microorganisms in sea ice can be posed: are bacteria metabolically active and growing *in situ?*

That this question arises stems from three independent observations: (1) microorganisms (microalgae) may be concentrated in sea ice by a purely physical means involving ice nucleation and ice scavenging;[27] (2) previously held views that polar oceans contained few microorganisms and that ambient low temperatures precluded their growth;[4,5] and (3) the lack of experimental evidence in the literature to suggest that *in situ* bacterial metabolism and growth actually did occur in the ice.

A. Physical Concentration

Ackley[27] and Clarke and Ackley[28] made observations in the Weddell Sea during February 1980 that indicated that newly formed (<1-day-old and <20 cm thick) sea ice contained algae at concentrations several times the concentration in the adjacent seawater. The short time of formation argued against a purely biological increase in algal number in the newly formed ice. They also reported microscopic observations of sea ice through sections in which algal cells could be observed both within individual frazil ice crystals and between ice crystals. Ackley[27] proposed two physical mechanisms by which algae could be concentrated in sea ice. The first is ice nucleation whereby a frazil ice crystal nucleates on an algal cell leading to a cell within an ice crystal. The second mechanism, concentration by scavenging, would occur as the result of preformed ice crystals in the water column contacting algal cells as they travel upwards towards the water surface. Cells adhering to the "sticky" crystal surface would accumulate between adjacent ice crystals in the newly formed ice. Ackley[27] argued that scavenging is likely to be the more effective mechanism because several cells can be accommodated on each crystal surface. Further evidence for these ideas was recently presented by Garrison et al.[29] It should be noted that physical concentration mechanisms may be unique to frazil ice such as occurs in the Weddell Sea. Columnar ice formation, typical of congelation ice, apparently proceeds by a different mechanism[30] and physical accumulation of microbial cells is not as likely. While these reports deal with algal cells, I believe that similar accumulation mechanisms might be envisioned for bacterial cells. The role of bacteria as ice nucleants will be discussed later in this chapter.

B. Biological Increase

The first experimental evidence to suggest an active bacterial metabolism in sea ice resulted from Horner and Alexander's[18] study of heterotrophy in Arctic sea ice algae. To test for heterotrophic activity associated with sea ice algae, they incubated ice meltwater samples with radiolabeled ^{14}C-glycine, ^{14}C-glucose, and ^{3}H-acetate and found that 4 of 15 samples tested by kinetic analysis showed Michaelis-Menton behavior. Estimates of (V_{max}) and K + S were underestimated because respiration was not measured. The turnover time, calculated according to Hobbie,[31] was 200 hr for ^{14}C-glycine. Little evidence of uptake by algae was found by examination of autoradiographs. Other cells described as "small particles, most likely bacteria, but possibly small squashed flagellates" were found to have dark spots (developed silver grains) associated with them. This work did not support the premise that heterotrophy was a major energy source for diatoms in sea ice and most of the radioisotope uptake was attributed to bacteria and small flagellates.[18]

Sullivan et al.[21] and Sullivan and Palmisano[12] used light and electron microscopy to examine bacteria from the lower 0.2 m of annual sea ice in McMurdo Sound, Antarctica, for evidence of *in situ* growth and metabolism. The bacteria in the lower portions of the ice were both more numerous and larger in size than those in the upper 1.5 to 2 m of ice. The ice bacteria were robust compared with the bacterioplankton of the underlying water column. Large cells with dimensions >1 μm by >0.5 μm were most common and were estimated to have larger biovolumes than bacterioplankton in the water column (Figure 1). Such large cells often are indicative of bacteria growing in high nutrient environments. A high frequency of dividing cells was found when samples fixed *in situ* were subsequently examined in the electron microscope. One morphologically distinct cell type was estimated to have a division frequency $>16\%$. Such frequencies of dividing cells are indicative of high population growth rates.[32] Microautoradiographic evidence of ^{14}C-L-serine uptake by ice bacteria under simulated *in situ* conditions provided additional evidence that the population was capable of active metabolism at $-1.8°C$.[21]

McConville and Wetherbee[23] reported the uptake of a ^{3}H-labeled amino acid mixture in laboratory incubations by the sea ice microbial community collected near the Australian Antarctic station of Casey. Turnover times of the amino acid were reported to vary from 15 to 154 hr. The highest bacterial numbers and shortest turnover times coincided with communities having the highest primary production, but these authors failed to identify which component of the ice community was responsible for the uptake (bacteria or microalgae). When the community was fractionated on 2 μm pore size Nucleopore filters, only 1 to 6% of the heterotrophic activity was found in the less than 2 μm size fraction. McConville and Wetherbee[23] suggested that free-living bacteria (<2 μm size fraction) in the ice play only a minor role in recycling of organic matter in the sea ice community.[23]

Evidence for bacterial growth in the ice was recently provided by Sullivan et al.[21] and

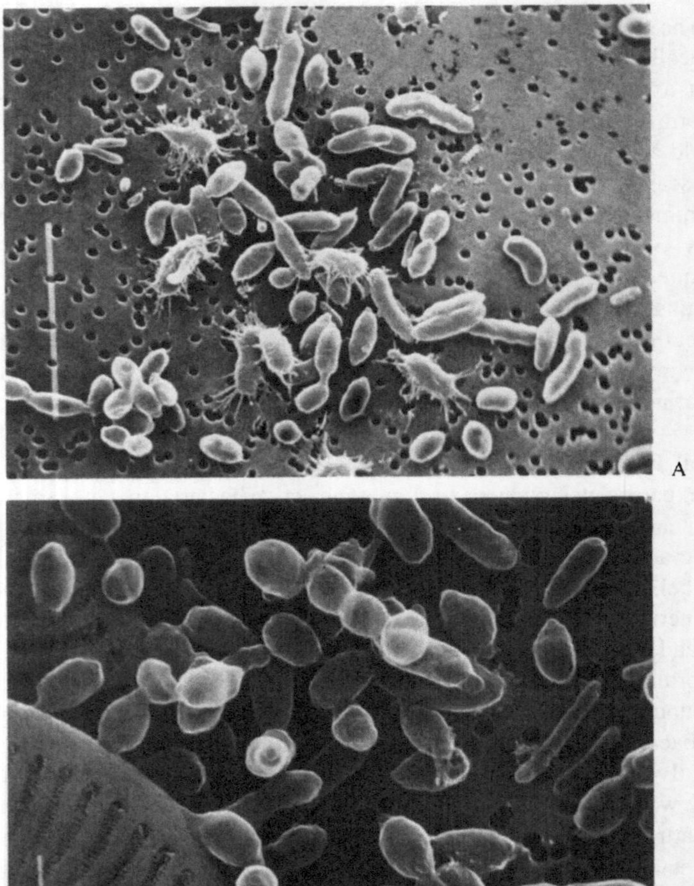

FIGURE 1. Scanning electron micrographs of sea ice bacteria collected in 2m fast ice from McMurdo Sound, Antarctica, 77°51′S, 166°40′E (A and B), and from 1.5 m pack ice in the Weddell Sea, 61°59.6′S, 36°27.3′W. A and B are from samples collected from the bottom 20 cm of ice; C was collected at a depth of 0.6 to 0.9 m from an ice floe 1.8 m thick. Note evidence of exopolymeric substance covering the surface of some cells in A and C, and also the predominance of large cell biovolumes. White bar represented (A) 5 μm, (B and C) 0.5 μm.

Grossi et al.[13] who conducted a light perturbation experiment on the annual sea ice of McMurdo Sound near Cape Armitage. During October to December 1981, two experimental quadrats of 100 m^2 were constructed by covering one with 70 cm of snow, while the second was left undisturbed with natural snow 7 cm deep. During the study period, both algae (cell numbers and chlorophyll *a*) and bacterial cell numbers increased in the unperturbed quadrat which had underice irradiance of 0.20 to 2.90 $\mu E\ m^{-2}\ sec^{-1}$.[33] Algal cell numbers increased approximately 15-fold and bacterial numbers 8-fold during the 45-day study. Neither algal nor bacterial cell numbers increased significantly in the experimentally darkened quadrat (0.06 to 0.11 $\mu E\ m^{-2}\ sec^{-1}$). Increases in algal cell numbers at the low irradiances of the unperturbed quadrat were attributed to a shade-adapted algal community with a low compensation intensity for photosynthesis. The lack of algal growth in the perturbed quadrat suggested that 0.11 $\mu E\ m^{-2}\ sec^{-1}$ was below compensation intensity for the community. The bacterial community revealed a similar growth pattern, yet they were shown not to be phototrophic. Some of the ice bacteria appeared to be physically attached to algal cell surfaces and the fraction of attached bacteria increased during the season. The authors interpreted their data to suggest that bacteria were physically and chemically coupled to the ice algae and grew in the unperturbed quadrat at the expense of reduced organic material provided by algal photosynthesis.

In further support of the idea of active bacterial growth in the ice, Sullivan et al.[21] showed that the attached bacteria took up ^{3}H-serine, ^{3}H-glucose, and ^{3}H-thymidine when small cubes (ca. 1 cm^3) of sea ice were perfused with filtered seawater containing radiolabeled compounds. After incubation under simulated *in situ* conditions in a chamber designed to mimic the under-ice conditions,[34] microautoradiographs were prepared and bacteria were observed to contain high densities of exposed silver grains for all radiolabels tested. This demonstrated not only intermediary metabolism of an amino acid and sugars by ice bacteria, but also showed that net DNA synthesis occurred by incorporation of ^{3}H-thymidine into macromolecules. Net DNA synthesis is a prerequisite for growth and cell division. Thus, three lines of evidence suggest net growth of ice bacteria *in situ:* DNA synthesis; high frequency of dividing cells; and net increase in bacterial cell concentration in the ice over time. Sinks for these organisms have not been investigated but bactivorus grazers associated with sea ice are likely consumers of bacterial biomass.

VI. ICE NUCLEATION ACTIVE BACTERIA

Freezing nuclei are defined as particles capable of nucleating ice in supercooled water. If no distinction is made between freezing and deposition the nuclei are referred to as ice nuclei.[35] Ice nuclei, both from terrestrial and marine sources, have been studied by atmospheric scientists for their role in precipitation processes (rain, snow, hail, and fog). Biotic and abiotic ice nuclei have been reported.[36-38]

Precipitation generally is believed to be closely related to the initiation and growth of ice crystals in supercooled clouds. Nuclei upon which such ice crystals form include soil particles, particularly kaolin and other clays, as well as extraterrestrial meteoritic materials.[39,40] Biogenic sources of ice nuclei have also been identified. Soulage[41] first reported that bacteria could become centers of ice crystals in cloud chambers. Later, Fresh[42] demonstrated the specific role of bacteria from natural materials in producing freezing nuclei.

Bacterial-derived nuclei have been observed to initiate ice in supercooled water at temperatures warmer than 3.0°C and some nucleation events have been observed at −1.3°C.[43-45] Schnell and Vali[46] showed that activity depended on an intact bacterial cell wall (living or dead), age, and growth environment of the bacterium.[43] Occasional transient appearances of freezing nuclei active at temperatures as warm as −1.3°C were observed during the earlier stages of leaf decay.[46] This activity was shown to depend on intact bacterial cell walls of *Pseudomonas syringae*.

Very few species of terrestrial bacteria appear to act as ice nuclei. Those associated with the initiation of ice formation have been in the *Pseudomonadaceae* group, e.g., *Pseudomonas syringae*, *P. fluorescence*, and *Erwinia herbicola*. The Pseudomonadaceae are well known to be an abundant, if not dominant, culturable group of freshwater and marine bacteria.

One species of epiphytic bacterium, *Pseudomonas syringae*, has been identified as a principal natural ice nucleant initiating frost damage in leaves of frost-prone North American plants at temperatures warmer than −5°C.[46] This bacterium was also found in clouds over the Arctic Ocean.[47] Bulk suspensions of *P. syringae* have shown threshold nucleation temperatures of −1.5 to −2.5°C.[45] Microorganisms with similar threshold nucleation temperatures may be relevant to ice formation in seawater. These nucleation events might result in conjunction with supercooling of ocean waters such as those associated with seawater under ice shelves. More likely, supercooling events may occur in the atmosphere near the sea surface when breaking waves loft water droplets containing ice nucleation active bacteria into the air. Air temperatures in the range −2 to −20°C are common in polar regions. Ice crystals thus nucleated might fall to the sea surface and contribute to the formation of grease ice, the nascent form of sea ice.

The source, composition, and number of small particles (ice nuclei) that may initiate freezing events during the formation of sea ice are not understood. The inorganic constituents of seawater have been shown to be poor ice nucleators.[49] The organic component of seawater has been observed to contain ice nuclei at temperatures as warm as −5°C.[50] These ocean-derived nuclei have also been captured in clean air and in fog above and down wind of marine sources.[51]

Samples collected in the Atlantic Ocean off Bedford, Nova Scotia, had considerably greater freezing nuclei concentrations than from other seawaters tested. A phytoplankton bloom occurred in these waters just prior to sample collection and it was suggested that the high concentrations of active freezing nuclei might be related to the phytoplankton event.[38] Some of the nuclei from seawater are active at temperatures around −4°C and concentrations reach up to 10^7 to 10^8 nuclei per gram of plankton at −10°C. However, no information was reported as to how many cells were present per gram plankton.

Marine bacteria isolated into pure culture have also been observed to nucleate ice formation. A marine bacterium living in close association with the temperate dinoflagellate *Cachonina nei* Loeblich, in culture at Scripps Institution of Oceanography, was shown to be an active ice nucleant at temperatures approaching −1°C.[52] Some bacteria isolated on nutrient agar plates from fog water were observed to initiate ice at −1.5°C. Fogs generally contain more ice nuclei per unit volume than do most seawaters.[38] Until recently, sea ice has not been examined for ice nucleation active bacteria.

Frazil ice is one of three types of sea ice: the other two commonly seen are congelation and snow ice.[30] Frazil ice is usually formed rapidly under dynamic and turbulent conditions in the water column and is composed of small (~1 mm) cylindrical disks of ice.[53] These small disks when driven by the wind can quickly aggregate into pieces of substantial size. The rapid (~hr) production of frazil ice with thickness >50 cm has been predicted by Bauer and Martin.[54] Because of the unique fashion by which frazil sea ice is formed, Ackley[27] predicted that organic material might be involved in its nucleation.

Ackley[27] reported viewing algal cells within individual ice crystals suggesting nucleation and cells at grain boundaries between crystals indicating scavenging. Garrison et al.[29] argued that the principal mechanism for incorporation is scavenging based on the number of crystals per volume of ice and the lack of preferential concentration of any one species. These authors[29] concluded that a mechanism which was independent of cell physiochemical characteristics such as scavenging was more likely than one with implicit dependence on cellular physiochemical characteristics such as nucleation.

The presence of large numbers of bacteria associated with the entire ice profile of landfast[11,12] and floating pack ice[24] suggested the possibility that bacteria rather than algae might be the responsible organisms for biogenic nucleation processes in sea ice.[55] Parker et al.[56] tested pure cultures of Antarctic sea ice diatoms and found that none were ice nucleation active by the freezing drop technique. However, several ice bacteria isolated from McMurdo Sound were active at $-10°C$ and one psychrophilic strain was active with an ice nucleation temperature of -2.0 to $-2.8°C$.

It should be emphasized that no direct connection between the existence of biologically derived nuclei and atmospheric precipitation processes or sea ice formation has been demonstrated. The potential is clearly there and if demonstrated it would represent a unique example of biosphere-atmosphere interaction.

VII. TRACE GASES IN SEA ICE

Hemmingsen[57] reported the relatively impermeable character of freshwater ice. On the other hand, Gosink et al.[58] found sea ice to be 10^4 to 10^5 times more permeable to gases. They postulated that sea ice should play a significant role in the exchange of gases across the sea-ice-air boundaries.

Trace gases such as methane (CH_4), carbon monoxide (CO), carbon dioxide (CO_2), hydrogen (H_2), and nitrous oxide (N_2O) were shown to be emitted at the ice-atmosphere interface in Arctic[58,59] and Antarctic[60] regions. These gases show a slow, but significant, exchange through the relatively porous sea ice which may be important to understanding trace gas composition of the polar atmosphere, aerosol formation, and climate. Microbial aerobic and anaerobic metabolism, nutrient regeneration processes, and photosynthetic activity within sea ice as well as under the ice in the water column and sediments probably produce these gases. It was postulated that pathways for gas exchange are the large grain boundaries between congelation ice crystals and brine channels in annual sea ice.

Gosink and Kelley[61] found rapidly decaying (June) coastal sea ice near Barrow, Alaska, contained surface bubbles with 5 to 20 ppm CO. Atmospheric CO in the Alaskan region is usually 0.05 ppm or less. The enrichment of CO in sea ice is suspected to be of biological origin, since winter studies revealed only a small enrichment compared with late spring values. Although no direct measurements of sea ice microbial community biomass or metabolism have been made in conjunction with atmospheric scientists studying trace gases in sea ice, it is clear that such interdisciplinary investigations are likely to be fruitful.

VIII. FUTURE STUDIES

While the presence of algae and protozoans in sea ice has been recognized for more than a century, evidence of bacteria associated with ice date back only 1 decade. The study of bacteria and indeed the role they play in the biology of sea ice microbial communities in ice-covered oceans, is yet in its infancy. Many important areas of research remain to be explored and they span temporal scales which vary from changes occurring during diurnal cycles to geological times and spatial scales from submicron cell-cell interactions to the global dimensions of the World Ocean.

In addition to ecological studies, the ice-associated bacteria may provide source material for unique microorganisms useful in biotechnology. Psychrophilic organisms from this permanently cold environment are physiologically adapted (at the molecular level) to growth at $-2°C$. They harbor enzymes that are catalytically active at low temperatures and possess enzymes that are easily inactivated at moderate temperatures. These are features that can be useful for a variety of industrial or scientific purposes.[62,63] Of the bacterial strains isolated from Antarctic seawaters by Kobori et al.,[62] 98% produced active alkaline phosphatases (APase). Antarctic bacteria apparently synthesize large quantities of APases constitutively and 26 strains of the 139 examined had heat-labile APases. Only one heat-sensitive APase was purified to homogeneity. It was inactivated by incubation at 55°C for 10 min. The remarkable temperature sensitivity of the APase was demonstrated to be a useful feature for 5′-end-labeling of endonuclease fragments of DNA because endonucleolytic cleavage, dephosphorylation, and subsequent end-labeling could be performed sequentially in one reaction mixture. Because of the rarity of heat-labile APase reported in the literature,[64] Antarctic bacteria are valuable as a source for further isolation of heat-labile enzymes with useful characteristics.

Kobori et al.[63] tested 155 strains of psychrophilic (23%) and psychrotrophic (77%) bacteria isolated from Antarctic sea ice, seawater, sediments, and ice-associated animals for the presence of plasmids. Bacteria isolated from sediments had the highest plasmid incidence (42%), those from seawater the lowest (20%), while 30% of sea ice bacteria harbored plasmids. Plasmids were carried by 16% of psychrophilic bacteria and 35% of psychrotrophic bacteria and yet an explanation for these differences was not apparent.

The frequency of plasmid incidence differed significantly depending on the medium used for the initial isolation. Plasmids were more frequent (46%) in strains isolated on low nutrient agar compared to high nutrient strains (25%). The authors[63] speculated that plasmids may encode traits that permit higher growth rates under conditions where nutrients are at very low concentrations as in seawater. Bacteria harboring such plasmids would have a selective advantage and these strains would be expressed faster as colonies than plasmidless strains.

This study suggests that plasmids are ubiquitous in natural microbial populations of the pristine marine ecosystem of Antarctica including sea ice. It provides preliminary information to microbial ecologists who are interested in examining the functions and evolution of plasmids in natural environments that are essentially free of selective pressures due to anthropogenic materials. Future studies must include the isolation and characterization of these plasmids and a determination of the proteins for which they code. Such investigations may lead to a better understanding of the role of plasmids in natural microbial populations.

At least one strain of an Antarctic sea ice bacterium has been shown to be ice nucleation active at relatively warm temperatures,[56] even after cultivation on laboratory media for several years. Sea ice may be a prime source for new strains of ice nucleation active bacteria since only a very few species have been isolated from terrestrial sources. For an excellent review of ice nucleation characteristics of economically important terrestrial bacteria see Lindow.[65]

Several aspects of the physiology, molecular biology, and ecology of ice bacteria are worthy of consideration for future studies.

ACKNOWLEDGMENTS

The author wishes to thank the NSF Division of Polar Programs through grants DPP-81-17237 and DPP-82-18752 for support during the writing of this chapter. I would also like to thank Dr. A. C. Palmisano and S. McGrath Grossi for critically reading the manuscript.

REFERENCES

1. **Maykut, G. A.,** The ice environment, in *Sea Ice Biota,* Horner, R., Ed., CRC Press, Boca Raton, Fla., 1985, chap. 2.
2. **Palmisano, A. C. and Sullivan, C. W.,** Growth, metabolism, and dark survival in sea ice microalgae, in *Sea Ice Biota,* Horner, R., Ed., CRC Press, Boca Raton, Fla., 1985, chap. 5.
3. **Morita, R. Y.,** Psychrophilic bacteria, *Bacteriol. Rev.,* 29, 144, 1975.
4. **Sorokin, Yu. I.,** On the role of bacteria in the productivity of tropical ocean waters, *Int. Rev. Gesamten Hydrobiol.,* 56, 1, 1971.
5. **Pomeroy, L. R., Wiebe, W. J., Frankenberg, D., Hendricks, C., and Layton, W. L.,** Metabolism of total water columns, *Antarct. J. U.S.,* 4, 188, 1969.

6. **Fuhrman, J. A. and Azam, F.,** Bacterioplankton secondary production estimates for coastal waters of British Columbia, Antarctica, and California, *Appl. Environ. Microbiol.*, 39, 1085, 1980.
7. **Hodson, R. E., Azam, F., Carlucci, A. F., Fuhrman, J. A., Karl, D. M., and Holm-Hansen, O.,** Microbial uptake of dissolved organic matter in McMurdo Sound, Antarctica, *Mar. Biol.*, 61, 89, 1981.
8. **Hansen, R. B., Shafer, D., Ryan, T., Pope, D. H., and Kenneth, H.,** Bacterioplankton in Antarctic Ocean waters during late austral winter: abundance, frequency of dividing cells, and estimates of production, *Appl. Environ. Microbiol.*, 45, 1622, 1983.
9. **Kaneko, T., Hauxhurst, J., Krichevsky, M. I., and Atlas, R. M.,** Numerical taxonomic studies of bacteria isolated from Arctic and subarctic marine environments, in *Microbial Ecology*, Loutit, M. W. and Miles, J. A. R., Eds., Springer-Verlag, Basel, 1978, 26.
10. **Kaneko, T., Krichevsky, M. I., and Atlas, R. M.,** Numerical taxonomy of bacteria from the Beaufort Sea, *J. Gen. Microbiol.*, 110, 111, 1979.
11. **Sullivan, C. W. and Palmisano, A. C.,** Sea ice microbial communities in McMurdo Sound, Antarctica, *Antarct. J. U.S.*, 16, 126, 1981.
12. **Sullivan, C. W. and Palmisano, A. C.,** Sea ice microbial communities. II. Distribution, abundance and diversity of ice bacteria in McMurdo Sound, Antarctica in 1980, *Appl. Environ. Microbiol.*, 47, 788, 1984.
13. **Grossi, S. M., Kottmeier, S. T., and Sullivan, C. W.,** Sea ice microbial communities. III. Seasonal abundance of microalgae and associated bacteria, McMurdo Sound, Antarctica, *Microb. Ecol.*, 10, 231, 1984.
14. **Kaneko, T., Atlas, R. M., and Krichevsky, M. I.,** Diversity of bacterial populations in the Beaufort Sea, *Nature (London)*, 270, 596, 1977.
15. **Fisher, R. A., Corbet, A. S., and Williams, C. B.,** The relation between the number of species and the number of individuals in a random sample of an animal population, *J. Anim. Ecol.*, 12, 42, 1943.
16. **Shannon, C. E. and Weaver, W.,** *The Mathematical Theory of Communication*, University of Illinois Press, Urbana, 1949.
17. **Horner, R.,** Ecology of sea ice microalgae, in *Sea Ice Biota*, Horner, R., Ed., CRC Press, Boca Raton, Fla., 1985, chap. 3.
18. **Horner, R. and Alexander, V.,** Algal populations in Arctic sea ice: an investigation of heterotrophy, *Limnol. Oceanogr.*, 17, 454, 1972.
19. **Griffiths, R. P., Hayasaka, S. S., McNamara, T. M., and Morita, R. Y.,** Relative microbial activity and bacterial concentrations in water and sediment samples taken in the Beaufort Sea, *Can. J. Microbiol.*, 24, 1217, 1978.
20. **Dahlback, B., Gunnarsson, L. A. H., Hermanson, M., and Kjellenberg, S.,** Microbial investigations of surface microlayers, water column, ice and sediment in the Arctic Ocean, *Mar. Ecol. Prog. Ser.*, 9, 101, 1982.
21. **Sullivan, C. W., Palmisano, A. C., Kottmeier, S. T., Grossi, S. M., and Moe, R.,** Influence of light on growth and development of the sea ice microbial community in McMurdo Sound in, *Fourth SCAR Symposium on Antarctic Biology: Food Webs and Nutrient Cycles*, Siegfried, R., Condy, P. R., and Laws, R. H., Eds., Springer-Verlag, Basel, 1985.
22. **Marra, J., Burckle, L., and Ducklow, H.,** Sea ice and water column plankton distributions in the Weddell Sea in late winter, *Antarct. J. U.S.*, 27, 111, 1982.
23. **McConville, M. J. and Wetherbee, R.,** The bottom-ice microalgal community from annual sea ice in the inshore waters of east Antarctica, *J. Phycol.*, 29, 431, 1983.
24. **Miller, M. M., Krempin, D., Manahan, D., and Sullivan, C. W.,** AMERIEZ 1983: growth rates, distribution and abundance of bacteria in the ice edge zone of the Weddell and Scotia seas, Antarctica, *Antarct. J. U.S.*, 19, 103, 1984.
25. **Weeks, W. F.,** Understanding the variations of the physical properties of sea ice, in *Symposium on Antarctic Oceanography*, Currie, R. I., Ed., W. Heffer & Sons Ltd., Cambridge, 1968, 173.
26. **Sullivan, C. W.,** Unpublished data, 1983.
27. **Ackley, S. F.,** Ice scavenging and nucleation: two mechanisms for incorporation of algae into newly forming sea ice, *EOS*, 63, 54, 1982.
28. **Clarke, D. B. and Ackley, S. F.,** Sea ice structure and biological activity in the Antarctic marginal ice zone, *J. Geophys. Res.*, 89, 2087, 1984.
29. **Garrison, D. L., Ackley, S. F., and Buck, K. R.,** A physical mechanism for establishing algal populations in frazil ice, *Nature (London)*, 306, 363, 1983.
30. **Weeks, W. F. and Ackley, S. F.,** The growth, structure and properties of sea ice, Monograph 82-1, Cold Regions Research and Engineering Laboratory, Hanover, 1982.
31. **Hobbie, J. E.,** A method for studying heterotrophic bacteria, in *A Manual on Methods for Measuring Primary Production in Aquatic Environments*, Vollenweider, R. A., Ed., IBP Handbook 12, Blackwell Scientific, Oxford, 1969, 146.

32. **Hagstrom, A., Larsson, U., Harstedt, P., and Normark, S.,** Frequency of dividing cells, a new approach to the determination of bacterial growth rates in aquatic environments, *Appl. Environ. Microbiol.*, 45, 1622, 1983.
33. **Palmisano, A. C., Kottmeier, S. T., Moe, R. L., and Sullivan, C. W.,** Sea ice microbial communities. IV. The effect of light perturbation on microalgae at the ice-seawater interface in McMurdo Sound, *Antarctica, Mar. Ecol. Prog. Ser.*, 21, 37, 1984.
34. **Palmisano, A. C. and Sullivan, C. W.,** Physiology of sea ice diatoms. I. Response of three polar diatoms to a simulated summer-winter transition, *J. Phycol.*, 18, 489, 1982.
35. **Schnell, R. C. and Vali, G.,** World-wide source of leaf derived freezing nuclei, *Nature (London)*, 246, 212, 1973.
36. **Bigg, E. K.,** Ice nucleus measurements in remote areas, *J. Atmos. Sci.*, 30, 1153, 1973.
37. **Brier, G. W. and Kline, D. B.,** Ocean water as a source of ice nuclei, *Science*, 30, 717, 1959.
38. **Schnell, R. C.,** Ice nuclei in seawater, fog water, and marine air of the coast of Nova Scotia: summer 1975, *J. Atmos. Sci.*, 34, 2399, 1977.
39. **Isono, K.,** The nature and origin of ice nuclei in the atmosphere, *J. Meteorol. Soc. Jpn.*, 37, 211, 1959.
40. **Kumani, M.,** Snow crystals and the identification of the nuclei in the northern United States of America, *J. Meteorol.*, 18, 139, 1961.
41. **Soulage, G.,** Les noyaux de congelation de l'atmosphere, *Ann. Geophys.*, 13, 103, 1957.
42. **Fresh, R. W.,** Microbial production of freezing nuclei from decomposing tree leaves, Report AR106, Department of Atmospheric Resources, University of Wyoming, Laramie, 1973.
43. **Maki, L. R., Galyan, E. L., Chien, M. C., and Caldwell, D. C.,** Ice nucleation induced by *Pseudomonas syringae*, *Appl. Microbiol.*, 38, 456, 1974.
44. **Lindow, S., Borchet, R. W., and Upper, C. D.,** The relationship between populations of bacteria active in ice nucleation and frost sensitivity in herbaceous plants, *EOS*, 56, 994, 1975.
45. **Vali, G., Christensen, M., Fresh, R. W., Galyan, E. L., Maki, L. R., and Schnell, R. C.,** Biogenic ice nuclei. II. Bacterial sources, *J. Atmos. Sci.*, 33, 1565, 1976.
46. **Schnell, R. C. and Vali, G.,** Atmospheric ice nuclei from decomposing vegetation, *Nature (London)*, 236, 163, 1972.
47. **Lindow, S. E., Arny, D. C., Upper, C. D., and Borchet, W. R.,** The role of bacterial ice nuclei in frost injury to sensitive plants, in *Plant Cold-hardiness and Freezing Stress Mechanisms and Crop Implications*, Li, P. and Sakai, A., Eds., Academic Press, New York, 1978, 244.
48. **Flanagan, P. W. and Jayaweera, K.,** Evidence for microbial ice nucleation over the Arctic Ocean, *EOS*, 61, 968, 1980.
49. **Pruppacher, H. R. and Klett, J. D.,** *Microphysics of Clouds and Precipitation*, D. Reidel, Boston, 1978, 1.
50. **Schnell, R. C. and Vali, G.,** Biogenic sources of ice nuclei. Part I. Terrestrial and marine sources, *J. Atmos. Sci.*, 33, 1554, 1976.
51. **Schnell, R. C.,** Naturally occurring biological ice nucleants: a review, Int. Conf. on Cloud Physics, *Am. Meteorol. Soc.*, Boston, 1976, 57.
52. **Fall, R. and Schnell, R. C.,** Isolation of ice nucleation-active bacteria from marine sources, Second American Conference on Ice Nucleating Bacteria, Northern Arizona University, Flagstaff, 1984.
53. **Martin, S.,** Frazil ice in rivers and oceans, *Annu. Rev. Fluid Mech.*, 13, 379, 1981.
54. **Bauer, J. and Martin, S.,** A model of grease ice growth in small leads, *J. Geophys. Res.*, 88, 2917, 1983.
55. **Sullivan, C. W.,** Bacteria associated with Antarctic sea ice, Second American Conference on Ice Nucleating Bacteria, Northern Arizona University, Flagstaff, 1984.
56. **Parker, L. V., Sullivan, C. W., Forest, T. W., and Ackley, S. F.,** Ice nucleation activity of Antarctic marine bacteria, Second American Conference on Ice Nucleating Bacteria, Northern Arizona University, Flagstaff, 1984.
57. **Hemmingsen, E.,** Permeation of gases through ice, *Tellus*, 11, 355, 1959.
58. **Gosink, T. A., Pearson, J. G., and Kelley, J. J.,** Gas movement through sea ice, *Nature (London)*, 263, 41, 1976.
59. **Gosink, T. A. and Kelley, J. J.,** Gases in sea ice, Annual report to The Office of Naval Research, Code 461, Institute of Marine Science, University of Alaska, Fairbanks, 1977.
60. **Gosink, T. A.,** Atmospheric trace gases in association with sea ice. *Antarct. J. U.S.*, 15(5), 82, 1980.
61. **Gosink, T. A. and Kelley, J. J.,** Gases in sea ice, Annual report to The Office of Naval Research, Code 461, Institute of Marine Science, University of Alaska, Fairbanks, 1979.
62. **Kobori, H., Sullivan, C. W., and Shizuya, H. S.,** Heat labile alkaline phosphatase from Antarctic bacteria: rapid end labeling of nucleic acids, *Proc. Natl. Acad. Sci. U.S.A.*, 81, 6691, 1984.
63. **Kobori, H., Sullivan, C. W., and Shizuya, H. S.,** Bacterial plasmids in Antarctic natural microbial assemblages, *Appl. Environ. Microbiol.*, 48, 515, 1984.

64. **McComb, R. B., Bowers, G. N., Jr., and Posen, S., Eds.,** *Alkaline Phosphatase*, Plenum Press, New York, 1979, 404.

65. **Lindow, S. E.,** The role of bacterial ice nucleation in frost injury to plants, *Annu. Rev. Phytopathol.*, 21, 363, 1983.

Chapter 8

MARINE ICE FAUNA: ARCTIC

Andrew G. Carey, Jr.

TABLE OF CONTENTS

I. INTRODUCTION

A. General

For all or varying portions of the year, sea ice is present as a unique environment providing a matrix or solid surface in the upper layers of the northern and southern oceans. A diverse group of animals ranging in size from protozoans[1] to marine mammals[2] utilize polar sea ice.[3] An extensive food web is supported during the growing season by algae on the ice undersurface. These food chains extend beyond the ice itself to include fish and zooplankton in the water layer beneath the ice,[4] birds,[5] and mammals.[2,6] A habitat for marine animals, sea ice has been described as a floor,[7] a sand-like substrate for ciliate protozoans, a ceiling for zooplankton,[8] and as an upside-down benthic environment for sea floor invertebrates.[9] The seawater-ice interface is a unique and extensive upper boundary in Arctic, Antarctic, and surrounding seas which is directly affected by changes in insolation and temperature. In the marginal seas this ice cover is seasonal with regular melting, breakup, and freezing cycles.

In the seasonal sea ice zone at the southern boundaries of the Arctic Basin and adjacent seas, the ice cover melts and breaks up generally in the early summer. The ice fauna, whether originating partly from the benthos in shallow water or from the plankton in deeper environments, appears to survive well in the absence of sea ice. During the winter and spring months animals colonize the lower layers of the ice. Other species living in the water column may respond to the presence of the ice cover by adjusting their vertical position, while still others may not be affected at all by the sea ice.[10]

The majority of data available on the animals associated with the undersurface of the ice environment is on the macrofauna, animals generally greater than 0.5 mm in size. These studies have been conducted independently in diverse environments, often without knowledge of previous studies and their results. The data obtained and conclusions reached have often been influenced by the scientific orientation of the researcher and the physical environment in which the work was accomplished. For example, phytoplankton and zooplankton researchers have viewed the ice community as part of the pelagic environment, while benthic workers have stressed interaction of ice with the sea floor in shallow water. The differing environments of fast ice, pack ice, inner continental shelf, outer shelf, and protected waters also undoubtedly affect the range of data collected and the interpretation of the results. The diverse data in the young scientific field of ice fauna ecology have not yet been summarized nor compared on a large scale. This paper provides a summary and some insight into the structure and the role of the ice community in the Arctic and subarctic ecosystems.

B. Definitions

The species composition and ecology of animals associated with the undersurface of the sea ice cover are not yet entirely known and terminology remains confused in the literature. The first descriptions of the bottom ice biota were primarily oriented toward the algae and were often general qualitative descriptions, e.g., "very dirty and often brownish color" or "reddish brown."[11] Early sealers called ice floes with reddish brown concentrations of algae on the undersurface "seal-ice". Seals prefer to lie on this type of ice presumably because increased primary production causes concentrations of associated crustaceans which seals utilize. Andriashev[4] described the overall appearance of the ice biota as "a coating of greenish-brown mud underneath". Antarctic ice algal concentrations were first termed epontic algae by Bunt and Wood[12] because attached species similar to estuarine epontic microalgae were present in addition to some phytoplanktonic species.

Andriashev[4] concluded that there are two faunal assemblages associated with Antarctic fast ice: *ice fauna* that lives at least temporarily within the lower layer of the sea ice and the *sub-ice fauna* that is not in physical association with the ice, but as part of the food

Table 1
MAJOR SYMPAGIC FAUNAL HABITATS IN THE ARCTIC AND SUBARCTIC

- Protected lagoon/embayment fast ice
  - First-year ice sheet
  - Multi-year ice sheet
  - Coastal tide cracks
  - Rough ice
- Shallow water oceanic fast ice
  - First-year ice
  - Multi-year ice
  - Tidal cracks
  - Pressure ridge keels and rough ice
- Deep water fast ice
  - First-year ice
  - Multi-year ice
  - Pressure ridge keels and rough ice
- Pack ice
  - First-year ice
  - Multi-year ice
  - Pressure ridge keels and rough ice
  - Edge ice

web, is connected to varying degrees with the ice fauna assemblage. The ice fauna proper are those animals living directly on the ice undersurface, within the soft lower ice layer, or within the more solid congelation ice in brine channels or pockets. The sub-ice fauna in the water column beneath the ice is more difficult to define as a distinct group and are probably analogous to the pelagobenthos of the benthic environment. Higher light levels found just beneath the ice may attract meroplanktonic and holoplanktonic animals, but they encounter the ice as a confining ceiling and accumulate in the waters beneath. Data suggest that some vertically migrating zooplankton species may concentrate under the ice cover at night[13] and may form a faunal assemblage not directly linked energetically to the ice algal food web.

Because of the international roots to the study of ice fauna and because in many studies the oceanographic conditions were unique, there is, at present, no universally recognized terminology. The meiofauna and macrofauna of the true ice community[4] include organisms from both the benthos and the plankton; descriptive terms for this biotic assemblage should be broadly inclusive. Of those that have been used in the literature, *sympagic,* meaning "with ice",[14] seems to be the most appropriate. This term can be used to refer to both benthic and pelagic fauna of the inner shelf fast ice community as well as the offshore ice communities populated more by zooplankton. *Cryopelagic* can be used to refer to a subset of the sympagic community and is descriptive of the ice community composed of pelagic organisms over deeper water. *Epontic,*[12] meaning "out of the sea",[14] seems less appropriate than terminology oriented specifically toward the unique ice habitat. Epontic may describe the pennate diatom component of the ice biota (also see Chapter 3), but the term is not a suitable description for vagile, benthic animals such as gammarid amphipods or species of benthic meiofauna otherwise generally associated with sedimentary substrates.

As an environment, sea ice is heterogeneous and patchy, and its characteristics largely depend on its age and the stresses to which it has been subjected.[15] The different ice types and the dissimilar microhabitats must be considered when studying the ecology of the ice fauna. Whitaker[14] defined a number of ice habitats important to the Antarctic fauna during his research. A similar scheme is developed here for the Arctic environment to facilitate ecological comparisons (Table 1). Ice age is important because the central Arctic Basin has a permanent ice cover in dynamic equilibrium. Pressure ridges and bottom fast ice are prevalent because of the large oceanic regions with shallow depths and are also important as ice types.

Depth of water determines the extent of the fast, unmoving sea ice sheet that is anchored to the shore line and to the sea floor by the deep draft keels caused by the piling up of ice in pressure ridges.[16] It also determines whether vagile benthic animals can effectively migrate between the two surfaces or whether small organisms can be advected up to the overlying ice ceiling. Therefore, water depth is a necessary criterion to use in defining several of the inner continental shelf sympagic environments.

During most of the year in oceanic waters, arctic ice generally exists as a mixture of old and new ice. Because of the dynamic forces involved, newly formed thin pack ice can exist adjacent to multi-year ice floes and pressure ridges tens of meters thick.[15] As age and thickness of the ice increase, its crystalline structure, salinity, and light transmissivity change. Age and deformation of ice are other characteristics necessary for consideration in defining ice habitats.

II. HISTORY

Detailed quantitative studies focused on Arctic and subarctic ice fauna have been undertaken only in the last decade. Explorers and scientists on early expeditions to the Arctic observed amphipods and other organisms associated with ice and ice edge environments, but these observations were often incidental and vague.[11] Early taxonomic works[17,18] and more recent quantitative and functional studies on ice algae[19,20] also noted the types of fauna associated with ice or living on the undersurface of the ice. Research on these faunal assemblages was first undertaken in the Arctic during the summer and fall of 1970 and 1973 by Golikov and Averincev,[21] along the coasts of Heis Island, Alexandra Land, Rudolph Island, and Victoria Island in the Franz Josef Land Archipelago in the Barents Sea. This extensive work utilized scuba diving to study the benthic and ice faunal assemblages.

Because Russian and Polish scientists were among the early workers in Antarctic ice faunal research,[22,23] it is not surprising that they also pioneered ice faunal research in the Arctic. Research on ice fauna began in Canadian and U.S. waters in 1979 when Cross[24] and Bradstreet and Cross[5] undertook extensive studies in Pond Inlet, Baffin Bay, in the eastern Canadian Arctic, and Carey and Montagna[25] initiated ice faunal research in Stefansson Sound near Prudhoe Bay, Alaska, in the western Beaufort Sea. Additional research was later undertaken off Narwhal Island in the Prudhoe Bay region.[26-28] Research on the animals associated with the undersurface of sea ice has also been done in McKinley Bay in the southern Beaufort Sea,[29] in Frobisher Bay,[10] and at the pack ice edge in the Barents Sea.[13] A more southern study was undertaken by Dunbar and Acreman[30] in the Gulf of St. Lawrence; though the research was oriented toward the ice algae, observations were made on the fauna. Dunbar had worked on Arctic and subarctic zooplankton earlier and had summarized data on amphipod species known to associate with the sympagic environment.[31,32]

Some of the earliest observations on animals associated with sea ice came from MacGinitie's[18] extensive systematic and natural history research on marine invertebrates in the Point Barrow, Alaska, region from 1948 to 1950. He reported large numbers of the amphipod *Apherusa glacialis* living under ice floes during the summer in the northeast Chukchi Sea in the vicinity of the U.S. Naval Arctic Research Laboratory.

Other early observations on ice fauna came from workers involved in the multidisciplinary research conducted on the floating ice island research station T-3[9,19,33-35] and from ice algal studies. Barnard[34] reported on numerous amphipods captured in baited traps beneath pack ice adjacent to the ice island, but no direct observations were made on their behavior and ecology at the bottom surface of the ice. Data on absolute abundance, feeding behavior, or the thigmotactic effects of ice were not available from these early studies. Apollonio[20] observed abundant gammarid amphipods that were "undoubtedly feeding on algae" while he was making collections of the bottom ice layer near Devon Island in the Canadian Arctic for determinations of chlorophyll concentrations. Mohr and Tibbs,[9] however, did not observe amphipods grazing on the ice algae. Horner and Alexander,[36] Clasby et al.,[37] and Horner[38] reported flagellates, heliozoans, hypotrichous ciliates, abundant nematodes, polychaete larvae, turbellarians, harpacticoid copepods, and amphipods as part of the ice community near Barrow, Alaska. The amphipods were observed living on the undersurface of nearshore ice and their fecal pellets contained fragments of ice diatoms.

Once scuba diving in polar ice-covered waters was proven to be feasible,[39] quantitative studies on the structure and function of the sea ice communities became possible. The field rapidly developed[22,40] as the use of this diving technology permitted direct observation, quantitative sampling, and experimentation. A number of specialized sampling devices have been utilized to sample the ice undersurface from above through holes augered in the ice.[10,22,23] The bottom of the ice at the ice-water interface generally consists of soft ice and fragile, open, crystalline structures[41] and cannot be effectively sampled or studied from above with standard coring devices.[40] The study of the Arctic ice faunal community has developed so rapidly that much of the information collected remains unpublished and is available only in a variety of data reports, particularly those compiled for environmental impact assessment research oriented toward the oil industry.

III. RECENT ARCTIC RESEARCH

A. Ice Fauna

1. Macrofauna

The true ice macrofauna (<0.5 mm) is almost entirely composed of gammarid amphipods[27,42] (Table 2). Cross[24] reported 10 species of macroinvertebrates collected from the undersurface of the sea ice in Pond Inlet, including eight gammarid amphipods, one hyperiid amphipod, and one mysid. The greatest number of species found at any one station was five to seven in the fast ice and in rough ice at the ice edge. However, two species, *Apherusa glacialis* and *Ischyrocerus anguipes*, together made up about 95% of the macrofaunal abundance and approximately 90% of the total biomass of the sympagic fauna. The dominant amphipods in the ice community in the Barents Sea were *A. glacialis* and *Gammarus wilkitzkii.*[13] This ice assemblage was correlated with the thicker, older ice floes. *G. wilkitzkii* appears to be generally associated with multi-year sea ice.[43] Distribution data for gammarid amphipod species derived from alcid bird stomachs indicate that in water depths greater than 100 m *Gammarus setosus, G. wilkitzkii,* and *Gammaracanthus loricatus* live in the under-ice habitat.[44] Further to the east, Golikov and Averincev[21] reported that *A. glacialis* and *G. wilkitzkii* were present at the lower surface of the pack ice near Franz Josef Land in the Barents Sea and DeLong Islands in the East Siberian Sea. In the central Arctic Basin, George and Paul[35] reported that during October to November, *G. wilkitzkii* was the predominant sympagic species, but in December, *Pseudalibrotus nanseni* increased in relative abundance at the ice undersurface to account for 90% of the macrofaunal population. It is possible that a seasonal change in the population structure of the two species caused this change in dominance.

In contrast to the species composition of the ice macrofauna in deeper and offshore waters, whether in fast ice or pack ice, the gammarid amphipod *Pseudalibrotus* (= *onisimus*) *litoralis* dominates the ice macrofaunal community on the inner shelf in water depths of about 10 m.[27,45] In Resolute Bay, Cornwallis Island, Northwest Territories (74°41'N, 95°W), collections were made of the ice fauna in December and June, 1974[45] and *P. litoralis* was found in high numbers. Off Narwhal Island in the western Beaufort Sea (70°24'N, 147°31.1'W), *P. litoralis* was the dominant member of the sympagic community from April through early June, 1980, in a water depth of 9 m.[27] A total of five amphipod species was collected from the ice. In the southern Beaufort Sea at 19 m depth near the edge of the landfast ice in a

Table 2
SYMPAGIC FAUNAL COMMUNITIES

Protected lagoon/embayment fast ice community
- Meiofauna
  - Nematoda
  - Rotifera
  - Polychaeta (larvae)
  - Crustacea
    - Nauplii
    - Copepoda
      - **Halectinosoma neglectum* (G. Sars, 1904)
      - *Pseudobradya* sp. C
      - *Cyclopina gracilis* (Claus, 1863)
- Macrofauna
  - Crustacea
    - Amphipoda
      - *Onisimus glacialis* (G. Sars, 1900)
      - *Boeckosimus (= Onisimus) affinis* (Hansen, 1886)
      - *Lagunogammarus setosus* (Dementieva, 1931)
      - **Gammaracanthus loricatus* (Sabine, 1821)
      - *Apherusa glacialis* (Hansen, 1887)

Shallow water fast ice community
- Meiofauna
  - Nematoda
  - Turbellaria
  - Copepoda

**Cyclopina gracilis* (Claus, 1863)
**Harpacticus* sp.
      - *Oncaea* sp.
      - *Dactylopodia signata* (Willey, 1920)
      - *Halectinosoma* sp.
- Macrofauna
  - Amphipoda
    - **Pseudalibrotus (= Onisimus) litoralis* (Kroeyer, 1845)
    - *Weyprechtia pingius* (Kroeyer, 1845)
    - *Lagunogammarus setosus* (Dementieva, 1931)
    - *Apherusa glacialis* (Hansen, 1887)
    - *Halirages mixtus* Stephenson, 1931

Deep water fast ice community
- Meiofauna (<1.0 mm)
  - *Nematoda
  - *Crustacea
    - Copepoda
      - *Calanus* sp. nauplii
      - **Harpacticus superflexus* (Willey, 1920)
      - **Ectinosoma finmarchicum* T. (Scott, 1903)
      - **Cyclopina schneideri* T. (Scott, 1903)
      - *Tisbe furcata* (Baird, 1837)
      - *Oithona similis* (Claus, 1866)
      - *Harpacticus uniremis* (Kroeyer, 1842)
      - *Oncaea borealis* (G. Sars, 1918)
  - Polychaeta
- Macrofauna
- Crustacea
  - Amphipoda
    - **Apherusa glacialis* (Hansen, 1887)
    - **Ischyrocerus anguipes* (Kroeyer, 1838)
    - *Onisimus glacialis* (G. Sars, 1900)
    - *Pseudalibrotus (= Onisimus) litoralis* (Kroeyer, 1845)

Table 2 (continued)
SYMPAGIC FAUNAL COMMUNITIES

Onisimus nanseni (G. Sars, 1900)
Gammarus wilkitzkii (Birula, 1897)
Gammaracanthus (juveniles)
Eusirus holmi (Hansen, 1887)
Parathemisto libellula (Mandt, 1822)
Mysidacea
Mysis polaris (Holmquist, 1959)
Multi-year ice community
Meiofauna
Crustacea
Copepoda
Calanoid spp.
Macrofauna
Crustacea
Amphipoda
**Apherusa glacialis* (Hansen, 1887)
**Gammarus wilktzkii* (Birula, 1897)
**Parathemisto libellula* (Mandt, 1822)
Gammaracanthus loricatus (Sabine, 1821)
**Pseudalibrotus nanseni* (G. Sars, 1900)
Pseudalibrotus sp. (spp.?)

Note: *Indicates dominant species within the ice communities.

transition zone to the moving pack ice, Pett et al.[29] reported that *Onisimus* sp. and *Apherusa* sp. were common amphipods in the ice. In this dynamic environment near the Issungnak gravel oil drilling island, the ice is generally thinner than stable landfast ice. In McKinley Bay, a semiprotected embayment on the Tuktoyaktuk Peninsula with an average water depth of 6 m, the gammarid amphipod *Gammaracanthus loricatus* was present. *G. loricatus* is circumpolar in distribution and is often found in low salinity environments.[46] *Boeckosimus* (= *Onisimus*) *affinis* was also reported during the spring in large numbers at one ice-water interface in the low salinity environment of the Eskimo Lakes, a protected embayment east of Tuktoyaktuk Peninsula in the southern Beaufort Sea.[47]

There appear to be four main macrofaunal sympagic communities:

1. The protected lagoon/bay fast ice community dominated by *Gammaracanthus loricatus*
2. The shallow water coastal fast ice community dominated by *Pseudalibrotus litoralis*
3. The deeper water seasonal sea ice zone fast ice community dominated by *Apherusa glacialis*
4. The multi-year ice community dominated by *A. glacialis* and *Gammarus wilkitzkii* (Tables 2, 3)

The *Apherusa* community over deeper water contained larger numbers of amphipod individuals. The shallow water fast ice sympagic community in the Beaufort Sea had a mean of 10.4 amphipods per m^{-2},[27,28] while the deeper water fast ice community in the Canadian Arctic had a mean of 30.9 amphipod individuals per m^{-2}.[24] In the multi-year ice community characterized by the amphipod species *A. glacialis* and *G. wilkitzkii* and studied in the Barents Sea by Gulliksen,[13] both numbers and biomass of the ice amphipods decreased significantly from the ice edge into the pack ice.

Amphipods appear to congregate at the ice undersurface not only to feed, but to release young. In the shallow western Beaufort Sea, the number of juvenile *Pseudalibrotus litoralis* rapidly increased at the end of May and early June. Preliminary data indicated that upon

Table 3
SYMPAGIC GAMMARID AMPHIPOD ABUNDANCES (MEAN NUMBER PER m^{-2}) IN THE ARCTIC

Species	Western Beaufort Sea[28]	Canadian Arctic[24]	Frobisher Bay[10]	Barents Sea[13]
Pseudalibrotus litoralis	9.1	<0.1		
Halirages mixtus	0.5			
Lagunogammarus setosus	0.4			
Apherusa glacialis	0.3	23.5		16.6
Weyprechtia pinguis	0.1			
Ischyrocerus anguipes		6.9		
Onisimus glacialis		0.5	62.8	
Onisimus nanseni		<0.1		
Gammarus wilkitzkii		<0.1	000	2.8

Note: All collections by ice net.

release from the females' brood pouches, the young of this normally benthic species were feeding on the ice algae.[28] In the Canadian Arctic, Cross[24] reported large numbers of young of the amphipods *Onisimus glacialis* and *Apherusa glacialis* in May, 1979, though only one cohort of *Ischyrocerus anguipes* was present. It is likely that the bottom ice layer acts as a nursery ground for a number of gammarid amphipod species.

2. Meiofauna

Both permanent and temporary meiofaunal organisms in the size range 62 to 500 μm, occur within the bottom layers of the sea ice habitat. Larvae of benthic polychaetes, pelecypods, gastropods, tunicates, turbellarians, and cirripedes have been reported within the bottom layers of sea ice in the Arctic Basin.[10,29,38] Pett et al.[29] reported spionid polychaete larvae in Canadian Beaufort Sea ice. Carey and Montagna[25] found hesionid polychaete larvae in western Beaufort Sea ice, including nectochaete larval stages that are probably capable of feeding on diatoms.[48]

Permanent meiofauna, including nematodes, harpacticoid and cyclopoid copepods, calanoid copepods, and rotifers, have been reported as members of the sympagic community.[24-26,29] These organisms often occur in large numbers in most environments except the protected lagoonal ice.[25] In Stefansson Sound, meiofaunal densities ranged from 4500 to 8000 m^{-2}, whereas offshore in the fast ice the range was 36,000 to 320,000 m^{-2}.[26]

Nematodes are the dominant members of the ice meiofauna. In Pond Inlet, they accounted for 58.8% of the meiofauna (<1.0 mm) by number.[24] In the shallow Beaufort Sea their numerical density exhibited an increase through the spring of 1980 and they ranged from 3590 m^{-2} in April to 62,360 m^{-2} in June. Nematodes averaged 47.0% of the meiofauna during the study.[26] There is little information available on the systematics biology and ecology of the sympagic Nematoda, but the strong trend in increasing population size through the spring suggests that these organisms are viable components of the ice community.

Copepods, mostly harpacticoids and cyclopoids, were the second most abundant group within the meiofauna in the sympagic environment. In the western Beaufort Sea, these copepods comprised 28% of the meiofauna population by number,[26] while in the eastern Canadian Arctic, they accounted for 39.7%.[24] In the shallow Beaufort Sea, the cyclopoid *Cyclopina gracilis* appeared to reproduce continuously throughout the spring; adults and young copepodites were always present. Initially, the harpacticoid *Harpacticus* sp. in the ice in the same area consisted entirely of juveniles that then grew as a single cohort. Though clasping adults were present in the collections, no gravid females appeared in the population associated with the ice.

3. Microfauna

Little is known of the sympagic microfauna, (<62 μm) though the list of major taxa represented continues to grow as more research is undertaken on the Arctic sea ice faunal community. Usachev[17] identified protozoans associated with ice floes in the Kara Sea, including the ciliates *Euplotes caudata, E. truncata, Probosidium armatum, Cephalotrichium tonsuratum, Amphorella vitrea,* and *Difflugia* sp. Heliozoans, hypotrichous ciliates, unpigmented flagellates, and other protozoans have been reported.[10,11,49]

4. Relationship to Ice Habitats

Scuba observations and sectioned ice core samples indicate that the sympagic amphipod species occupy a variety of subhabitats. Generally, the macrofauna occupy the lower 5 to 10 cm of ice,[29] though individuals were observed living in brine channels[24,50-52] and partially or fully embedded within the ice.[13] Green and Steele[45] reported that *Gammaracanthus loricatus* was the only amphipod species associated with ice stalactites and may at times be frozen within them. On occasion *Apherusa glacialis* occupied the brine channels, though they were mostly at the ice-water interface[24] and distribution tended to be patchy with aggregations of juveniles concentrated on clumps and strands of ice algae. Cross[24] also noted that the amphipod *Ischyrocerus anguipes* was abundant at only one station where it was evenly distributed on the bottom surface of the ice in Pond Inlet. Some of the *I. anguipes* had their posterior ends embedded within the ice, while others were on the bottom of the ice. *Onisimus glacialis* was rarely at the ice-water interface, but inhabited small pockets or brine channels in rough ice. Gulliksen[13] observed both *Gammarus wilkitzkii* and *Apherusa glacialis* in small pockets within the ice. *Pseudalibrotus litoralis* lives at the ice-water interface and moves along the ice surface.[13,28] *Parathemisto libellula,* a hyperiid amphipod, swims near the sympagic environment and is often found in patches in areas beneath the ice where there is greater light transmission.[13] Perhaps the role of this zooplankter in the under-ice community is similar to that of vagile epibenthic forms on the sea floor.

In correlation with the ice algal bloom, total ice fauna in the Canadian Beaufort Sea increased in numerical density through the spring to a maximum of about 100,000 m^{-2} in mid-June, 1982.[29] The abundance peak was reached several weeks after the ice algal bloom began. Large temporal changes in ice fauna species composition were correlated with the development of the ice algal bloom, changes in salinity at the ice undersurface, and the presence of meroplanktonic larvae.[26,29]

During August, 1982, more crustaceans with a total higher biomass were found near the ice edge in the Barents Sea than further northward into the pack ice.[13] Calanoid copepods were particularly abundant near the ice edge, while the amphipod *Apherusa glacialis* was relatively common and was the dominant member of the ice faunal community within the pack ice some 60 nautical miles north of the ice edge. *Gammarus wilkitzkii* was less abundant and its distribution pattern relative to the ice edge was less clear cut. Very few ice algae were present in August and it appears that oceanographic and biological processes at the ice edge have an effect on the biomass of the sympagic fauna present. Perhaps there is an ice edge effect on phytoplankton production[53] that supports more grazers at the ice edge than further within the pack ice.

Gammarid amphipods were attracted to turbid, sediment-laden ice within Stefansson Sound, Beaufort Sea.[54] They were observed feeding on the ice undersurface and their guts were empty. Five gammarid species (*Gammaracanthus loricatus, Gammarus setosus, Melita formosa, Pseudalibrotus litoralis,* and *Weyprechtia pinguis*), one polychaete species (*Antinoella sarsi*), and one fish species (*Boreogadus saida*) were present in this turbid ice habitat. The polychaete and two of the amphipods were gravid. The relationship between the fauna and the turbid ice is not understood at this time.

B. Sub-Ice Fauna

According to the ecological groups described by Andriashev,[4] Arctic cod (*Boreogadus saida*), glacial cod (*Arctogadus glacialis*), and sand lance (*Ammodytes hexapterus*) are fish living in the sub-ice environment that are trophically involved with the sympagic fauna, feeding extensively on *Apherusa glacialis*.[55-57] The Arctic cod also feed on the hyperiid amphipod *Parathemisto libellula* that is found at times beneath the ice. *P. libellula* can be correlated with the presence of ice, but it does not appear to be a member of the ice algal food web. The ice cover may act as a ceiling for this and a number of other pelagic zooplankton species, particularly those that may be vertically migrating diurnally.

C. Relationship to the Pelagic and Benthic Fauna

Though information is sketchy, it is becoming apparent that the ice biota can interact with both the benthic and pelagic environments. Water depth and the length of the ice-covered season appear to affect the source and fate of the ice fauna and the fate of the detrital materials produced by the community.

It appears that within at least the 10 m depth contour, the source of the ice fauna is from both the bottom sediments and the water column.[27,42,43,58] The ice meiofauna in the lagoonal and the shallow fast ice environments is comprised of benthic groups such as nematode worms and harpacticoid copepods,[24-26] and meroplanktonic larval forms of benthic polychaetes and cirripedes.[24,25] In shallow water these smaller benthic organisms may be advected upward by storms.[59] Pett et al.,[29] using near-surface particle traps, collected ice diatoms that had been resuspended from the sediments by early fall gales. On the other hand, some benthic meiofauna such as harpacticoid copepods may actively migrate upwards from the sediments[60] and either reach the ice directly or become more susceptible to turbulent transport.

Though there were only two species of harpacticoid copepods in the impoverished lagoonal ice community in Stefansson Sound compared with 28 species in the sediments, Carey and Montagna[25] reported that both of these sympagic species (*Halectinosoma neglectum* and *Pseudobradya* sp. C) were in common between the ice and sediment communities. A cyclopoid copepod, *Cyclopina gracilis*, thought to be an epiphytic form,[61] was found in the ice, but not in the sediments. Nematoda, normally a benthic group, were found in the ice, but species comparisons are not available. Though sympagic meiofauna were numerous, ranging from 200 to 320,000 individuals per m^{-2} (Table 4), their numerical density in the sediments of Stefansson Sound was 2 to 5 orders of magnitude greater.[25] As the ice assemblage early in the spring was dominated by larval polychaetes and crustacean nauplii, the sources of the sympagic meiofauna in the protected lagoon environment are from both the pelagic and benthic environments.

In shallow water on the inner shelf much of the macrofauna is derived from the benthic assemblage. In the Beaufort Sea lagoonal systems, Griffiths and Dillinger[42] reported benthic gammarid amphipods and mysids associated with the undersurface of the ice cover. In February, *Gammaracanthus loricatus* was the dominant species of the ice macrofauna, though densities were low (5.5 m^{-2}). Pett et al.[29] also reported the gammarid *G. loricatus* as a member of the ice community in shallow McKinley Bay in May. By late winter and spring, epibenthic and sympagic species compositions change in the nearshore lagoons. Griffiths and Dillinger[42] suggested that the lagoonal fauna changes through the season as species migrate in and out of the lagoons and to and from the shallow shelf in the nearshore western Beaufort Sea. More marine species move onshore and include the amphipods *Onisimus glacialis* and *Boeckosimus affinis*. During the winter large numbers of the gammarid amphipod *Onisimus affinis* collect at the ice-water interface in the Eskimo Lakes, where they "presumably" graze on the ice diatoms.[47] Part of the *O. affinis* population remains in the sediment with few at intermediate depths. By late May when the salinity had dropped in the uppermost water layer, the sympagic population had disappeared. Percy[47] concluded that the brackish water layer functioned as a barrier to the amphipod during the melt season.

Table 4
MEIOFAUNA DENSITIES IN ARCTIC FAST ICE

Locality	Number ($\bar{x}$ m^{-2})	Ref.
Beaufort Sea		
Stefansson Sound	4,500—8,000	25
Narwhal Island	36,000—320,000	26
Canadian Arctic		
Baffin Bay	300—53,000	24
Frobisher Bay	200—323,297	10

Under these conditions *O. affinis* becomes totally benthic and feeds on the sediments as a detritivore.

In the shallow oceanic environment, the fast ice macrofauna are dominated by benthic gammarid amphipods.[27,28,45] In the western Beaufort Sea, *Pseudalibrotus litoralis* was found to be benthic during the ice-free summer months with its population center at 5 to 20 m depth, though its distribution extended to the 100 m isobath.[62] From April through early June, *P. litoralis* was more abundant in the under-ice environment than in the benthos,[27] feeding on the ice algae and releasing its young there.

Further offshore in the deep fast ice community, the sympagic macrofaunal species are pelagic in origin. The gammaridean Amphipoda appear to be well-adapted as a group to live in this habitat. Dunbar and Acreman[30] comment that "it is remarkable" that the gammarid amphipod *Gammarus wilkitzkii,* a member of a benthic genus, dominates the offshore multi-year ice community rather than the much more abundant pelagic hyperiid amphipod *Parathemisto libellula*. The distribution of *G. wilkitzkii* appears to be independent of water depth.[63]

D. Food Web

Many authors have suggested that the ice algal bloom is an important food source in Arctic waters.[20,64,65] The data accumulated to date document that an extended food web is directly supported during the spring months by these algae. Gammarid amphipods appear to be the major grazers in all Arctic sympagic environments.[24,28,66] Ice diatoms have been noted in the gastrointestinal tracts and in the fecal pellets of sympagic amphipods.[38,58]

The functional role of the sympagic meiofauna in the ice algae-based food web is unknown. The large numbers of organisms present on the ice undersurface and observations of copepod growth and reproduction indicate that at least a portion of the fauna must derive an energy source from the ice environment. It seems likely that they are feeding on the algae, bacteria, microfauna, and detritus derived from the ice community or combinations of these sources. It is also likely that the macrofaunal browsers feed on the meiofauna, although whether or not this is selective predation is not yet known. As the ice amphipod *Pseudalibrotus litoralis* is a scavenger, it probably feeds on the smaller fauna in addition to the pennate diatoms present in the ice.

Amphipods appear to be the pivotal prey species in the ice faunal assemblage during the spring in all Arctic and subarctic waters. They graze on the ice algae that are concentrated in a narrow zone in the lower layer of sea ice.[5,10,24,28] In turn, the amphipods are fed upon by fish,[57] seals,[2] gulls, murres, and guillemots.[5,44,67,68] Some of the sympagic amphipods such as *Pseudalibrotus litoralis, Gammarus wilkitzkii,* and *Apherusa glacialis,* are active scavengers as well as grazers.[28,34] *Boeckosimus* (=*Onisimus*) *affinis* are attracted to bait from as far as 30 m downstream of traps.[69] The importance of the alternate trophic links to amphipods via the meiofauna and microfauna in the simplified sympagic food web is not known at this time.

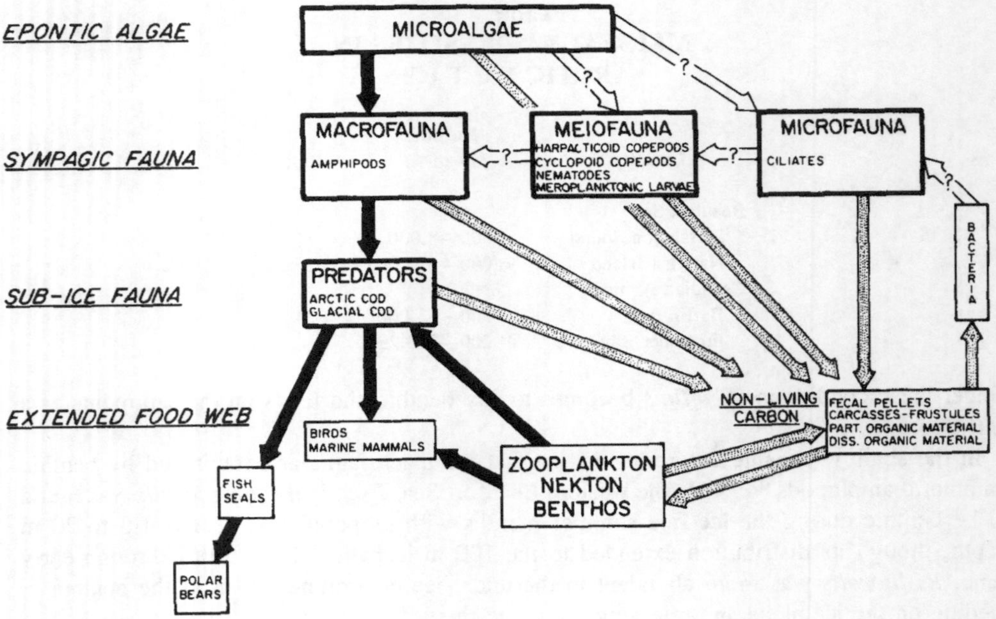

FIGURE 1. A simplified food web to illustrate the major trophic relationships of the ice biota. The energetic linkages are designated at 3 levels of information: (1) solid arrows = observed, (2) stippled arrows = inferred from other data, and (3) open arrows = hypothesized.

There is little quantitative data available on the ice-based food web, and spatial and temporal coverage is scant. Figure 1 summarizes the major links within the food web, but the scattered information available precludes a presentation of more than general structure. *Boreogadus saida* is a major energetic link in the ice-based food web at a higher trophic level.[70,71] During the spring and summer months these fish feed on amphipods associated with the ice undersurface[57] and their mouth morphology appears to be adapted to feeding on an overlying ice surface.[56] The cod are prey for a diverse group of predators including birds, seals, and other marine mammals.

It has been conjectured that the downward transport of organic detritus, fecal pellets, crustacean molts, and pennate diatom cells from the ice provides a food source for the pelagic and benthic faunas.[5,7,24,49,56,72-75] The organic matter not consumed during its sinking through the water column would be available to the benthos as food. The depth of water and the temporal pattern and flux rate of detrital sinking influence the availability of the organics to the 2 faunas.[76] The supposition has been stated that during the melt season beginning in late May to early June, particles and algae slough off the bottom surface of the ice. In shallow water this sudden flux provides a pulse of food materials to the sea floor as well as to the water column. In deep water less utilizable organic matter reaches the benthos. If the flux from the sympagic community takes place over an extended period of time, more food could be utilized by zooplankton on its way down to the bottom than if there was a sudden influx from the ice during accelerated ice melt in early summer.

Particles from the sympagic community reach the sea floor in shallow water throughout the winter and spring. In the Canadian Beaufort Sea, pennate diatoms from the ice algal assemblages were captured in near-bottom sediment traps in 6 to 19 m of water on the shelf offshore of the Mackenzie River and in McKinley Bay, a shallow, coastal embayment.[29] Algal cells increased from 4×10^6 cells m^{-2} day^{-1} before the ice algal bloom to 28×10^6 cells m^{-2} day^{-1} in May to June. The total particulate flux of carbon and nitrogen do not seem to be affected by the accelerated loss of the ice algal community from the lower layers of ice in early June.[29,58] The carbon flux to the sediments through a 7 m water column in the western Beaufort Sea off Narwhal Island varied between 29 and 48 mg C m^{-2} day^{-1} with no significant trends of correlations with environmental events. Nitrogen flux was low, 1.5 to 6.2 mg N m^{-2} day^{-1} and erratic throughout the spring season. Fecal pellets from the grazing sympagic amphipod *Pseudalibrotus litoralis* contained high concentrations of pennate ice diatom frustules; the flux of pellets to the bottom increased from April through May, 1980, and then fell almost to zero in early June. The carbon and nitrogen flux data indicate that there is no sudden influx of organic materials to the benthic environment during the melting period, but rather there is generally a high and continual rain of organic matter to the bottom throughout the spring growing season. Horner[38] also noted that amphipods grazing on the lower ice surface during the spring produced fecal pellets packed with the remains of ice diatoms.

IV. DISCUSSION

The development of the bottom-ice biotic community lessens the impact of marked seasonality and restricted summer primary production in both the Arctic[64] and Antarctic.[4] The primary production season begins earlier because of the ice algae and may extend later if fall blooms of ice algae are significant. Dunbar[77] described the ice community as a system of high local production embedded within a low-productivity system, similar to coral reefs and salt marshes. Because the ice algal bloom develops before significant seasonal primary productivity rates by benthic diatoms in shallow sediments and by phytoplankton in the upper layers of the water column,[78] ice diatom production can provide an early food source for the ice fauna. Although a study in the shallow Beaufort Sea by Horner and Schrader[75] measured relatively low rates of ice algal primary productivity, this community still provided the earliest new carbon source to the sympagic fauna.

Although a number of gammarid amphipods have been shown to feed on the ice algae,[5,10,24,28] the food sources of other animal species observed and collected from the ice-water interface and the water layer directly beneath the ice remain unknown. They may be located near this substrate for reasons other than utilizing the ice algae as a food source.[25] Zooplankters such as *Parathemisto libellula* may be concentrated by the ice barrier and others may be attracted by the greater light intensities adjacent to the lower surface of the ice. Meguro et al.[79] noted that at times the structure of the lower ice layers restricted the grazing of zooplankton on the ice algae. The benthic invertebrate larvae close to metamorphosis may be aggregating on the more solid ice substrate.[25]

Depth has proven to be a controlling factor in the species composition and structure of sympagic communities on the inner continental shelf. In protected bays and on the shallow shelf in the Arctic, the normally benthic amphipods *Boeckosimus affinis* and *Pseudalibrotus litoralis* are dominant species of the ice macrofauna during the winter and spring.[47,58] At these same depths, 2 to 21 m, the benthic amphipod species *Paramoera walkeri* lives in the ice community during the winter months in Antarctic waters.[23] During the ice-free summer months, this amphipod is restricted to sandy sediments that lie within a very narrow depth zone in the Alasheyev Bay,[80] though individuals have been collected out to depths of 250m and from the water column.[81] There is a strong parallelism between the Arctic and Antarctic inner shelf communities, but because the Antarctic continental shelf is deep and often rocky, the shallow community is not as apparent. The Antarctic sympagic communities appear to be almost completely dominated by zooplankton[77] because of the greater proportion of deep shelf environments. In contrast, the Arctic Basin and surrounding seas contain large areas of shallow, benthic-oriented water, permitting greater interaction between the faunas of the sediments and ice surfaces.

As a physical substrate, sea ice has effects on both the pelagic and benthic fauna. Some of these effects may be an adaptive interaction such as the use of the ice by the microfauna and meiofauna as a more solid substrate for living than the surrounding water.[1,24,26] Young Arctic cod (*Boreogadus saida*) are closely associated with ice and live in holes and crevices particularly in the melting ice during the summer.[45,55,82] Young sand lance (*Ammodytes dubius*) have also been observed in cavities within melting ice in summer off the Greenland west coast.[83] This behavior may provide the young fish with a refuge from predaceous birds, e.g., thick-billed murres and alcids.

V. RESEARCH NEEDS

Additional information on the species composition and abundance is necessary to define the variability in space and time. The geographic extent of the sympagic communities is not known. Further correlations with the oceanographic and geological environments and with the pelagic and benthic faunas are needed to provide an understanding of the physical factors controlling the community structure. Data on the effects of ice microhabitats on the small-scale distribution of the ice fauna are few; such detailed studies would probably explain some of the variability in published results.

More basic quantitative ecological information on the ice fauna will provide insights into the interactions of the fauna with each other and with their environment. Food web studies, particularly of the smaller-size categories of ice organisms, will define the roles of the meiofauna and microfauna. The effects of grazing and predation on the structures of the ice assemblage should be investigated and the feeding behavior of the sub-ice fauna needs to be clarified. Year-round comparative studies will provide basic information on the relative importance of the ice habitat in the ecology of the benthic and pelagic faunas and on the source and fate of the sympagic fauna during the summer in the seasonal sea ice zone. Comparative studies between the faunal assemblage in annual and multi-year ice should be accomplished to determine if sympagic communities exist at other times of the year. It is unknown whether there are significant summertime sympagic assemblages on the under-surface of multi-year ice and whether the summer populations are ecologically important.

Data on life histories and reproduction are also few. Such information would allow evaluation of the role of the ice substrate as a temporary, but food-rich, spawning, and nursery ground for a number of benthic and pelagic species. The dynamics and particularly the energetics of the sympagic species populations are virtually unknown. Information on trophic energy transfer, secondary production, and metabolism would allow more firm conclusions to be drawn about the relative importance of the sympagic faunal assemblages and the ice algae in the Arctic ecosystem. Physiological studies would provide more insight into possible adaptations to the cold temperatures and to the restricted primary productivity in the Arctic environments. Further comparisons of the Arctic sympagic fauna and ice biotic community structure and energetics with those from the Antarctic would provide insights into the interactions of this assemblage with the biota and the oceanographic environments surrounding them.

Information on the impact of pollution on the under-ice biotic communities is also lacking. Oil-related activities in Arctic waters could impact these communities. Oil itself, resuspended sediments, and other disturbances may adversely affect the functioning of these productive assemblages during critical spawning periods for the amphipods and other animals.

VI. SUMMARY

It is evident that significant research progress has been made in the last 2 decades in the recent field of Arctic ice faunal ecology. Enough is known of the species composition to define the ice habitats and their faunas. On the inner shelf and within protected lagoons, depth and the hydrographic conditions appear to be important environmental factors. There also appear to be effects on the fauna by the relative age of the ice and the proximity of the ice field edge.

The sympagic microfauna is dominated by ciliate protozoans, the meiofauna by nematode worms, copepods, and sometimes by meroplankton, and the macrofauna by gammarid amphipods. Arctic cod are primary members of the sub-ice fauna trophically linked to the sympagic community. The amphipods and cod are the primary links in the food web based on the ice algae.

Interactions of the ice biotic community with the pelagic and benthic communities consist of a direct exchange of fauna as well as an indirect input of organic material by vertical flux from the ice. Species may actively migrate to the ice environment seasonally or diurnally to feed there, or they may be advected to this environment by currents and storm turbulence. It is likely that both benthic and pelagic gammarid amphipod species release young at the sympagic environment and utilize the ice as a nursery ground. Plagic larval stages of normally benthic invertebrates may interact with the ice biota. Members of the sub-ice faunal assemblage may or may not be trophically linked to the true-ice community. During the spring months organic detritus is formed by the ice algae and sympagic fauna and is probably available to both the pelagic and benthic animal assemblages.

ACKNOWLEDGMENTS

Special thanks go to R. E. Ruff for his investigative literature search and his editorial support and to R. Horner for her patience as an editor. Fruitful discussions with B. Gulliksen about ice fauna nomenclature and ecology were most helpful. The unpublished data came from research supported by the Bureau of Land Management through an interagency agreement with the National Oceanic and Atmospheric Administration as part of the Alaskan Outer Continental Shelf Environmental Assessment Program.

REFERENCES

1. **Fenchel, T. and Lee, C. C.,** Studies on ciliates associated with sea ice from Antarctica, *Arch. Protistenkd.*, 114, 231, 1972.
2. **Burns, J. J., Shapiro, L. H., and Fay, F. H.,** Ice as marine mammal habitat in the Bering Sea, in *The Eastern Bering Sea Shelf: Oceanography and Resources*, Vol. 2, Hood, D. W. and Calder, J. A., Eds., University of Washington Press, Seattle, 1981, 781.
3. **El-Sayed, S. Z.,** Biological aspects of the pack ice ecosystem, in *Proc. Symposium on Antarctic Ice and Water Masses*, Deacon, Sir. G., Ed., Scientific Committee on Antarctic Research, W. Heffer & Sons Ltd., Cambridge, 1971, 35.
4. **Andriashev, A. P.,** The problem of the life community associated with the Antarctic fast ice, in *Symposium on Antarctic Oceanography*, Currie, R. I., Ed., W. Heffer & Sons Ltd., Cambridge, 1968, 147.
5. **Bradstreet, M. S. and Cross, W. E.,** Trophic relationships at high Arctic ice edges, *Arctic*, 35, 1, 1982.
6. **Fay, F. H.,** The role of ice in the ecology of marine mammals of the Bering Sea, in *Oceanography of the Bering Sea with Emphasis on Renewable Resources*, Hood, D. W. and Kelley, E. J., Eds., Institute of Marine Science, University of Alaska, Fairbanks, 1974, 383.
7. **Alexander, V.,** Interrelationships between the seasonal sea ice and biological regimes, *Cold Regions Sci. Tech.*, 2, 157, 1980.
8. **Dunbar, M. J.,** Biological production in the Gulf of St. Lawrence, in *Marine Production Mechanisms*, Dunbar, M. J., Ed., Cambridge University Press, Cambridge, 1979, 151.
9. **Mohr, J. L. and Tibbs, J.,** Ecology of ice substrates, in *Proc. Arctic Basin Symposium*, Tidewater Publ., Centreville, 1963, 245.
10. **Grainger, E. H. and Hsiao, S. I. C.,** A study of the ice biota of Frobisher Bay, Baffin Island 1979—1981, *Can. Manuscr. Rep. Fish. Aquat. Sci.*, No. 1647, Ottawa, 1982.
11. **Nansen, F.,** Protozoa on the ice-floes of the North Polar Sea, *Sci. Res. Norw. N. Polar Exped.*, 5(16), 1, 1906.

12. **Bunt, J. S. and Wood, E. J. F.,** Microalgae and Antarctic sea-ice, *Nature (London)*, 199, 1254, 1963.
13. **Gulliksen, B.,** Cryopelagic fauna from Svalbard waters, *Sarsia*, 69, 17, 1984.
14. **Whitaker, T. M.,** Sea ice habitats of Signy Island (South Orkneys) and their primary productivity, in *Adaptations within Antarctic Ecosystems*, Llano, G. A., Ed., Gulf Publ., Houston, 1977, 75.
15. **Zubov, N. N.,** *L'dy Arktiki* (*Arctic Ice*), Izdatel'stvo Glavsevmorputi, Moscow, 1945, 1. (Translated from Russian by U.S. Navy Oceanographic Office and The American Meteorological Society.)
16. **Reimnitz, E., Toimil, L., and Barnes, P.,** Arctic continental shelf morphology related to sea-ice zonation, Beaufort Sea, Alaska, *Mar. Geol.*, 28, 179, 1978.
17. **Usachev, P. I.,** The microflora of polar ice, *Tr. Inst. Okeanol., Akad. Nauk SSSR*, 3, 216, 1949 (in Russian).
18. **MacGinitie, G. E.,** Distribution and ecology of the marine invertebrates of Point Barrow, Alaska, *Smithson. Misc. Collect.*, 128, 1, 1955.
19. **English, T. S.,** Some biological oceanographic observations in the central North Polar Sea, Drift Station Alpha, 1957—1958, Sci. Rep. No. 15, The Arctic Institute of North America, Washington, D.C., 1961.
20. **Apollonio, S.,** Chlorophyll in Arctic sea ice, *Arctic*, 18, 118, 1965.
21. **Golikov, A. N. and Averincev, A. G.,** Distribution patterns of benthic and ice biocenoses in the high latitudes of the Polar Basin and their part in the biological structure of the World Ocean, in *Polar Oceans*, Dunbar, M. J., Ed., The Arctic Institute of North America, Calgary, 1977, 331.
22. **Gruzov, Ye., Propp, M. V., and Pushkin, A. F.,** Biological associations of coastal areas of the Davis Sea (based on the observations of divers), *Sov. Antarct. Exped. Inf. Bull.*, 6, 523, 1968.
23. **Rakusa-Suszczewski, S.,** The biology of *Paramoera walkeri* Stebbing (Amphipoda) and the Antarctic sub-fast ice community, *Pol. Arch. Hydrobiol.*, 19, 11, 1972.
24. **Cross, W. E.,** Under-ice biota at the Pond Inlet ice edge and adjacent fast ice areas during spring, *Arctic*, 35, 13, 1982.
25. **Carey, A. G., Jr. and Montagna, P. A.,** Arctic sea ice faunal assemblages: first approach to description and source of the under-ice meiofauna, *Mar. Ecol. Prog. Ser.*, 8, 1, 1982.
26. **Kern, J. C. and Carey, A. G., Jr.,** The faunal assemblage inhabiting seasonal sea ice in the nearshore Arctic Ocean with emphasis on copepods, *Mar. Ecol. Prog. Ser.*, 10, 159, 1983.
27. **Carey, A. G., Jr.,** Shallow southwest Beaufort Sea macrofauna: interactions with the epibenthic community, *EOS*, 63, 949, 1982.
28. **Boudrias, M. and Carey, A. G., Jr.,** unpublished data, 1984.
29. **Pett, R. J., Vickers, G. P., Acreman, J. C., and Ethier, A. G.,** A seasonal study of the epontic and planktonic communities near the Issungnak artificial island and McKinley Bay in the Canadian Beaufort Sea, Contract ISZ81-00026 for Department of Supply and Services Canada by Arctic Laboratories Ltd., Sydney, 1983.
30. **Dunbar, M. J. and Acreman, J. C.,** Standing crops and species composition of diatoms in sea ice from Robeson Channel to the Gulf of St. Lawrence, *Ophelia*, 19, 61, 1980.
31. **Dunbar, M. J.,** On *Themisto libellula* in Baffin Island coastal water, *J. Fish. Res. Board Can.*, 6, 419, 1946.
32. **Dunbar, M. J.,** Euphausiids and pelagic amphipods, distribution in North Atlantic and Arctic waters, *Serial Atlas of the Marine Environment*, Folio 6, Am. Geogr. Soc., New York, 1964, 1.
33. **Mohr, J. L.,** Marine biological work, *Geophys. Res. Pap.*, 63, 82, 1959.
34. **Barnard, J. L.,** Epipelagic and under-ice Amphipoda of the central Arctic Basin, *Geophys. Res. Pap.*, 63, 115, 1959.
35. **George, R. Y. and Paul, A. Z.,** USC-FSU biological investigations from the Fletcher's Ice Island T-3 on deep-sea and under-ice benthos of the Arctic Ocean, Tech. Rep. No. 1, University of Southern California Arctic Project, Los Angeles, 1970, 1.
36. **Horner, R. and Alexander, V.,** Algal populations in Arctic sea ice: an investigation of heterotrophy, *Limnol. Oceanogr.*, 17, 454, 1972.
37. **Clasby, R. C., Alexander, V., and Horner, R.,** Primary productivity of sea-ice algae, in *Assessment of the Arctic Marine Environment: Selected Topics*, Hood, D. W. and Burrell, D. C., Eds., Institute of Marine Science, University of Alaska, Fairbanks, 1976, 289.
38. **Horner, R.,** History and recent advances in the study of ice biota, in *Polar Oceans*, Dunbar, M. J., Ed., The Arctic Institute of North America, Calgary, 1977, 269.
39. **Peckham, V.,** Year-round SCUBA diving in the Antarctic, *Polar Rec.*, 12, 143, 1964.
40. **Clasby, R. C., Horner, R., and Alexander, V.,** An in situ method for measuring primary productivity of Arctic sea ice algae, *J. Fish. Res. Board Can.*, 30, 835, 1973.
41. **Kovacs, A. and Mellor, M.,** Sea ice morphology and ice as a geologic agent in the southern Beaufort Sea, in *The Coast and Shelf of the Beaufort Sea*, Reed, J. C. and Sater, J. E., Eds., The Arctic Institute of North America, Calgary, 1974, 113.
42. **Griffiths, W. B. and Dillinger, R. E.,** Beaufort Sea barrier island-lagoon ecological process studies: final report, Simpson Lagoon, pt. 5, invertebrates, *OCSEAP, NOAA*, 8, 1, 1981.

43. **Steele, D. H. and Steele, V. J.,** The biology of *Gammarus* (Crustacea, Amphipoda) in the northwestern Atlantic. VIII. Geographic distribution of the northern species, *Can. J. Zool.*, 52, 1115, 1974.
44. **Bradstreet, M. S. W.,** Thick-billed murres and black guillemots in the Barrow Strait area, NWT, during spring: diets and food availability along ice edges, *Can. J. Zool.*, 58, 2120, 1980.
45. **Green, J. M. and Steele, D. H.,** Observations on marine life beneath sea ice, Resolute Bay, N.W.T., in *Proc. Circumpolar Conference on Northern Ecology*, Vol. 2, National Research Council of Canada, Ottawa, 1975, 77.
46. **Steele, D. H. and Steele, V. J.,** Some aspects of the biology of *Gammaracanthus loricatus* (Sabine) (Crustacea, Amphipoda), *Astarte*, 8, 69, 1976.
47. **Percy, J. A.,** Ecological physiology of Arctic marine invertebrates. Temperature and salinity relationships of the amphipod *Onisimus affinis* H. J. Hansen, *J. Exp. Mar. Biol. Ecol.*, 20, 99, 1975.
48. **Blake, J. A.,** The larval development of Polychaeta from the Northern California coast. III. Eighteen species of Errantia, *Ophelia*, 14, 23, 1975.
49. **Horner, R.,** Sea ice organisms, *Oceanogr. Mar. Biol. Annu. Rev.*, 14, 167, 1976.
50. **Welch, H. E. and Kalff, J.,** Marine metabolism at Resolute Bay, Northwest Territories, in *Proc. Circumpolar Conference on Northern Ecology*, Vol. 2, National Research Council of Canada, Ottawa, 1975, 69.
51. **Lewis, E. L. and Milne, A. R.,** Underwater ice formations, in *Polar Oceans*, Dunbar, M. J., Ed., The Arctic Institute of North America, Calgary, 1977, 230.
52. **Newbury, T. K.,** Under landfast ice, *Arctic*, 36, 328, 1983.
53. **Niebauer, H. J., Alexander, V., and Cooney, R. T.,** Primary production at the eastern Bering Sea ice edge: the physical and biological regimes, in *The Eastern Bering Sea Shelf: Oceanography and Resources*, Vol. 2, Hood, D. W. and Calder, J. A., Eds., University of Washington Press, Seattle, 1981, 763.
54. **Dunton, K. H., Reimnitz, E., and Schonberg, S.,** An Arctic kelp community in the Alaskan Beaufort Sea, *Arctic*, 35, 465, 1982.
55. **McAllister, D. E.,** Ecology of the marine fishes of Arctic Canada, in *Proc. Circumpolar Conference on Northern Ecology*, Vol. 2, National Research Council of Canada, Ottawa, 1975, 51.
56. **Alexander, V.,** Interrelationships between the seasonal sea ice and biological regimes, *Cold Regions Sci. Tech.*, 2, 157, 1980.
57. **Lowry, L. F. and Frost, K. J.,** Distribution, growth and food of Arctic cod (*Boreogadus saida*) in the Bering, Chukchi, and Beaufort seas, *Can. Field Nat.*, 95, 186, 1981.
58. **Carey, A. G., Jr.,** unpublished data, 1984.
59. **Hagerman, G. M., Jr. and Rieger, R. M.,** Dispersal of benthic meiofauna by wave and current action in Bogue Sound, North Carolina, USA, *Mar. Ecol.*, 2, 245, 1981.
60. **Hauspie, P. and Polk, R. H.,** Swimming behavior patterns in certain benthic harpacticoids, *Crustaceana*, 25, 95, 1973.
61. **Sars, G. O.,** *An Account of the Crustacea of Norway. Vol. VI. Copepoda Cyclopoida*, Bergen Museum, Alb. Cammermeyers Forlag, Christiania, 1918, 1.
62. **Carey, A. G., Jr.,** The distribution, abundance, composition and variability of the western Beaufort Sea benthos, *OCSEAP, NOAA* , 2, 27, 1981.
63. **Tencati, J. R. and Geiger, S. R.,** Pelagic amphipods of the slope waters of northeast Greenland, *J. Fish. Res. Board Can.*, 25, 1637, 1968.
64. **Dunbar, M. J.,** *Ecological Development in Polar Regions*, Prentice-Hall, Englewood Cliffs, N.J., 1968, 1.
65. **Alexander, V.,** Primary productivity regimes of the nearshore Beaufort Sea, with reference to the potential roles of the ice biota, in *The Coast and Shelf of the Beaufort Sea*, Reed, J. C. and Sater, J. E., Eds., Arctic Institute of North America, Calgary, 1974, 609.
66. **Apollonio, S.,** The chlorophyll content of Arctic sea ice, *Arctic*, 14, 197, 1961.
67. **Divoky, G. J.,** The pelagic feeding habits of ivory and Ross' gulls, *Condor*, 78, 85, 1976.
68. **Divoky, G. J.,** Birds and the ice-edge ecosystem in the Bering Sea, in *The Eastern Bering Sea Shelf: Oceanography and Resources*, Vol. 2, Hood, D. W. and Calder, J. A., Eds., University of Washington Press, Seattle, 1981, 799.
69. **Busdosh, M., Robilliard, G. A., Tarbox, K., and Beehlez, C. L.,** Chemoreception in an arctic amphipod crustacean: a field study, *J. Exp. Mar. Biol. Ecol.*, 62, 261, 1982.
70. **Klumov, S. K.,** Polar cod (*Boreogadus saida* Lepech.) and its importance for certain life processes in the Arctic, *Izv. Akad. Nauk SSSR, Ser. Biol.*, 1, 175, 1937 (in Russian, French summary).
71. **Andriashev, A. P.,** Cryopelagic fishes of the Arctic and Antarctic and their significance in polar ecosystems, in *Antarctic Ecology*, Vol. 1, Holdgate, M. W., Ed., Academic Press, New York, 1970, 297.
72. **Golikov, A. N. and Scarlato, O. A.,** Comparative characteristics of some ecosystems of the upper regions of the shelf in tropical, temperate and arctic waters, *Helgol. Wiss. Meeresunters.*, 24, 219, 1973.
73. **Bradford, J. M.,** Sea ice organisms and their importance to the Antarctic ecosystem (review), *N.Z. Antarct. Rec.*, 1(2), 43, 1978.

74. **Alexander, V. and Chapman, T.,** The role of epontic algal communities in Bering Sea ice, in *The Eastern Bering Sea Shelf: Oceanography and Resources,* Vol. 2, Hood, D. W. and Calder, J. A., Eds., University of Washington Press, Seattle, 1981, 773.
75. **Horner, R. and Schrader, G. C.,** Relative contributions of ice algae, phytoplankton, and benthic microalgae to primary production in nearshore regions of the Beaufort Sea, *Arctic,* 35, 485, 1982.
76. **Green, K. A.,** Simulation of the pelagic ecosystem of the Ross Sea, Antarctica: a time varying compartmental model, Ph.D. thesis, University Microfilms International, Ann Arbor, 1976.
77. **Dunbar, M. J.,** The evolution of polar ecosystems, in *Adaptations within Antarctic Ecosystems,* Llano, G. A., Ed., Gulf Publ., Houston, 1977, 1063.
78. **Matheke, G. E. M. and Horner, R.,** Primary productivity of the benthic microalgae in the Chukchi Sea near Barrow, Alaska, *J. Fish. Res. Board Can.,* 31, 1779, 1974.
79. **Meguro, H., Ito, K., and Fukushima, H.,** Ice flora (bottom type): a mechanism of primary production in polar seas and the growth of diatoms in sea ice, *Arctic,* 20, 114, 1967.
80. **Gruzov, Ye. N.,** Seasonal alterations in coastal communities in the Davis Sea, in *Adaptations within Antarctic Ecosystems,* Llano, G. A., Ed., Gulf Publ., Houston, 1977, 263.
81. **Barnard, K. H.,** Crustacea, Part XI — Amphipoda, *British Antarctic ("Terra Nova") Expedition, 1910, Natural History Report,* 8, 307, 1930.
82. **Andriashev, A. P.,** Fishes of the northern seas of the USSR, *Izdatel'stvo Akademii Nauk SSSR, Moskva-Leningrad, 1954, 1. (Translated from Russian by Israel Program for Scientific Translations, Jerusalem, 1964.)*
83. **Holmquist, C.,** An observation on young specimens of *Ammodytes dubius, Medd. Groenl.,* 159(2), 9, 1958.

APPENDIX

ALGAL SPECIES REPORTED FROM SEA ICE

This is a partial list of algal species reported from sea ice. The nomenclature follows Van Landingham[1] for diatoms; Schiller[2] and Balech[3] for dinoflagellates; and Parke and Dixon[4] for most other groups. Names not found in these publications are from the references cited for the individual species. Diatom species marked with an asterisk (*) are of questionable validity according to Van Landingham. Species marked with a dagger (†) are of freshwater origin, but may be found in the marine environment. Other abbreviations: I = ice; IE = ice edge; PI = pack ice; W = water column; N = Arctic; S = Antarctic.

Species	Region	Habitat	Ref.
Bacillariophyceae			
Achnanthes delicatula (Kützing) Grunow	N	I	5
Achnanthes taeniata Grunow	N	I	5, 6
Achnanthes spp.	N	I	5, 7, 8
**Actinoptychus antarcticus* (Castracane) Heiden	S	I	10
Actinoptychus undulatus (Bailey) Ralfs	S	I	9
Amphipleura micans (Lyngbye) Cleve	S	I	10
Amphipleura rutilans (Trentepohl) Cleve	N, S	I	5, 12
A. rutilans var. *antarctica* (Harvey) Cleve	S	I	11
A. rutilans var. *obtusa* (Greville) Hustedt	S	I	10
Amphipleura sp.	S	I	13
Amphiprora acuta M. Peragallo	S	I, W	10, 26
**Amphiprora concilians* Cleve (*Entomoneis concilians* (Cleve) Poulin et Cardinal)	N	I	5, 14, 15
Amphiprora gigantea var. *septentrionalis* (Grunow) Cleve (*Entomoneis gigantea* var. *septentrionalis* (Grunow) Poulin et Cardinal)	N	I	5, 6, 15
Amphiprora kjellmanii Cleve	N, S	I	5, 6, 16—18
A. kjellmanii var. *kariana* (Grunow) Cleve (*Entomoneis kjellmanii* var. *kariana* (Grunow) Poulin et Cardinal)	N	I	5, 6, 15
A. kjellmanii var. *striolata* (Grunow) Cleve	N, S	I	5, 19
A. kjellmanii var. *subtilis* (Grunow) Cleve (*Entomoneis kjellmanii* var. *subtilis* (Grunow) Poulin et Cardinal)	N	I	15
A. kjellmanii var. *subtilissima* Van Heurck	S	I, W	10, 19
Amphiprora kryophila Cleve	N	I	5, 20
Amphiprora kufferathii Manguin	S	I, W	11
Amphiprora lata Greville	S	I, W	10
Amphiprora oestrupii Van Heurck	S	I, W	10, 16, 19, 22, 23
A. oestrupii var. *minor* Van Heurck	S	I, W	10
Amphiprora paludosa var. *hyperborea* (Grunow) Cleve (*Entomoneis paludosa* var. *hyperborea* (Grunow) Poulin et Cardinal)	N, S	I	5—7, 14, 15, 19
Amphiprora paludosa var. *punctulata* Grunow	N	I	6, 14, 23
Amphiprora spp.	S	I	24—26
Amphora cymbelloides Grunow	S	I, W	10
Amphora exsecta Grunow	N	I	5, 8
Amphora laevis Gregory	N	I	8
A. laevis var. *laevissima* (Gregory) Cleve	N	I	5, 8, 15
**Amphora laevissima* var. *minuta* Cleve	N	I	5, 6
Amphora ocellata Donkin	N	I	7
Amphora proteus Gregory	N	I	5
Amphora spp.	N	I	8
Arachnoidiscus ehrenbergii Bailey	S	I	13
Asterionella formosa Hassall	N	I	8
Asteromphalus beaumontii Ehrenberg	S	I	9
Asteromphalus brookei Bailey	N	I	5
Asteromphalus heptactis (Brébisson) Ralfs	S	PI, W	10
Asteromphalus hookeri Ehrenberg	N, S	I, W	5, 9, 10, 13, 23
Asteromphalus parvulus Karsten	S	IE	27
Bacillaria paradoxa Gmelin	N	I	5
Biddulphia aurita (Lyngbye) Brébisson	N	I	5
B. aurita var. *obtusa* (Kützing) Hustedt	S	I	13
Biddulphia litigiosa Van Heurck	S	I	13
Biddulphia punctata Greville	S	I	17
Biddulphia striata Karsten	S	I	26
Biddulphia weissflogii Grunow	S	I	10, 13, 16, 28
Biddulphia spp.	S	I	12, 25
Caloneis brevis (Gregory) Cleve	N	I	5

Species	Region	Habitat	Ref.
Caloneis kryophila (Cleve) Cleve	N	I	5, 8
Caloneis liber (Wm. Smith) Cleve	N	I	5
†*Caloneis obtusa* (Wm. Smith) Cleve	N	I	5
Caloneis semiinflata (Østrup) Boyer	N	I	5
Chaetoceros atlanticus Cleve	N	I	5, 6, 23
C. atlanticus var. *skeleton* (Schütt) Hustedt	S	I	19
Chaetoceros borealis Bailey	N	I, W	5, 6, 8
Chaetoceros bulbosum (Ehrenberg) Heiden	S	I, W	10, 13, 18
Chaetoceros ceratosporum Ostenfeld	N	I	8
Chaetoceros compressus Lauder	N	I	8
Chaetoceros concavicornis Mangin	N	I	6
Chaetoceros curvatum Castracane	S	I	19
Chaetoceros debilis Cleve	S	I	28
Chaetoceros decipiens Cleve	N	I	5, 8, 23
Chaetoceros dichaeta Ehrenberg	S	I, W	10, 18, 19, 22, 26
Chaetoceros fragilis Meunier	N, S	I	5, 8, 29
Chaetoceros furcellatus Bailey	N	I	5, 6, 14
Chaetoceros karianus Grunow	N	I, W	8
Chaetoceros neglectus Karsten	S	IE	27
Chaetoceros neogracile Van Landingham	N, S	I, W	14, 18, 28
Chaetoceros pelagicus Cleve	S	I	30
Chaetoceros pendulus Karsten	S	I	26
Chaetoceros peruvianus Brightwell	S	I	19
Chaetoceros radiculus Castracane	S	I	19
Chaetoceros schimperianus Karsten	S	I	17
Chaetoceros septentrionalis Østrup	N	I	5, 6, 8, 23
Chaetoceros simplex Ostenfeld	S	I	30
Chaetoceros socialis Lauder	N, S	I, IE	6, 27
Chaetoceros spp.	N, S	I, W	8, 12, 14, 25, 28
Charcotia actinochilus (Ehrenberg) Hustedt	S	I	9, 13, 21, 30
Charcotia apollinis (Ehrenberg) M. Peragallo	S	I	9
Charcotia australis (Karsten) M. Peragallo	S	I, W	11
Charcotia bifrons (Castracane) M. Peragallo	S	I	10, 22
Charcotia janus (Castracane) M. Peragallo	S	I, W	10
Charcotia sp.	S	I	12
Chuniella sp.	S	I	25
Climacosphenia sp.	S	I	25
Cocconeis costata Gregory	N	I	5, 8
Cocconeis fasciolata (Ehrenberg) Brown	S	I	13, 30
Cocconeis gautieri Van Heurck	S	I	13
Cocconeis placentula var. *euglypta* (Ehrenberg) Grunow	N	I	5
Cocconeis scutellum Ehrenberg	N, S	I	5, 6, 13
C. scutellum var. *parva* (Grunow) Cleve	N	I	5
C. scutellum var. *stauroneiformis* Wm. Smith	N	I	5
Corethron criophilum Castracane	S	PI, I, W	18, 26, 27, 30
Corethron pinnatum (Grunow) Ostenfeld	S	I	19
Corethron valdiviae Karsten	S	PI, I, W	10, 19
Coscinodiscus anomalus Van Landingham	S	I, W	10
Coscinodiscus antarcticus (Grunow) Rattray	S	I	10, 13
Coscinodiscus curvatulus Grunow	N	I	6
Coscinodiscus denarius A. Schmidt	S	I, W	10
Coscinodiscus excentricus Ehrenberg	S	I	9
Coscinodiscus furcatus Karsten	S	I	11, 28, 30, 31
Coscinodiscus gemmatulus Castracane	S	PI	10
Coscinodiscus gerlachii Van Heurck	S	PI	10
Coscinodiscus gigas Ehrenberg	S	I	13
Coscinodiscus gracilis Karsten	S	I	13, 21
Coscinodiscus impolitus Rattray	S	I, W	10

Species	Region	Habitat	Ref.
Coscinodiscus kützingii A. Schmidt	S	I, W	10
Coscinodiscus kützingii var. *glacialis* Grunow	N	I	5
Coscinodiscus lentiginosus Janisch	S	I, W	10, 13
C. lentiginosus var. *confusa* Van Heurck	S	I, W	10
**C. lentiginosus* var. *minor* Heiden et Kolbe	S	I	10
Coscinodiscus lineatus Ehrenberg	S	I	9, 10
Coscinodiscus lunae Ehrenberg	S	I	9
Coscinodiscus oculus-iridis Ehrenberg	S	I	9
Coscinodiscus pacificus (Grunow) Rattray	S	PI, I, W	10
Coscinodiscus polyacanthus Grunow	N	I	5
Coscinodiscus pseudodenticulatus Hustedt	S	I	21
Coscinodiscus pyrenoidophorus Karsten	S	I	30
**Coscinodiscus radiolatus* Ehrenberg	S	I	9
Coscinodiscus stellaris Roper	N, S	I	6
C. stellaris var. *symbolophorus* (Grunow) Jørgensen	S	I	9, 10, 13
Coscinodiscus subtilis Ehrenberg	S	I	9, 16
Coscinodiscus suspectus Janisch	S	I, W	10
Coscinodiscus tabularis Grunow	S	PI, I	10, 13
Coscinodiscus tumidus Janisch	S	I, W	10, 13
**C. tumidus* var. *fasciculata* Rattray	S	I	13
Coscinodiscus varians var. *major* Karsten	S	I	13
Coscinodiscus velatus Ehrenberg	S	I	9
Coscinodiscus spp.	N, S	I	8, 13, 23, 25
Coscinosira antarctica Kozlova	S	I	13
Cyclotella comta (Ehrenberg) Kützing	S	I	13
Cylindrotheca closterium (Ehrenberg) Reimann et Lewin	N, S	I, W	5, 7, 8, 10, 12, 14, 17, 18, 19, 23, 27, 28, 30
Cymbella cistula (Hemprich) Kirchner	S	I	19
Dactyliosolen antarcticus Castracane	S	I, W	10, 18, 28
D. antarcticus var. *laevis* (Karsten) Heiden et Kolbe	S	I	10, 13, 19
Dactyliosolen areolatus Heiden et Kolbe	S	PI, I, W	10
Detonula confervacea (Cleve) Gran	N	I	5, 23
Diatoma vulgare var. *ehrenbergii* Kützing) Grunow	S	I	19
Diatoma sp.	S	I	25
Diploneis didyma (Ehrenberg) Ehrenberg	N	I	5
Diploneis incurvata (Gregory) Cleve	N	I	5
Diploneis lineata (Donkin) Cleve	N	I	5, 14
Diploneis litoralis (Donkin) Cleve	N	I	5, 6, 8, 23
D. litoralis var. *arctica* Cleve	N	I	5, 8, 32
D. litoralis var. *clathrata* (Østrup) Cleve	N	I	5, 8, 32
Diploneis sejuncta (A. Schmidt) Jørgensen	S	I, W	10
Diploneis smithii (Brébisson) Cleve	N	I	5, 6
D. smithii var. *recta* Peragallo	N	I	32
Diploneis vacillans (A. Schmidt) Cleve	N	I	5, 6
Eucampia balaustium Castracane	S	IE, I, W	10, 11, 13, 16, 27
Eucampia groenlandica Cleve	N	I, W	5, 8, 23
Eucampia zodiacus Ehrenberg	N, S	I	8, 26
†*Eunotia lunaris* (Ehrenberg) Brébisson	S	I	21
Eunotia sp.	N	I	7
Fragilaria islandica Grunow	N	I	6
F. islandica var. *adeliae* Manguin	S	I	24
Fragilaria pinnata Ehrenberg	N	I	5
Fragilaria striatula Lyngbye	N, S	I	6, 19
Fragilaria spp.	N, S	I	7, 25
Frustulia rhomboides var. *crassinervis* (Brébisson) Ross	S	I	19
†*Gomphonema clevei* Fricke	S	I	21
Gomphonema exiguum Kützing	N	I, W	5, 8, 14
G. exiguum var. *arcticum* (Grunow) Cleve	N	I	6, 7, 20

Species	Region	Habitat	Ref.
G. exiguum var. *pachycladum* (Brébisson) Cleve	N	I	5, 6
G. groenlandicum Østrup	N	I	5, 15
Gomphonema kamtschaticum Grunow	N	I	6
Gomphonema septentrionale Østrup	N	I	15
Grammatophora angulosa Ehrenberg	N	I	5, 8
Grammatophora arctica Cleve	N	I	5
Grammatophora cf. *charcotii* M. Peragallo	S	I	13
Grammatophora hamulifera Kützing	N	I	5
Guinardia blavyana H. Peragallo	S	I	19
Gyrosigma compactum (Greville) Cleve	S	I	19
Gyrosigma fasciola (Ehrenberg) Griffith et Henfrey	N	I	5, 7
Gyrosigma hudsonii Poulin et Cardinal	N	I	32
Gyrosigma tenuissimum var. *hyperborea* (Grunow) Cleve	N	I	32
Gyro-Pleurosigma spp.	N	I	7
Gyrosigma spp.	S	I	26
Hantzschia weyprechtii Grunow	N	I	5, 6
Lauderia annulata Cleve	S	PI, I, W	10
Lauderia antarctica Heiden et Kolbe	S	PI, I, W	10
Licmophora antarctica var. *minor* Frenguelli	S	I	13
Licmophora dalmatica (Kützing) Grunow	N	I	5
Licmophora decora Heiden et Kolbe	S	I	13
Licmophora gracilis (Ehrenberg) Grunow	N	I	5
Licmophora gracilis var. *anglica* (Kützing) Peragallo et Peragallo	N	I	5
Licmophora hyalina (Kützing) Grunow	N	I	5, 6
Licmophora tenuis (Kützing) Grunow	N	I	14
Licmophora spp.	N, S	I	7, 8, 12
Melosira arctica Dickie	N	I	5, 6, 8, 23
Melosira distans (Ehrenberg) (Kützing)	S	I	13
Melosira sol (Ehrenberg) Kützing	S	I	13
Melosira sp.	S	I	13
Navicula algida Grunow	N	I	5, 8, 20, 33
**Navicula antarctica* Frenguelli	S	I	13
Navicula bolleana (Grunow) Cleve	N	I	7
Navicula cancellata Donkin	N	I	5
Navicula cluthensis Gregory	N	I	5, 23
N. cluthensis var. *pagophila* Grunow	N	I	5
Navicula corymbosa (Agardh) Cleve	S	I	22
Navicula crassirostris Grunow	N	I	5, 8
**Navicula criophila* (Castracane) DeToni	S	I	13, 19
Navicula crucigeroides Hustedt	N	I	5, 8, 20, 33
Navicula decipiens O'Meara	N	I	5
Navicula detersa (Grunow) Gran	N	I	6
Navicula digitoradiata (Gregory) Ralfs	N	I	5
Navicula directa (Wm. Smith) Ralfs	N, S	I	5—8, 20, 23, 30, 33
N. directa var. *constricta* Heiden et Kolbe	S	I	10
N. directa var. *javanica* Cleve	N, S	I	5, 10, 33
N. directa var. *rostrata* Heiden et Kolbe	S	I	10
N. directa var. *subtilis* (Gregory) Cleve	N, S	I	5, 13
Navicula frequens Van Heurck	S	I, W	10, 19
Navicula forcipata Greville	N	I	5, 8
N. forcipata var. *densestriata* A. Schmidt	N	I	33
Navicula gastrum (Ehrenberg) Kützing	N	I	5, 6, 8
Navicula gelida Grunow	N	I	5, 6, 8
N. gelida var. *radissonii* Poulin et Cardinal	N	I	33
Navicula glacialis (Cleve) Grunow	N	I	5, 33
N. glacialis var. *hudsonii* Poulin et Cardinal	N	I	33
Navicula glaciei Van Heurck	N, S	I	17, 19, 21, 24, 30, 33

Species	Region	Habitat	Ref.
Navicula gordonii M. Peragallo	S	I	13
Navicula gracilis Ehrenberg	S	I	21
N. gracilis var. *inaequalis* Østrup	N	I	20
Navicula granii (Jørgensen) Gran	N	I	5
Navicula imperfecta Cleve	N	I	5, 33
N. imperfecta var. *antarctica* Heiden et Kolbe	S	I, W	10
Navicula implexa Hustedt	N	I	33
Navicula jejunoides Van Heurck	S	I, W	10, 19
N. jejunoides var. *longissima* Van Heurck	S	I	13
Navicula kariana Grunow	N	I	5
N. kariana var. *detersa* Grunow	N	I	5, 8, 33
N. kariana var. *frigida* (Grunow) Cleve	N	I	33
Navicula kepesii Grunow	N	I	6
Navicula kjellmanii (Cleve) Cleve	N	I	5—7, 20, 23, 33
Navicula kryokonites Cleve	N	I	33
N. kryokonites var. *subprotracta* Cleve	N, S	I, W	10, 33
Navicula kryophila (Cleve) Poulin et Cardinal	N	I	33
Navicula kuripanensis var. *borealis* Poulin et Cardinal	N	I	33
Navicula libellus Gregory	S	I	13
Navicula lineola Grunow	N	I	33
N. lineola var. *perlepida* (Grunow) Cleve	N	I	33
Navicula ludloviana var. *unilateralis* Poulin et Cardinal	N	I	33
Navicula lyra var. *atlantica* A. Schmidt	N	I	5
Navicula maculosa Donkin	N	I	7
Navicula marina Ralfs	N	I	5, 7, 14
Navicula mauriciana Van Heurck	S	I	19
Navicula membranacea Cleve	N	I	8
Navicula muticopsis Van Heurck	S	I	19
Navicula novadecipiens Hustedt	N	I	5, 8, 33
**Navicula obtusa* (Cleve) Cleve	N	I	5, 20
Navicula oestrupi Cleve	N	I	5, 33
Navicula pagophila var. *manitounukensis* Poulin et Cardinal	N	I	33
Navicula pelagica Cleve	N	I	6—8, 33
**Navicula pellucida* Karsten	N	I	5
Navicula pellucidula var. *contracta* (Østrup) Hustedt	N	I	33
Navicula pennata A. Schmidt	S	I, W	10
Navicula peregrina (Ehrenberg) Kützing	N	I	5
Navicula perlucens Østrup	N	I	5
Navicula praetexta var. *antarctica* Van Heurck	S	I	13
†*Navicula pygmaea* Kützing	N	I	5, 8
Navicula recurvata Gran	N	I	5, 6, 8
†*Navicula rhynchocephala* Kützing	N	I	8
†*Navicula salinarum* Grunow	N	I	5, 8
Navicula scopulorum Brébisson	N	I	8
Navicula siberica (Grunow) Cleve	N	I	5, 6, 14, 23, 33
Navicula solitaria Cleve	N	I	5, 14, 33
Navicula spicula (Hickie) Cleve	N	I	5, 7, 8, 33
**Navicula stuxbergii* var. *subglabra* Østrup	N	I	5
Navicula subantarctica Frenguelli (as *N. gelida* var. *antarctica* Heiden et Kolbe)	S	I, W	10
Navicula subdelicata var. *arctica* Poulin et Cardinal	N	I	33
Navicula subinflata Grunow	N	I	5, 23, 33
Navicula superba Cleve	N	I	5, 6, 8, 33
N. superba var. *crassa* (Østrup) Gran	N	I	5
N. superba var. *elliptica* Cleve	N	I	5, 33
N. superba var. *subacuta* Gran	N	I	5, 33
Navicula transfuga Grunow	N	I	5

Species	Region	Habitat	Ref.
N. transfuga var. *plagiostoma* (Grunow) Cleve	N	I	5
N. transfuga var. *septentrionalis* Østrup	N	I	5
Navicula transitans Cleve	N	I	5—8, 14, 20
N. transitans var. *asymmetrica* (Cleve) Cleve	N	I	33
N. transitans var. *derasa* (Grunow) Cleve	N	I	5, 6, 8, 14, 20, 23, 33
N. transitans var. *erosa* (Cleve) Cleve	N	I	5, 20
N. transitans var. *incudiformis* (Grunow) Cleve	N	I	5, 8, 33
**Navicula trigonocephala* Cleve	N	I	5, 7, 8, 14, 20
**N. trigonocephala* var. *contracta* Østrup	N	I	5
**N. trigonocephala* var. *depressa* Østrup	N	I	5, 6, 8, 33
Navicula triplex Van Landingham (as *N. subpolaris* (M. Peragallo) Hendey)	S	I	22
Navicula trompii Cleve	S	I, W	10, 13
N. trompii var. *major* Heiden	S	I	10, 13
Navicula ungavae Poulin et Cardinal	N	I	33
Navicula valida Cleve	N	I	5, 6, 8, 14, 20
N. valida var. *minuta* Cleve	N	I	5, 7, 8, 33
**Navicula variabilis* (Heiden et Kolbe) Mills	S	I, W	10
Navicula vitrea Grunow	N	I	33
Navicula sp. A	N	I	7
Navicula spp.	N, S	I	8, 12, 13, 23, 25, 26
Nitzschia acicularis (Kützing) Wm. Smith	N, S	I, W	5, 6, 10, 19, 21
N. acicularis f. *angustior* O. Müller	S	I	10
Nitzschia acuminata (Wm. Smith) Grunow	N	I	5
Nitzschia angularis Wm. Smith	N, S	I, W	5—8, 10, 15, 19
N. angularis var. *tennuistriata* Van Heurck	S	I	21
Nitzschia angulata Hasle (as *Fragilaria rhombica* (O'Meara) Heiden, *Fragilariopsis rhombica* (O'Meara) Hustedt)	N, S	I, W	5, 10, 11, 18, 21, 28, 30
Nitzschia angustissima Van Heurck	S	I	19, 21
Nitzschia barbieri var. *minor* M. Peragallo	S	I	22
Nitzschia brebissoni var. *borealis* Grunow	N	I	5, 8, 15
Nitzschia castracanei Hasle (*Fragilaria linearis* Castracane, *Fragilariopsis linearis* (Castracane) Hustedt, *Nitzschia lineata* Hasle)	S	I	10, 12, 16, 17, 21, 22, 30
Nitzschia curta (Van Heurck) Hasle (*Fragilaria curta* Van Heurck, *Fragilariopsis curta* (Van Heurck) Hustedt)	N, S	I, W	5, 11, 17, 18, 19, 21, 22, 26, 28, 30
Nitzschia cylindrus (Grunow) Hasle (*Fragilaria cylindrus* Grunow, *Fragilariopsis cylindrus* (Grunow) Helmcke et Krieger)	N, S	I	5—8, 10, 11, 15, 18, 19, 21, 28, 30, 34
Nitzschia debilis (Arnott) Grunow	S	I	19
Nitzschia dissipata Kützing	N	I	8
Nitzschia distans Gregory	N, S	I	8, 10
Nitzschia frigida Grunow	N, S	I, W	5—8, 10, 12, 14, 15, 23
Nitzschia gelida Cleve et Grunow	N	I	5
N. gelida var. *manitounukensis* Poulin et Cardinal	N	I	15
Nitzschia gracillima Heiden et Kolbe	S	I, W	10
Nitzschia gruendleri Grunow	N, S	I	5, 11
Nitzschia grunowii Hasle (*Fragilaria arctica* Grunow, *Fragilaria oceanica* Cleve, *Fragilariopsis oceanica* (Cleve) Hasle)	N, S	I	5, 6, 8, 14, 19
Nitzschia heimii (Manguin) Hasle	S	I	30
Nitzschia hudsonii Poulin et Cardinal	N	I	15
Nitzschia hybrida Grunow	N, S	I, W	5, 8, 10, 15
Nitzschia kerguelensis (O'Meara) Hasle (*Fragilariopsis antarctica* (Castracane) Hustedt)	S	I, W	10, 11, 18, 19, 21, 22, 27, 30, 35
Nitzschia kryophila Cleve	N	I	15
Nitzschia laevissima Grunow	N, S	I, W	5, 6, 8, 10, 15
Nitzschia lanceolata var. *pygmaea* Cleve	N	I	5, 15

Species	Region	Habitat	Ref.
Nitzschia lecointei Van Heurck	N, S	I	5, 15, 19, 30
†*Nitzschia linearis* (Agardh) Wm. Smith	N	I	5, 8
Nitzschia lineata Grunow	N	I	5
Nitzschia longissima (Brébisson) Grunow	N	I	5, 8, 23
Nitzschia marginulata Grunow	N	I	5
Nitzschia martiana (Agardh) Van Heurck	S	I	16
Nitzschia neglecta Hustedt	S	I	18
Nitzschia obliquecostata (Van Heurck) Hasle (*Fragilaria obliquecostata* Van Heurck, *Fragilariopsis obliquecostata* (Van Heurck) Heiden)	N, S	I	5, 10, 11, 12, 19
Nitzschia ostenfeldii Van Heurck	S	I	19
Nitzschia polaris Grunow	N, S	I	5, 6, 8, 15, 19, 30
Nitzschia ritscheri (Hustedt) Hasle (*Fragilariopsis ritscheri* Hustedt)	S	I	12
Nitzschia semigibbosa Van Heurck	S	I	19
Nitzschia seriata Cleve	N, S	I, W	5—7, 16, 19, 23, 26, 35
Nitzschia sicula var. *migrans* (Cleve) Hasle	S	I, W	10, 19
Nitzschia sigma (Kützing) Wm. Smith	N, S	I, W	7, 8, 10
N. sigma var. *rigida* (Kützing) Grunow	N	I	6
Nitzschia sigmoidea (Nitzsch) Wm. Smith	N	I	7
Nitzschia stellata Manguin	S	I, W	11, 24, 36
Nitzschia subcurvata Hasle	S	I	18, 28
Nitzschia sublineata Hasle (*Fragilaria sublinearis* Van Heurck, *Fragilariopsis sublinearis* (Van Heurck) Heiden et Kolbe)	S	I, W	10, 18, 19, 21, 28, 29
Nitzschia tergestina (Kützing) Ralfs	N	I	5
†*Nitzschia vitrea* Norman	S	I	19
Nitzschia sp. A	N	I	7
Nitzschia sp. B	N	I	7
Nitzschia spp.	N, S	I	7, 8, 12, 17—19, 23, 25
Pinnularia ambigua Cleve	N	I	8, 32
Pinnularia polaris Heiden	N	I	32
Pinnularia quadratarea (A. Schmidt) Cleve	N, S	I	5, 7, 8, 14, 20, 21, 32
P. quadratarea var. *bicontracta* (Østrup) Heiden	N	I	5, 8, 20, 32
P. quadratarea var. *bicuneata* Heiden et Kolbe	N, S	I, W	5, 10, 11, 32
P. quadratarea var. *capitata* Heiden	N	I	20, 32
P. quadratarea var. *constricta* (Østrup) Heiden	N, S	I, W	5, 6, 8, 10, 20, 32
P. quadratarea var. *constricta* f. *interrupta* Heiden	N	I	32
P. quadratarea var. *cuneata* Østrup	N	I	5, 32
P. quadratarea var. *densestriata* Cleve	N	I	5
P. quadratarea var. *dubia* Heiden	N	I	32
P. quadratarea var. *kerguelensis* (Cleve et Grunow) Cleve	N	I	7
P. quadratarea var. *leptostauron* (Cleve) Cleve	N	I	5, 32
P. quadratarea var. *maxima* (Østrup) Boyer	N	I	5, 32
P. quadratarea var. *minima* (Østrup) Boyer	N	I	5
P. quadratarea var. *minor* (Østrup) Heiden	N	I	32
P. quadratarea var. *patrickae* Poulin et Cardinal	N	I	32
P. quadratarea var. *stuxbergii* (Cleve) Cleve	N	I	5, 6, 20, 32
P. quadratarea var. *subconstricta* (Østrup) Heiden	N	I	5
P. quadratarea var. *subcontinua* (Cleve) Cleve	N	I	5
P. quadratarea var. *subglabra* (Østrup) Poulin et Cardinal	N	I	32
P. quadratarea var. *theelii* (Cleve) Cleve	N	I	7
Pinnularia semiinflata (Østrup) Gran	N	I	6, 8, 20, 32
P. semiinflata var. *decipiens* (Cleve) Gran	N	I	20, 32
Pinnularia spp.	N	I	8, 23
Plagiogramma staurophorum (Gregory) Heiberg	N	I	5
Plagiotropis lepidoptera (Gregory) Poulin et Cardinal	N	I	15
Plagiotropis maxima (Gregory) Poulin et Cardinal	N	I	15

Species	Region	Habitat	Ref.
Pleurosigma angulatum (Quekett) Wm. Smith	N	I	5, 8, 23
**Pleurosigma antarcticum* Heiden et Kolbe	N, S	I, W	5, 10, 16
Pleurosigma clevei Grunow	N	I	5, 6, 32
P. clevei var. *manitounukensis* Poulin et Cardinal	N	I	32
**Pleurosigma cuspidatum* (Cleve) H. Peragallo	N	I	5
Pleurosigma elongatum Wm. Smith	S	I	13
P. elongatum var. *kariana* (Grunow) Cleve	N	I	5, 8
Pleurosigma interrmedium Wm. Smith	S	I	13
Pleurosigma marinum Donkin	N	I	5
Pleurosigma stuxbergii Cleve et Grunow	N	I	5, 6, 14, 20, 32
P. stuxbergii var. *minor* Grunow	N	I	5
P. stuxbergii var. *rhomboides* (Cleve) H. Peragallo	N	I	5, 20, 32
Pleurosigma spp.	N, S	I, W	8, 11, 12, 23
Porosira glacialis (Grunow) Jørgensen	N, S	I	5, 6, 30
Porosira pseudodenticulata (Hustedt) Jousé	S	I	13, 28, 30, 31
Pseudonitzschia delicatissima (Cleve) Heiden (Nitzschia *delicatissima* Cleve)	N, S	I	5—8, 14, 23, 28
Rhabdonema arcuatum (Lyngbye) Kützing	N	I	5
Rhabdonema minutum Kützing	N	I	5
Rhizosolenia alata Brightwell	N, S	I	5, 16, 19
Rhizosolenia antennata (Ehrenberg) Brown	S	I	9
Rhizosolenia fragilissima Bergon	S	PI	10
Rhizosolenia hebatata Bailey	N	I	5, 23
R. hebatata f. *semispina* (Hensen) Gran	S	PI, W	10
R. hebatata f. *subacuta* Grunow	N	I	5
Rhizosolenia inermis Castracane	S	I	19
R. inermis f. *rostrata* Heiden	S	PI, I, W	10
Rhizosolenia styliformis Brightwell	S	I	9
R. styliformis var. *latissima* Brightwell	S	I	9, 10, 19
Rhizosolenia truncata Karsten	S	I	19, 26, 30
Rhizosolenia spp.	N	I	8
**Stauroneis amphioxys* Gregory	S	I	37
**Stauroneis glacialis* Heiden et Kolbe	S	I	10
**Stauroneis pacifica* var. *minor* Van Heurck	S	I	19
Stauroneis perpusilla Grunow	S	I	10
Stauroneis quadripedis (Cleve-Euler) Hendey	N	I	5, 7, 8
Stauroneis radissonii Poulin et Cardinal	N	I	32
Stauroneis septentrionalis Grunow	N	I	6
Stauroneis sp.	S	I	13
Stenoneis inconspicua (Gregory) Cleve	N	I	20
S. inconspicua var. *baculus* (Cleve) Cleve	N	I	5, 8
†*Stephanodiscus astraea* (Ehrenberg) Grunow	S	I	13
Stephanopyxis sp.	S	I	12
**Surirella japonica* A. Schmidt	N	I	5
Synedra camtschatica var. *finnmarchica* Cleve et Grunow	N	I	5
Synedra formosa Hantszch	S	I	13
Synedra hyperborea Grunow	N	I	5, 6
S. hyperborea var. *rostellata* Grunow	N	I	5
†*Synedra pulchella* (Ralfs) Kützing	N	I	5
Synedra reinboldii Van Heurck	S	I	10, 13, 19
Synedra tabulata (Agardh) Kützing	N	I	5, 6
S. tabulata var. *fasciculata* (Lyngbye) Hustedt	N	I	5
†*S. tabulata* var. *obtusa* (Pantocsek) Hustedt	N	I	5
Synedra spp.	N, S	I	7, 12, 13, 28
Thalassiosira anguste-lineata (A. Schmidt) Fryxell et Hasle (*Coscinosira polychorda* (Gran) Gran)	N	I	6
Thalassiosira antarctica Comber	S	I, IE, W	30, 35
Thalassiosira bioculata (Grunow) Ostenfeld	N	I	5, 6
T. bioculata var. *exigua* (Grunow) Hustedt	N	I	5

Species	Region	Habitat	Ref.
Thalassiosira decipiens (Grunow) Jørgensen	N, S	I, W	5, 6, 10, 19
Thalassiosira expecta Van Landingham (*T. delicatula* Hustedt)	S	I	13
Thalassiosira dubia Kozlova	S	I	21
Thalassiosira fallax Kozlova	S	I	13
Thalassiosira gravida Cleve	N, S	I, W	5—7, 10, 19, 23
Thalassiosira hyalina (Grunow) Gran	S	I, W	10
Thalassiosira kozlovii (Kozlova) Makarova	S	I	13
Thalassiosira lacustris (Grunow) Hasle (*Coscinodiscus lacustris* Grunow)	N	I, W	8, 38
T. lacustris var. *septentrionalis* (Grunow) Rattray	N	I	5
Thalassiosira leptopus (Grunow) Hasle et Fryxell (*Coscinodiscus lineatus* Ehrenberg)	N	I	14
Thalassiosira nordenskioeldii Cleve	N	I	5, 6, 8, 14
Thalassiosira obica Kozlova	S	I	13
Thalassiosira oestrupii (Ostenfeld) Proschkina-Lavrenko *ex* Hasle	N	I	5, 6
Thalassiosira ritscherii (Hustedt) Hasle	S	I	30
Thalassiosira tumida (Janisch) Hasle	S	I	17
Thalassiosira sp.	S	I	18
Thalassiothrix antarctica Schimper	S	IE, I, W	13, 35
Thalassiothrix longissima Cleve et Grunow	N	I	5, 23
Thalassiothrix spp.	N	I	7, 14
Trachyneis aspera (Ehrenberg) Cleve	N, S	I	5, 13
Triceratium arcticum Brightwell	N	I	5
Triceratium formosum Brightwell	S	I	13
Triceratium nobile Witt	N	I	5
Tropidoneis belgicae (Van Heurck) Heiden	S	I, W	10, 22
T. belgicae var. *major* Van Heurck	S	I, W	10
Tropidoneis fusuformis Manguin	S	I	18
Tropidoneis gausii Heiden	S	I	10, 13, 18
Tropidoneis glacialis Heiden et Kolbe	S	I	10
T. glacialis var. *constricta* Heiden et Kolbe	S	I	10
Tropidoneis maxima (Gregory) Cleve	N	I	5
T. maxima var. *dubia* (Cleve et Grunow) Cleve	N	I	5
Tropidoneis vanheurckii (Grunow) Cleve	S	I	30
Tropidoneis sp.	S	I	26
Vanhoeffenus antarcticus Heiden et Kolbe	S	I	10
Unidentified pennate diatoms	N	I	7, 8, 14
Chlorophyceae			
Carteria sp.	N	I, W	8
Chlamydomonas ballenyana Kol et Flint	N	I	5
Chlamydomonas sp.	S	I	12
Platymonas sp.	N	I	14
Green flagellates	S	I	26
Chrysophyceae			
Chloridella glacialis Kol	N	I	5
Dinobryon balticum (Schütt) Lemmermann	N	I, W	8
Paraphysomonas imperforatus Lucas	S	I	12
Archaeomonads, unidentified cysts	S	I	30, 39
Craspedophyceae			
Acanthoecopsis spiculifera Norris	S	I, W	40
Bicosta antennigera Moestrup	S	I, W	40, 41
Bicosta spinifera Throndsen	S	I, W	40
Calliacantha multispina Manton et Oates	S	I, W	40
Calliacantha simplex Manton et Oates	S	I, W	40
Crinolina aperta (Leadbeater) Thomsen	S	I, W	40
Diaphanoeca multiannulata Buck	S	I, W	40
Parvicorbicula socialis (Meunier) Deflandre	S	I, W	40, 41

Species	Region	Habitat	Ref.
Saepicula leadbeateri Takahashi	S	I	41
Unidentified Craspedophyceae	S	I	41
Cryptophyceae			
Cryptomonas cryophila Taylor et Lee	S	I	12
Unidentified Cryptophyceae	N	I	7, 14
Dictyochophyceae			
Distephanus speculum (Ehrenberg) Haeckel	N, S	I	6, 9, 23, 27
Dinophyceae			
Amphidinium sp.	N	I	7
Dinophysis arctica Mereschkowsky	N	I	8
Discoasteromonas calciferus Bursa	N	I	42, 43
Exuviaella sp.	N	I	23
Gymnodinium lohmanni Paulsen	N	I	14
Gymnodinium spp.	N, S	I	7, 12, 23
Peridiniella catenata (Levander) Balech (*Goniaulax catenata* (Levander) Peters)	N	I	5
Pronoctiluca sp.	N	I	23
Prorocentrum rampii Sournia	N	I	5
Protoperidinium bipes (Paulsen) Balech (*Peridinium minusculum* Pavillard)	N	I	6
Protoperidinium depressum (Bailey) Balech	N	I	8
Protoperidinium grenlandicum (Woloszynska) Balech	N	I	7, 14
Protoperidinium pellucidum Bergh	N	I	6
Protoperidinium sp.	N	I	23
Unidentified Dinophyceae	N, S	I	7, 14, 26
Euglenophyceae			
Dinematomonas litorale (Perty) Silva	N	I	7, 14
†*Euglena geniculata* Dujardin	N	I	5
†*Euglena proxima* Dangeard	N	I	5
†*Euglena viridis* Ehrenberg	N	I	5
Eutreptiella sp.	N	I	7, 14
Urceolus sp.	N	I	7, 14
Haptophyceae (Prymnesiophyceae)			
Chrysochromulina birgeri Hällfors et Niemi	N	I	44
Phaeocystis antarctica Karsten	S	I	17
Phaeocystis pouchetii (Hariot) Lagerheim	N, S	I	5, 23, 26
Phaeocystis sp.	S	I	12, 28
Prasinophyceae			
Mantoniella squamata (Manton et Parke) Desikachary	S	I	12
Pyramimonas gelidicola McFadden, Moestrup, et Wetherbee	S	I	12
Unidentified Prasinophyceae	N	I	23

REFERENCES

1. **Van Landingham, S. L.,** *Catalogue of the Fossil and Recent Genera and Species of Diatoms and their Synonyms. A Revision of F. W. Mills' "An Index to the Genera and Species of the Diatomaceae and their Synonyms,"* Pt. 1-8, J. Cramer, Lehre, 1967-1979.
2. **Schiller, J.,** Dinoflagellatae (Peridineae), in *Kryptogamen-Flora von Deutschland, Österreich und der Schweiz,* Rabenhorst, L., Ed., Bd. 10, Pt. 1 and 2, 1933, 1937.
3. **Balech, E.,** El genero *"Protoperidinium"* Bergh, 1881 (*"Peridinium"* Ehrenberg, 1831, partim), *Rev. Mus. Argent. Cienc. Nat. "Bernardino Rivadavia" Inst. Nac. Invest. Cienc. Nat. Hidrobiol.*, 4, 1, 1974.
4. **Parke, M. and Dixon, P. S.,** Check-list of British marine algae — third revision, *J. Mar. Biol. Assoc. U.K.*, 56, 527, 1976.
5. **Hsiao, S. I. C.,** A checklist of marine phytoplankton and sea ice microalgae recorded from Arctic Canada, *Nova Hedwigia,* 37, 225, 1983.

6. **Usachev, P. I.,** The microflora of polar ice. *Tr. Inst. Okeanol., Akad. Nauk SSSR,* 3, 216, 1949 (in Russian).
7. **Horner, R. and Schrader, G. C.,** Relative contributions of ice algae, phytoplankton, and benthic microalgae to primary production in nearshore regions of the Beaufort Sea, *Arctic,* 35, 485, 1982.
8. **Pett, R. J., Vickers, G. P., Acreman, J. C., and Ethier, A. G.,** A seasonal study of the epontic and planktonic communities near the Issungnak artificial island and McKinley Bay in the Canadian Beaufort Sea, Contract ISZ81-00026 for Department of Supply and Services Canada by Arctic Laboratories Ltd., Sydney, 1983.
9. **Hooker, J. D.,** *The Botany of the Antarctic Voyage of H. M. Discovery Ships Erebus and Terror in the Years 1839-1943. I. Flora Antarctica,* J. Cramer, Weinheim, 1847 (reprint 1963), 503.
10. **Heiden, H. and Kolbe, R. W.,** Die marinen Diatomeen der deutschen Südpolar-Expedition 1901-1903, *Dtsch. Süd-Polar-Exped. 1901-1903,* 8(5), 450, 1928.
11. **Fukushima, H. and Meguro, H.,** The plankton ice as basic factor of the primary production in the Antarctic Ocean, *Nankyoku Shiryo,* 27, 99, 1966.
12. **McConville, M. J. and Wetherbee, R.,** The bottom-ice microalgal community from annual ice in the inshore waters of east Antarctica, *J. Phycol.,* 19, 431, 1983.
13. **Buinitsky, V. Kh., Nikolayev, V. A., and Nikolayev, S. Ye.,** Microscopic algae in fast ice of Alasheyev Bay, *Probl. Arktiki Antark.,* 49, 87, 1977 (in Russian).
14. **Alexander, V., Horner, R., and Clasby, R. C.,** Metabolism of Arctic sea ice organisms, Rep. R74-4, Institute of Marine Science, University of Alaska, Fairbanks, 1974.
15. **Poulin, M. and Cardinal, A.,** Sea ice diatoms from Manitounuk Sound, southeastern Hudson Bay, (Quebec, Canada). III. Cymbellaceae, Entomoneidaceae, Gomphonemataceae, and Nitzschiaceae, *Can. J. Bot.,* 61, 107, 1983.
16. **Bunt, J. S. and Wood, E. J. F.,** Microalgae and Antarctic sea-ice, *Nature (London),* 199, 1254, 1963.
17. **Whitaker, T. M.,** Sea ice habitats of Signy Island (South Orkneys) and their primary productivity, in *Adaptations within Antarctic Ecosystems,* Llano, G. A., Ed., Gulf Publ., Houston, 1977, 75.
18. **Garrison, D. L., Ackley, S. F., and Buck, K. R.,** A physical mechanism for establishing algal populations in frazil ice, *Nature (London),* 306, 363, 1983.
19. **Van Heurck, H.,** Diatomeés. Resultats du Voyage du S. Y. "Belgica" en 1897-9 sous le Commandement de A. de Gerlache dë Gomery. Rapports Scientifiques: Botaniques, *Exped. Antarct. Belg.,* 5(2), 3, 1909.
20. **Meguro, H., Ito, K., and Fukushima, H.,** Ice flora (bottom type): a mechanism of primary production in polar seas and the growth of diatoms in sea ice, *Arctic,* 20, 114, 1967.
21. **Buinitsky, V. Kh.,** The influence of microalgae on the structure and strength of Antarctic sea ice, *Oceanology (USSR),* 8, 771, 1968.
22. **Hendey, N. I.,** The plankton diatoms of the Southern Seas, *Discovery Rep.,* 16, 151, 1937.
23. **Booth, J. A.,** The epontic algal community of the ice edge zone and its significance to the Davis Strait ecosystem, *Arctic,* 37, 234, 1984.
24. **Grossi, S. M. and Sullivan, C. W.,** The vertical distribution of algae in bottom sea ice, *EOS,* 64, 1090, 1983.
25. **Hoshiai, T. and Kato, M.,** Ecological notes on the diatom community of the sea ice in Antarctica, *Bull. Mar. Biol. St. Asamushi, Tohoku Univ.,* 10, 221, 1961.
26. **Burkholder, P. R. and Mandelli, E. F.,** Productivity of microalgae in Antarctic sea ice, *Science,* 149, 872, 1965.
27. **Hart, T. J.,** Phytoplankton periodicity in Antarctic surface waters, *Discovery Rep.,* 21, 261, 1942.
28. **Ackley, S. F., Taguchi, S., and Buck, K. R.,** Primary productivity in sea ice of the Weddell Region, Report 78-19, Cold Regions Research and Engineering Laboratory, Hanover, 1978.
29. **Bunt, J. S. and Lee, C. C.,** Data on the composition and dark survival of four sea-ice microalgae, *Limnol. Oceanogr.,* 17, 458, 1972.
30. **Krebs, W. N.,** Ecology of neritic marine diatoms, Arthur Harbor, Antarctica, *Micropaleontology,* 29, 267, 1983.
31. **Watanabe, K.,** Centric diatom communities found in the Antarctic sea ice, *Nankyoku Shiryo,* 74, 119, 1982.
32. **Poulin, M. and Cardinal, A.,** Sea ice diatoms from Manitounuk Sound, southeastern Hudson Bay (Quebec, Canada). I. Family Naviculaceae, *Can. J. Bot.,* 60, 1263, 1982.
33. **Poulin, M. and Cardinal, A.,** Sea ice diatoms from Manitounuk Sound, southeastern Hudson Bay (Quebec, Canada). II. Naviculaceae, genus *Navicula, Can. J. Bot.,* 60, 2825, 1982.
34. **Hasle, G. R.,** An analysis of the phytoplankton of the Pacific Southern Ocean, *Hvalråd. Skr.,* 52, 1, 1969.
35. **Hart, T. J.,** On the phytoplankton of the south-west Atlantic and the Bellingshausen Sea, 1929-31, *Discovery Rep.,* 8, 1, 1934.
36. **Palmisano, A. C. and Sullivan, C. W.,** Sea ice microbial communities (SIMCO). I. Distribution, abundance, and primary production of ice microalgae in McMurdo Sound, Antarctica in 1980, *Polar Biol.,* 2, 171, 1983.

37. **Gillen, F. T., McFadden, G. I., Wetherbee, R., and Johns, R. B.,** Sterols and fatty acids of an Antarctic sea ice diatom, *Stauroneis amphioxys*, *Phytochemistry*, 20, 1935, 1981.
38. **Grant, W. S. and Horner, R.,** Growth responses to salinity variation in four Arctic ice diatoms, *J. Phycol.*, 12, 180, 1976.
39. **Mitchell, J. G. and Silver, M. W.,** Modern archaeomonads indicate sea-ice environments, *Nature (London)*, 296, 437, 1982.
40. **Buck, K. R.,** Morphology and distribution of the Acanthoecidae (Choanoflagellata) from the Weddell Sea during the austral summer, 1977, Report 80-16, Cold Regions Research and Engineering Laboratory, Hanover, 1980.
41. **Takahashi, E.,** Loricate and scale-bearing Protista from Lützow-Holm Bay, Antarctica. I. Species of the Acanthoecidae and the Centrohelida found at a site selected on the fast ice, *Nankyoku Shiryo*, 73, 1, 1981.
42. **Bursa, A. S.,** *Discoasteromonas calciferus* n. sp., an Arctic relict secreting *Discoaster* Tak [sic] Sin Hok 1927, *Grana Palynol.*, 6, 147, 1965.
43. **Horner, R.,** *Discoasteromonas calciferus* Bursa, an unusual organism occurring in Arctic sea ice, *J. Phycol.*, 8, (Suppl.) 7, 1972.
44. **Hällfors, G. and Niemi, Å.,** A *Chrysochromulina* (Haptophyceae) bloom under the ice in the Tvärminne Archipelago, southern coast of Finland, *Memo. Soc. Fauna Flora Fenn.*, 50, 89, 1974.

INDEX

A

C

F

G

H

I

J

K

L

M

N

O

P

R

U

V

W

X

Y

Z